Foundations of
Environmental
Engineering

Foundations of Environmental Engineering

C. David Cooper
John D. Dietz
and
Debra R. Reinhart

University of Central Florida

WAVELAND
PRESS, INC.
Prospect Heights, Illinois

For information about this book, write or call:
Waveland Press, Inc.
P.O. Box 400
Prospect Heights, Illinois 60070
(847) 634-0081

Cover photo credits:
upper left, ©1990 Greenpeace Images; *upper right,* © Thomas A. Schneider; *middle left,* Illinois Environmental Protection Agency; *lower left,* © Thomas A. Schneider.

Printed in the United States of America

7 6 5 4 3 2 1

The authors dedicate this book to our fathers:

Jack F. Cooper, P.E., J.D. (1921–1991)

Jess C. Dietz, P.E. (1914–)

Rex W. Hitchcock (1924–1998)

Contents

Chapter 3

Mathematical and Physical Foundations 79

Chapter 4

Biological Foundations 121

Preface

Environmental engineering is an important branch of engineering. Rarely can one pick up the newspaper without reading an article concerning the environment; for example, an old hazardous waste dump site that has been discovered, a need for a new wastewater treatment plant in a growing city, the increasing frequency of air pollution problems caused by industry or motor vehicles, an accidental release of a toxic chemical, environmental impacts from a large construction project, ozone depletion, or global climate change. All engineers can benefit from additional education in the field of environmental engineering.

This book has three major objectives. The first is to introduce the key concepts of material and energy balances and to make those principles an integral part of the student's thought processes. This approach, which has long been followed in educating chemical engineers, is helpful to all engineers. Because engineering is an interdisciplinary effort, the second major objective is to introduce the major topics of environmental engineering, including pollution prevention, to all engineering students. Fore-knowledge of possible environmental consequences should lead to better designs from engineers in all disciplines. The third goal of this book is to provide a strong foundation for further study by environmental engineering majors. In that regard, we believe the book strikes a proper balance between the voluminous and detailed presentation of knowledge seen in some texts, and the general and qualitative coverage given in others.

Pre-requisites for this course include college chemistry and two semesters of calculus. The units in this text are a mixture of English and metric units, reflecting the different sets of units that are used in the modern practice of engineering in the United States. Careful attention to units by engineering students therefore is essential. Several types of errors may be identified and diagnosed merely by examination of units for dimensional consistency. It is noted that one consequence of dual unit systems is the fact that, as part of building the new international space station, the shuttle must carry both metric and English tools and parts!

Some competence with computers is assumed. While many undergraduate students are familiar with several widely available applications software packages, they often lack a general ability to develop code (which we mean to include the writing of an original application in a spreadsheet) to solve specific problems. Therefore, we have chosen to introduce numerical solution techniques in this text. Computer solutions are presented in a "logic code" or "pseudocode" format because it better illustrates the thought processes and the sequence of computations. In some cases, we present the solution in a spreadsheet format because of the widespread use of this tool. Instructors should encourage students to convert certain pseudocode examples to spreadsheet solutions to develop those skills in students who lack them.

This book is organized as follows: initially, the basic foundations are laid by an introductory chapter followed by chapters that review chemistry, math (including a detailed treatment of material and energy balances), and biology. Next, three comprehensive chapters are presented covering environmental engineering applications to air resources, water resources, and solid/hazardous wastes. These three chapters can be taught in any order the instructor chooses. Finally, the book ends with case studies that illustrate the foundations taught in this book.

Chapter 1

Historical, Social, and Political Foundations

1.1 Introduction

This book is a first text on environmental engineering. However, it is intended for engineering students in all disciplines, and for others interested in obtaining a quantitative understanding of our air and water resources, environmental quality, pollution control processes, pollution prevention, and restoration of contaminated land areas. As a formal discipline, environmental engineering is not very old. However, the problems now being solved by environmental engineers have been faced by people in one form or another for centuries—controlling or preventing human pollution of the environment, making a more healthy environment for people, and (more recently) preserving and restoring environmental quality for its own sake. Thus, in a sense, environmental engineering has a long history. Even today, much environmental work is accomplished by civil, chemical, industrial, and mechanical engineers; chemists; biologists; and many others.

Environmental engineering can be defined as that branch of engineering which is concerned with (a) the protection of human society from adverse environmental factors, (b) the protection of ecosystems (local and global) from the potentially harmful effects of human activities, and (c) the improvement of environmental quality.

The common theme of environmental engineering is a basic understanding of the interactions of human activities and the environment, how emissions from human activities can damage the environment, what hazards pollution can present to people, and how both people and the environment can be protected from such effects. Environmental engineers not only design, operate, maintain, and manage facilities and systems for environmental protection, they also measure environmental quality and continually seek ways to improve it at a reasonable cost. They deal with atmospheric, aquatic, and

1

terrestrial environments, as well as interactions among these environments. Their work products include process designs, feasibility studies, environmental assessments and compliance reports, among other things. Modern practitioners of environmental engineering try to stay knowledgeable about all these areas, though most specialize in one field such as air pollution control, water or wastewater treatment, or solid and hazardous waste management.

Environmental engineers work in consulting firms; industrial corporations; local, state, and federal governments; private research organizations; and, in small but increasing numbers, with public-interest groups. In addition, some environmental engineers with doctoral degrees are on university faculties, although many doctoral-level environmental engineers are employed in the above-mentioned sectors. Regardless of where they work, many environmental engineers have a high level of job satisfaction and feel their work is challenging and has a direct impact on our quality of life.

All engineers must be "numerate"—that is, literate with numbers. Before progressing very far in their chosen curriculum, engineering students must be able to handle calculations involving many kinds of units and be able to convert values from one set of units to another. This skill can be improved with practice. Some of these simple calculations are illustrated in the following example problem; some common conversion factors are tabulated in appendix A.

Example Problem 1.1

(a) A track star runs a mile in 4 minutes, 16.2 seconds. What is her average speed in m/s?

Solution

$$\frac{1 \text{ mile}}{256.2 \text{ sec}} \times \frac{5280 \text{ ft}}{1 \text{ mile}} \times \frac{1 \text{ m}}{3.281 \text{ ft}} = 6.281 \text{ m/s}$$

(b) An industrial plant discharges a wastewater containing 0.50 mg/L of a pollutant. The wastewater flow averages 30 L/s. How many lb of pollutant are discharged in a year?

Solution

$$\frac{30 \text{ L}}{\text{s}} \times \frac{0.50 \text{ mg}}{\text{L}} \times \frac{86,400 \text{ s}}{1 \text{ day}} \times \frac{365 \text{ day}}{1 \text{ yr}} \times \frac{1 \text{ g}}{1000 \text{ mg}} \times \frac{1 \text{ lb}}{454 \text{ g}} = \frac{1040 \text{ lb}}{\text{yr}}$$

(c) A power plant discharges 0.60 lb of SO_2 per million Btu of heat input. It burns 1.20 tons/minute of coal with a heating value of 25,000 kJ/kg. How much SO_2 is discharged, in lb/day?

Solution

$$\text{Heat in} = \frac{1.2 \text{ tons}}{\text{min}} \times \frac{2000 \text{ lbs}}{1 \text{ ton}} \times \frac{1 \text{ kg}}{2.205 \text{ lb}} \times \frac{25{,}000 \text{ kJ}}{\text{kg}} \times \frac{1 \text{ Btu}}{1.055 \text{ kJ}} = 2.579 \, (10)^7 \, \frac{\text{Btu}}{\text{min}}$$

$$SO_2 \text{ out} = 2.579 \, (10)^7 \, \frac{\text{Btu}}{\text{min}} \times \frac{0.60 \text{ lb } SO_2}{(10)^6 \text{ Btu}} \times \frac{1440 \text{ min}}{1 \text{ day}} = \frac{22{,}300 \text{ lb } SO_2}{\text{day}}$$

Students should note that the answers to the questions in the preceding example problem were reported to only three or four significant figures. Even though your hand calculator will give an answer to part (a) of 6.281286 m/s, it should be obvious that the answer cannot be this precise. It is common in engineering to use three or four significant figures for all answers, except when more are needed and are fully justified by precise measurements of the data or exact values of conversion factors.

1.2   People and Pollution

Human society has evolved into a civilization that has a tremendous capacity to produce wastes. Everyone produces a small amount of personal waste (including excreta, inedible parts of food, worn out clothing, and so forth). In the scattered nomadic or agrarian societies of the distant past, these small amounts of organic materials were easily assimilated back into nature. When populations were small and dispersed, the impacts of humans on the environment were small. With the advance of civilization and the growth of urban societies, the types and amounts of wastes increased. During the 1700s and 1800s it was rare to find a city that did not have severe local pollution problems (with serious consequences for public health) due to improper disposal of wastes. Today, modern societies are producing hazardous and radioactive wastes as well as ever-increasing quantities of "traditional" wastes. Developed countries (DCs) produce much more waste per capita than less developed countries (LDCs), and thus can have a great impact on the environment. Recently, the per capita waste production trend in some LDCs has been rising. The consequences of these practices and trends are potentially global in scale and may continue to be felt for centuries to come.

Even though the formal discipline of environmental engineering is less than forty years old, there are numerous examples in history of environmental engineering. Public water supplies and waste disposal facilities have existed since the days of ancient Rome, which was supplied fresh water by nine aqueducts. Some of these aqueducts were up to 80 kilometers long and up to 15 meters across. There is a bridge across the Tiber River in Rome that today carries automobile traffic, yet was built 1900 years ago! The ancient structures in Bath, England, are another example of Roman construction of public works. In the city of Pompeii, buried by a volcanic eruption in A.D. 79, some homes had running water (albeit through lead pipes), but the sewage was carried away via the stone-paved streets.

In the Middle Ages, local city-states had to defend themselves against warlike neighbors. People who built machines of war became known as engineers, so the term took on a military context. In the late 1700s, John Smeaton, a nonmilitary builder of roads, buildings, and canals in England, decided he should more properly be called a civil engineer (Vesilind, Peirce, and Weiner, 1994). The title became widely adopted by other engineers around the world who designed and built these kinds of public works.

Prior to the early 1800s, it was common practice to discharge human wastes into the street where they might seep into the ground or flow slowly into a nearby ditch and then into a stream, polluting local groundwater wells and small streams and rivers. In addition to the aesthetic offense, such practices contaminated local drinking water supplies with pathogens, causing frequent outbreaks of deadly waterborne diseases. As cities grew, public water supplies often became grossly polluted, and private water supplies became extremely expensive.

Although it may seem obvious to us today that drinking polluted water can make you sick, it was not until the mid-1800s that the transmission of disease was linked conclusively to contaminated drinking water. John Snow traced the 1849 cholera epidemic in London to one drinking water well located on Broad Street, and curbed the epidemic by removing the handle from the pump. Through his work and that of others, water filtration for large public water supplies came into common practice by the late 1800s (Vesilind, Peirce, and Weiner, 1994). However, it took considerable time for more advanced technology to come into widespread use. In 1885, almost 90,000 people in Chicago died of typhoid or cholera when untreated sewage was accidentally discharged into and mixed with the public drinking water supply (Masters, 1991). By the early 1900s, the idea of treating drinking water to kill bacteria became more widely accepted, and today we have modern facilities that can produce clean drinking water from almost any raw water source.

Early engineers proposed sewer systems to separate and transport these waste discharges away from living areas. The first sewers emptied directly into the nearest river, lake, or ocean, without treatment. Wastewater treatment in the early 1900s attempted only to remove the offensive solids (primary treatment). Later, secondary treatment processes were developed to better protect human health. From 1956 to 1965, more than $7.5 billion was spent by the federal government on construction of new municipal wastewater treatment plants. However, it was not until 1972 that federal requirements were established for minimum sewage treatment levels. Around the same time, the idea that reducing damage to the environment was itself a worthwhile and necessary goal gained acceptance. Even after that, there were numerous cases of water pollution in lakes and rivers caused by industrial discharges and agricultural wastes in rainfall runoff. Fish kills—cases where thousands of fish die in a body of water with no obvious source of pollution—were ultimately traced to rainfall that washed pollutants off streets and/or backyards into nearby water bodies. (A fish kill is not a pretty sight; see Figure 1.1). Again, this kind of event is evidence that as we crowd more and more people into smaller and smaller areas, the effects of "normal" human activity can be harmful to the environment.

1.2.1 The Mathematics of Population Growth

Adverse effects of pollution are related both to the number of people and to their crowding into urban areas. Also, as the economic well-being of people improves, their habits become more wasteful. Human population growth (and the trend towards urbanization) from the year A.D. 1000 to present is depicted in Figure 1.2, and shows that world population has been rising very rapidly during the past two hundred years or so. Such growth can be modeled as exponential growth. Also, note from the figure the very recent and very significant shift of people from farms and rural communities into urban centers.

Figure 1.1
A fish kill in Lake County, Florida, in 1997. (*Julie Fletcher/The Orlando Sentinel*)

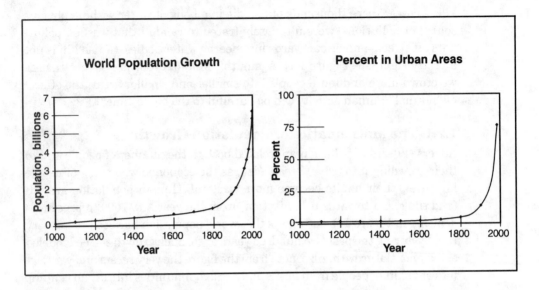

Figure 1.2
World population growth and urbanization in the last thousand years.

Exponential growth is deceiving in that it appears slow for a long time, and appears to "speed up" at the end. Exponential growth results when the quantity grows in proportion to how much is already there, as shown in equation (1.1).

$$\frac{dN}{dt} = rN \tag{1.1}$$

where:

    N = quantity present at time t
    t = time, in arbitrary units (days, years, etc.)
    r = growth rate, expressed as a fraction per unit time

Equation (1.1) has the solution:

$$N = N_0 e^{rt} \tag{1.2}$$

where: N_0 = the amount present initially, at the start of the exponential growth period.

The doubling time is a simple way of viewing exponential growth. **Doubling time** is the time it takes for a quantity, growing exponentially at a constant growth rate, r, to double from its present level. For example, the world's population increased from 1.0 billion in 1815 to 2.0 billion in 1930, yielding a doubling time of 115 years. But it reached 4.0 billion in 1975, a doubling time of only 45 years. If population were to grow at the same rate as from 1930 to 1975, the population would reach 8.0 billion by 2020. Mathematically, by setting N equal to $2N_0$ in equation (1.2), we find that

$$T_d = \frac{\ln2}{r} \cong \frac{70}{R} \qquad (1.3)$$

where:

T_d = doubling time

R = growth rate expressed as a percentage

Using the doubling time, equation (1.2) can be written as:

$$N = N_0 \, 2^{t/T_d} \qquad (1.4)$$

Example Problem 1.2

At one time, the rate of growth of kudzu (a fast-growing vine which completely covers trees, killing them) in a certain part of Georgia was 11% per month. If only 3% of the trees in this Georgia county were covered with kudzu in May 1995, and if exponential growth continued at the same rate, by what date would the entire county be covered with kudzu? Repeat the solution using the doubling time.

Solution

Rearrange equation (1.2) and then take the natural log of both sides.

$$\ln\left(N/N_0\right) = 0.11(t)$$

Set N/N_0 to 1.00/0.03, and solve for t.

$$t = 3.507 / 0.11 = 31.9 \text{ months (less than 3 years)}$$

Thus, by January of 1998, the county would be covered with kudzu.

Using equation (1.3), T_d = 70/11 = 6.36 months. Rearranging equation (1.4) and plugging in the numbers, we must solve the following for t:

$$1.00 = 0.03 \times 2^{t/T_d}$$

Thus, $t = T_d \times \ln(1.0/0.03)/\ln(2)$ or t = 32.2 months.

This is essentially the same answer as before.

1.2.2 The Implications of Population Growth

Of course, it should be understood that no biological population can continue to grow exponentially forever. Some limit—perhaps food supply, water supply, space available, or something else—will eventually stop the exponential growth. At that time either the population will stabilize at the sustainable limit (the carrying capacity), or will crash back down to a much smaller level.

As an example, consider the reindeer population in the following case study (Miller, 1996). In 1910, twenty-six reindeer were placed onto an island off the coast of Alaska. Food supplies were plentiful, and there were no wolves or other predators, so the reindeer population mushroomed to 2000 by 1935. Then came several very harsh winters which, combined with the overgrazing that had occurred, caused a population crash. By 1950, the herd had plummeted to only eight individuals; the reindeer population was reduced by more than 99%!

Of course, people are not reindeer; they have intelligence and free will, and can consciously choose to limit population, and/or take other actions to avoid a disastrous population crash. However, despite efforts of governments and individuals to educate and convince societies to reduce population growth rates, the system has a lot of inertia, and progress has been slow. The pressures of population growth continue through the present day. Not only is population growing, but so also is the tendency for inefficient and wasteful usage of energy and material resources, which contributes to environmental pollution. This will be addressed further in the next section.

The world population reached 1.0 billion people in 1815, 2.0 billion in 1930, and 4.0 billion in 1975. As of 1997, the world population was 5.8 billion people, with an overall growth rate of about 1.6% per year—which adds about 90 million people (equal to another Mexico) every year (Miller, 1996). Although this rate of growth has decreased in the last few years, the world's population has grown enormously during the twentieth century. It took from the beginning of human history to about 1930 to add 2 billion people to the planet, but it took only another 45 years to add another 2 billion (in 1975). The next two-billion addition will likely be reached in only 25 years (in the year 2000)!

If we investigate growth dynamics by regions, we find that the highest growth rates are in the least developed countries (in Africa, Latin America, and Asia)—the ones that can least afford it. Therefore, these countries are getting poorer while their populations are getting larger. Denmark, for example, has a stable population, while Nigeria, Mexico, and Indonesia are all growing faster than 2% per year. The United States is growing moderately at 1.1% per year (of the annual increases in the 1990s, about half were due to legal and illegal immigration).

Also, we note that the population distribution in the fastest-growing countries is heavily weighted towards young people (see Figure 1.3). This over-weighting of youth gives these countries a growth momentum that is difficult to appreciate and extremely hard to stop. That is, as children attain childbearing age, they then have more children. If each couple has only two children they simply "replace" themselves. This is called **replacement fertility**. (Actually, due to accidental deaths of children and other reasons, the replacement fertility rate averages about 2.1 children per couple in the more developed nations, and somewhat higher in the less developed countries.) If

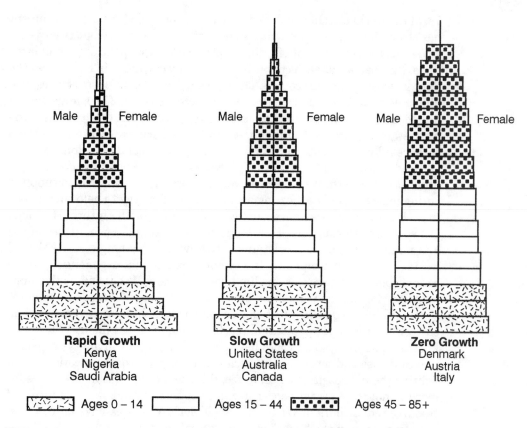

Figure 1.3
Population distributions in countries with different growth rates. (*Adapted from Miller, 1996*)

a nation averages more than 2.1 children per couple, its population continues to grow.

By the time this book is published, the world population will be about 6 billion. Even if childbearing-age couples all over the world were to *immediately* achieve replacement fertility rates, the momentum of the youthful distribution of population implies that the total population will continue to grow to about 8 billion before stabilizing by about 2075. Of course, people who grow up in large families tend to want to have large families. If it takes another forty years to change people's behavior and attain the goal of replacement fertility worldwide, then the stabilized population level will be significantly higher, perhaps about 12 billion. A world population twice as big as it is today would put enormous new strains on the global environment, and may well be beyond the earth's carrying capacity for human beings. Thus it is vital to change people's behavior and attitudes.

Such change of human behavior is difficult but not impossible. China now has 1.2 billion people, or about 21% of the world's population, with only

7% of the world's arable land. For obvious reasons, the Chinese government has strong incentives to control its country's population. About twenty-five years ago, the government of China implemented strict policies to limit birth rates. In urban areas each couple is allowed only one child; in rural areas the limit is two children per couple. Through education, political pressures, and economic and other penalties, the behavior and attitudes of people are changing. In fact, some Chinese women see this as an opportunity to overcome the old view of their role as nothing more than childbearers, and to allow them to more fully participate in work, societal, and political institutions.

Right now, human population growth is still on the rapidly rising leg of exponential growth. We have been very successful in creating food supplies, curing illnesses, and harnessing vast amounts of energy to help sustain our lives. But our successes have serious implications for the future. How long can such growth continue? How long can we continue to use up fossil fuels and mineral resources to sustain such growth? How long can we discard wastes into our environment? It was noted earlier that exponential growth cannot continue forever—some limit will be reached. What will be the limiting factor for human population on earth—water supply? food? air quality? What will be the *quality* of life for the majority of human beings living on earth when the world population reaches 8 billion or 10 billion? Within any known economic system, there have always been (and will always be) some rich countries and some poor countries. With modern communications highlighting these disparities for all to see, there will be increasing pressures for more immigration from poorer to richer countries, and/or other forms of resource redistribution. What effects will such pressures have on the relationships between the "have" and "have not" countries?

These are exceedingly difficult questions, but there is an urgent need to address them. If people are not successful in controlling world population and economic disparities very soon, the societal and environmental consequences may be disastrous. More efforts are needed by the "have" countries to help the "have nots" with agricultural and industrial development, and environmental preservation. The developed countries must also put much more effort into preventing pollution and reducing their consumptive use of energy and other resources.

Fortunately, world leaders are beginning to recognize and address some of the difficult societal and cultural problems identified in the preceding paragraphs. Meanwhile, we as engineers must also continue to address the technical problems of trying to increase production, reduce pollution, improve the distribution and utilization of resources, improve quality, and reduce costs in order to support and improve our global living conditions.

1.2.3 Energy Use, Resource Consumption, and Pollution Generation

The problem of population growth has imposed large demands on the earth's ability to sustain large societies. Population growth might well be considered

Table 1.1 Selected statistics for world population, food production, and energy and resource consumption rates.

Commodity	Units	Annual Production or Consumption, units/year		
		1970	1980	1990
People (cumulative)	10^9	3.67	4.43	5.25
Wheat	10^6 MT	318	446	560
Rice	10^6 MT	315	399	450
Fertilizers	10^6 MT	67	120	155
Coal	10^9 MT	2.14	2.73	3.31
Oil	10^9 B	16.5	19.6	24
Natural Gas	10^{12}m³	1.04	1.44	2.17
Iron Ore (Fe content)	10^6 MT	421	555	570
Copper Ore (Cu)	10^6 MT	6.5	7.9	8.8
Bauxite	10^6 MT	56.8	93.2	98
Passenger Cars	10^6	22.7	29.1	33

Notes: 1 MT (metric ton) of coal (1000 kg) contains about 25 million Btus (but can vary widely).
1 B (barrel) of oil (42 gal) contains about 6 million Btus.
1 m³ of natural gas contains about 35,000 Btus.

Source: Data compiled from several sources.

the single most important cause of environmental problems over the past two hundred years. A close second, however, has been the huge increases in the use of energy. Because of the industrial revolution, people are able to harness fossil fuels and other forms of energy to do vast amounts of work, and to alter their environment. Table 1.1 displays some statistics about world population, food production, and energy and resource consumption.

Human discards have expanded far beyond personal wastes to industrial, mining, petrochemical, and agricultural wastes as well. The ever-growing use of energy and materials to support modern lifestyles produces large quantities of gaseous, liquid, and solid wastes, which can pollute our air, water, and land resources. As an example, air pollution is closely and directly linked to the use of energy (combustion of fossil fuels). As early as A.D. 1285 there were air pollution problems in London, England, problems that have recurred from time to time up until the present. Even though we may try to use materials and energy efficiently, it is inevitable that something will end up as a waste. As humans gain economic wealth, they want more "stuff," which in turn requires the use of more materials and more energy. No other creatures on earth can leverage the use of energy and machinery to move, change, use, and discard such massive amounts of materials. From the data in Table 1.1, we can calculate growth rates over a twenty-year period. While population growth has averaged under 2% per year, energy growth has averaged almost twice that rate. It is sobering to consider that we are now capable, as a species, of permanently changing our environment. And we seem to be doing so in more and more places around the world.

Finally, it is noted that a waste need not be toxic, hazardous, or even unpleasant to cause concern. An overly large generation rate of any substance can create a serious problem. For example, consider the "greenhouse" gas carbon dioxide (more details about global warming will be given in chapter 5). Each person, in the simple act of breathing, emits about one-half kg of CO_2 into the atmosphere each day. If exhaled CO_2 were the only anthropogenic source of CO_2, the ecological balance between plants and animals could be maintained. Excess CO_2 emissions come mainly from burning fossil fuels at enormous rates. For instance, one gasoline-fueled car emits about 5 metric tons (MT) of CO_2 in a year (more than the weight of the car itself), and an average size coal-fired power plant emits about 10,000 MT of CO_2 per day! Furthermore, world demand for fossil fuels is projected to continue growing, as shown in Table 1.2.

Table 1.2 World energy consumption, recent past and near-term future.

	Energy Use, Quadrillion Btu/year			
Fuel	1970	1995	2010	2015
Oil	97	141	195	213
Natural Gas	36	78	129	145
Coal	59	93	123	135
Nuclear	1	23	25	23
Biomass/Renewables	12	30	42	46

Source: Oil & Gas Journal, 1997

Example Problem 1.3

One measure of the economic wealth (or opulence) of a society is the prevalence of cars. In 1995, there were about 280 million people in the United States and about 150 million cars. (We have so many cars because we mostly drive alone; that is, one person per car—a very wasteful habit.) In China, there were about 1.1 billion people, but only 5 million cars. How much CO_2 was emitted from people and from cars in each country in 1995? The Chinese economy is growing very rapidly. What if China had the same ratio of cars to people as the U.S.? How much CO_2 would be emitted from Chinese cars?

Solution

In the U. S., from people:

$$280\left(10\right)^6 \text{ people} \times 0.5 \text{ kg } CO_2 \text{ / person} - \text{day} \times 365 \text{ day / yr} \times 0.001 \text{ MT/kg}$$
$$= 5.1\left(10\right)^7 \text{ MT } CO_2 \text{ / year}$$

and from cars:

$$150\left(10\right)^6 \text{ cars} \times 5 \text{ MT } CO_2 \text{ / car} - \text{year} = 7.5\left(10\right)^8 \text{ MT } CO_2 \text{ / year}$$

In China, from people:

$$1.1\,(10)^9\,\text{people} \times 0.5\,\text{kg CO}_2\,/\,\text{person–day} \times 365\,\text{day/yr} \times 0.001\,\text{MT/kg}$$
$$= 2.0(10)^8\,\text{MT CO}_2/\text{year}$$

and from cars:

$$5\,(10)^6\,\text{cars} \times 5\,\text{MT CO}_2\,/\,\text{car–year} = 2.5\,(10)^7\,\text{MT CO}_2\,/\,\text{year}$$

For the "what if?" scenario for China,

$$1.1\,(10)^9\,\text{people} \times (150\,\text{cars}/280\,\text{people}) = 5.9(10)^8\,\text{cars}$$
$$5.9\,(10)^8\,\text{cars} \times 5\,\text{MT CO}_2\,/\,\text{car–year} = 3.0\,(10)^9\,\text{MT CO}_2\,/\,\text{year}$$

From the results, it seems clear that the Chinese must not obtain as many cars per person as the U.S. has now, and/or Americans must change their driving habits, and/or the world must find a better way to move people from place to place!

Example Problem 1.4

Assume that global energy use grows exponentially at 5% per year for the next 100 years. Estimate the present use from Table 1.2, and assume, for simplicity, that the energy demand that is met by fossil fuels (coal, oil, and gas) can be represented by oil only. Oil has an energy content of about 1,000,000 Btu/ft³.

(a) How much oil (in ft³) would be consumed during the next 100 years?

(b) It has been estimated that the world's reserves (discovered and undiscovered) of fossil fuels (coal, oil, and gas) are equivalent to a quadrillion barrels of oil. This amount of oil would cover the earth to a depth of one foot. How long would that amount of oil last?

Solution

(a) In 1995 fossil fuel use was 312 (10)¹⁵ Btu. Thus present "oil" demand (in 1997) is:

$$312\,(10)^{15}\,\text{Btu/yr} \times (1.05)^2 \times 1\,\text{ft}^3/10^6\,\text{Btu} = 3.44\,(10)^{11}\,\text{ft}^3/\text{yr}$$

In any future year, t, the oil demand is:

$$N_t = 3.44(10)^{11}e^{0.05t}$$

The cumulative demand over the next 100 years is:

$$D \sim \int_0^{100} 3.44\,(10)^{11}e^{0.05t}$$
$$= 3.44\,(10)^{11}\left[e^{0.05(100)} - e^0\right]/0.05$$
$$= 1.02\,(10)^{15}\,\text{ft}^3$$

(b) The volume of oil in a layer that covers the earth to a depth of one foot is the difference between the volumes of spheres of radii 4000 miles +1 ft and 4000 miles.

$$V = 4/3 \, \pi \left(R_1{}^3 - R_2{}^3 \right)$$

$$= 4/3 \, \pi \left[\left(4000 \times 5280 + 1 \right)^3 - \left(4000 \times 5280 \right)^3 \right]$$

$$= 5.6 \left(10 \right)^{15} \; ft^3$$

The time to use that oil is obtained by setting $D = 5.6 \, (10)^{15} \; ft^3$ and solving for t.

$$D = 5.6 \left(10 \right)^{15} = 3.44 \left(10 \right)^{11} \left[e^{0.05(t)} - e^0 \right] / 0.05$$

$$e^{0.05t} = 8.14 \left(10 \right)^2 + 1$$

$$0.05t = 6.70$$

$$t = 134 \; \text{years}$$

Under the assumption of continuous exponential growth, our fossil fuel reserves, even if they were so vast as to blanket the earth, would last only about another 134 years!

This example demonstrates the power of exponential growth (and the absurdity of expecting things to continue to grow exponentially forever). Long before we burned as much oil as implied by this example problem, we would have filled the atmosphere with carbon dioxide (and possibly used up a good portion of the oxygen in the air!). Obviously, we must find renewable alternatives to our energy needs!

1.3   The Development of an Environmental Ethic

A basic concept of science is that matter cannot be created or destroyed. When this simple concept is applied to the environmental field, it tells us that pollution does not simply "go away" when we discharge it to the air, water, or land. Unfortunately, it took decades, even centuries, of pollution to make people understand that this concept makes no exceptions, and that the earth itself is not a big enough place for us to just keep throwing things away.

During the industrial revolution, people lost sight of this basic concept because the world seemed so vast that when wastes were discharged into the air or water, they seemed to "disappear." Today, we are more aware than ever that material is cycled through the environment, sometimes with grave consequences. Matter can be changed in form through chemical reactions, and it can be "stored" for long periods of time. But it never just goes away.

For example, phosphorus is a key element in sustaining life. Phosphate ore (the skeletal remains of prehistoric marine life) is mined in central Florida to produce fertilizer; when the phosphate fertilizers are applied to crops, the phosphorus that was stored millions of years ago is absorbed into plant fiber and food, and finally returns to the environment. However, excessive applications of phosphate fertilizers to the lawns and gardens of homes and businesses near lakes can result in excess phosphorus being carried by rainwater runoff into the lakes. This can produce rapid growth of algae and water weeds, and may result in a fish kill such as the one pictured in Figure 1.1.

Another example involves a case in which a pollutant adversely affected human health—namely the mercury poisoning of the people living around Minamata Bay in Japan in the 1950s (Vesilind, Peirce, and Weiner, 1994). Mercury was discharged by nearby industries into the waters of the bay. That mercury accumulated in bottom sediments, was methylated (transformed into methyl mercury which is soluble) by aquatic microorganisms, and subsequently transferred up the food chain to fish and then to people. The people became ill due to high mercury levels that accumulated in their systems.

The history of the development of environmental awareness in the United States is a long story. Previously, we saw how the rudimentary science and common sense of the 1700s and 1800s developed into the engineering practices needed to protect public water supplies from contamination with human sewage. Early concepts of nature clearly influenced society's values about the environment. During the 1600s and 1700s as pioneers struggled for survival, nature was viewed as dangerous and capricious. It was filled with beasts, droughts, plagues, blizzards, and other things and events that caused hardship and death. Nature was something that had to be fought and tamed. During the 1800s, as the country expanded westward, nature was viewed as a commodity. It was still dangerous, but it was rich with resources (forests, animals, minerals) that could and should be exploited.

With the harnessing of energy and the development of machinery, people were able to extract and utilize many natural resources at rates never before imagined. These natural resource industries fueled the development of the United States into a great economic power by the early 1900s. Understandably, the resource development philosophy that was so successful in the 1800s persisted well into the twentieth century. However, there were visionaries even in the 1800s who foresaw that the domination and exploitation of nature would end up being self-defeating.

Early environmentalists like George Marsh in the 1870s and Gifford Pinchot in the 1890s championed the value of the environment for its own sake. With Pinchot as an advisor, President Teddy Roosevelt helped establish many national parks in the early 1900s. President Franklin Delano Roosevelt, in response to the Great Depression of the 1930s, created the Soil Conservation Service, the Civilian Conservation Corps, and other programs that benefited the environment.

In 1949, Aldo Leopold published *A Sand County Almanac*. This book introduced a new way of viewing the environment and was quite radical for its time. Leopold advocated a "land ethic" that espoused reverence for the land, and which obligated people to respect and care for nature. That is, we should be custodians or stewards of the land rather than owners (this effectively would limit the rights of property owners). However, it was not until the 1950s and 1960s that various consequences of severe environmental pollution became apparent and began to be widely publicized. It was then that a wide cross section of society began to realize that we could not continue to dispose of untreated or improperly treated municipal, industrial, and hazardous wastes without dire consequences for the environment and for all of us.

In 1962, Rachel Carson authored a book called *Silent Spring*, which may have had more to do with the general public becoming involved with and aware of the environment than any other single event of the 1950s and 1960s. The title of her book refers to a future springtime that is silent because all the birds have died, having been poisoned by environmental pollution. Interestingly, in a 1997 journal article, one chemical engineer reminisced about his first job in 1941, working at a cyanide plant in New Jersey. One of his recollections was that "when the HCN [hydrogen cyanide] in the stack gas became strong enough, birds innocently flying overhead would fall onto the ground dead at one's feet. Similarly poisoned rats would run out from under buildings and drop dead" (Connolly, 1997).

In the 1960s and 1970s, many people became outraged at the state of the environment. They joined forces in environmental groups such as the Sierra Club or Greenpeace, and demanded that their legislators do something about protecting the environment. Such vocal advocacy was what was needed at the time to provoke government and industry into action. The following excerpts from an EPA report (U.S. EPA, 1980) are provided to help students understand the conditions that prompted the public to push Congress for passage of environmental laws and regulations.

> During the 1950s and 1960s, toxic wastes were placed into drums and discarded in a landfill near the towns of Toone and Teague, Tennessee. When the landfill closed in 1972, the site held some 350,000 drums, many of them leaking pesticide wastes. The towns' water supply was made unusable in 1978 when water leaching from the landfill reached the drinking water zone; it was contaminated with a number of organic compounds. The towns no longer had access to uncontaminated groundwater, and had to pump water in from other locations.

> Another underground water supply near Denver was contaminated from disposal of pesticide waste in unlined disposal ponds. The wastes were produced from manufacturing activities of the U.S. Army and a chemical company during the 1943 to 1957 period.

> In 1978, a truck driver was killed as he discharged waste from his truck into one of the four open pits at a disposal site in Iberville Parish, Louisiana. He was asphyxiated by hydrogen sulfide produced when the liquid wastes mixed and reacted in the open pit.

In one of the most publicized environmental disasters of the twentieth century, the health of residents of Love Canal, near Niagara Falls, was seriously damaged by chemical wastes buried in the 1950s. As drums holding the wastes corroded, their contents percolated through the soil into yards and basements, forcing the evacuation of over two hundred families in 1978 and 1979. (A more detailed discussion of this case is presented in chapter 8).

By the late 1960s there were calls for action from many groups, and a number of federal legislative initiatives were begun. The 1970s became known as the "decade of the environment." As mentioned earlier, attitudes toward the environment had evolved to the point where many people believed we should adopt a custodial or stewardship role. That is, as a society we had a moral obligation to preserve and care for the land. This philosophy was often justified by enumerating the benefits of such custodial actions (increased fishing yields, or better enjoyment of forests, for example). There were calls for stopping all further land development and growth; however, in the face of growing populations, this course of action was unacceptable to most people.

Sustainable development is a recent concept defined as "development that meets the needs of the present without compromising the needs of the future." A simple example is selectively cutting and replanting "timber farms," rather than harvesting old-growth forests. Under this concept, growth and development are not halted but are modified to include recycling and restructuring. Technological progress must consider not only efficiency and profit, but also resource and energy conservation, and must adapt to a changing world. Many countries around the world, both developed and developing, recognize the need to encourage and embrace sustainable development. This concept will be discussed further in the next section.

Most recently, some individuals have begun advocating the inalienable rights of nature to exist, even if no observable benefits accrue to people. This concept includes not only the rights of plants and animals to live and enjoy their habitats, but also the rights of the habitat itself (including the rocks, rivers, and soil). This view may be the most enlightened one of all, but is certainly the most difficult to put into practice.

1.4   Environmental Laws, Regulations, and Agencies

Environmental protection in this country is supported by a complex web of laws and regulations, which are implemented and enforced by numerous individuals, agencies, lawmakers, and the courts.

Environmental laws (whether federal, state, or local) establish broad goals for environmental quality; set standards of behavior for individuals, industry, and governments; create agencies to monitor the environment and to oversee the actions of other groups; and establish penalties for failure to adhere to the laws. For example, the National Environmental Policy Act (signed into law by then-President Nixon on January 1, 1970) was a six-page document that set a new policy for the United States to help ensure that decision makers would consider all the environmental impacts of any major actions (such as building a new highway). It also led to the creation later that year of the Environmental Protection Agency (EPA), the main federal agency charged with monitoring environmental quality; creating specific regulations to protect the quality of our air, water, and land resources; and enforcing those regulations and laws. Figure 1.4 traces the growth in environmental legislation during the past one hundred years, while Table 1.3 describes some of the key pieces of federal environmental legislation that have been passed since 1970.

Environmental agencies are governmental units charged with monitoring and protecting the environment. There are federal, state, and local environmental agencies, and all must interact, each respecting the (sometimes differing) opinions, priorities, authorities, and responsibilities of the others in order to get things done for the ultimate benefit of the environment. In most states, the state environmental agencies have been given significant powers by the U.S. EPA. For example, the states have significant flexibility in interpreting and enforcing federal mandates and guidelines. An example of the federal/state/local agency organization in the Orlando, Florida area is the U.S. EPA, the state of Florida Department of Environmental Protection (FDEP), and the Orange County Environmental Protection Department (OCEPD).

Environmental agencies usually are organized into specific media-oriented groups (such as air quality or water quality divisions in the U.S. EPA) or mission-oriented groups (such as the regulation enforcement division of the U.S. EPA). Their staffs include technical, administrative, legal, and support personnel. Even nonenvironmental agencies often have large environmental sections (such as the Department of Transportation, the Department of Energy, and the U.S. Army). Of course, we cannot overlook our judicial system. In the United States, the most litigious country in the world, the courts have played a major role in directing the course of environmental protection.

Environmental regulations are not laws but are specific rules that have the power of law. Regulations are created by environmental agencies to protect existing environmental quality; to limit present or future discharges of pollutants from industry, municipalities, and individuals; and to provide the legal means and authority to monitor the actions of other groups and enforce compliance with the regulations. The regulations that have been created by the U.S. EPA alone take up tens of thousands of printed pages, and can be very confusing to read and understand.

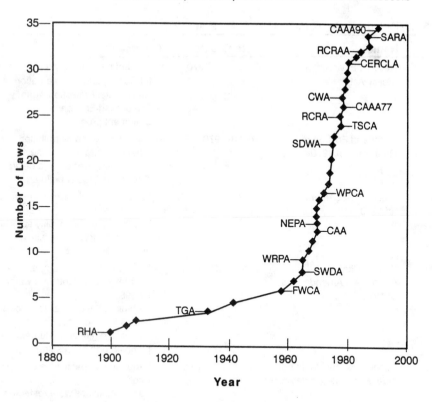

Key

RHA	Rivers and Harbors Act, 1899
TGA	Taylor Grazing Act, 1934
FWCA	Fish and Wildlife Coordination Act, 1958
SWDA	Solid Waste Disposal Act, 1965
WRPA	Water Resources Planning Act, 1965
NEPA	National Environmental Policy Act, 1970
CAA	Clean Air Act, 1970
WPCA	Water Pollution Control Act, 1972
SDWA	Safe Drinking Water Act, 1974
TSCA	Toxic Substances Control Act, 1976
RCRA	Resource Conservation and Recovery Act, 1976
CAAA77	Clean Air Act Amendments, 1977
CWA	Clean Water Act, 1977
CERCLA	"Superfund" Act, 1980
RCRAA	RCRA Amendments, 1984
SARA	Superfund Amendments and Reorganization Act, 1986
CAAA90	Clean Air Act Amendments, 1990

Note: Symbols without names represent other federal environmental laws in the United States; not all laws that were promulgated are named in this chart.

Figure 1.4
A graphical chronology of environmental laws in the United States.

Table 1.3 Major federal environmental legislation since 1970.

Acronym	Name	Date Enacted	Comments
NEPA	National Environmental Policy Act	January 1, 1970	—Created the Council on Environmental Quality, and led to creation of Environmental Protection Agency —Established Environmental Impact Statement process
WEQA	Water and Environmental Quality Act of 1970	April 3, 1970	—Concentrated on oil pollution and other discharges by vessels —Directed the President to designate "hazardous" water pollutants
CAAA–70	Clean Air Act Amendments of 1970	December 31, 1970	—Established emissions standards for many sources —Led to ambient air quality standards —Technology-forcing legislation led to automobile emissions controls
WPCA	Water Pollution Control Act Amendments of 1972	October 18, 1972	—Set a goal of eliminating all discharges of pollutants into waterways by 1985 —Prohibited the discharge of toxic pollutants in toxic amounts —Required regional waste treatment planning and processing
FIFRA	Federal Insecticide, Fungicide, and Rodenticide Act	October 21, 1972	—Required that pesticides be registered —Some pesticides were restricted —Commercial users must be certified to use restricted pesticides —Required that EPA establish procedures and regulations for disposal or storage of certain pesticides
ESECA	Energy Supply and Environmental Coordination Act	June 22, 1974	—Recognized the interdependence of energy and the environment —Allowed some delays in meeting emissions standards but basically said environment would not be sacrificed for energy
SDWA	Safe Drinking Water Act	December 16, 1974	—Provided for federal primary and secondary drinking water quality standards for public water supplies —Provided for regulation of public water systems —Provided for protection of underground drinking water sources —Regulated all underground injection
TSCA	Toxic Substances Control Act	October 11, 1976	—Provided pre-marketing review of all new chemicals —Provided for direct regulation of manufacture, use, and disposition of all chemicals —Required extensive testing of chemicals for hazardous effects

Acronym	Name	Date Enacted	Comments *(continued)*
RCRA	Resource Conservation and Recovery Act of 1976	October 21, 1976	—Mandated comprehensive regulation for disposal of all wastes (both by-products and consumer products) —Defined hazardous wastes; listed a number of specific hazardous wastes —Concentrated on "land" pollution rather than air or water pollution —Regulated interstate transportation of hazardous wastes as well as disposal, treatment and/or storage; a "cradle to grave" approach —Required detailed permit process for everyone handling hazardous waste
CAAA-77	Clean Air Act Amendments of 1977	August 7, 1977	—Provided for prevention of significant deterioration of "clean" air regions —Designated regulations for nonattainment areas —Delayed attainment of emissions standards for automobiles
CERCLA	Comprehensive Environmental Response Compensation and Liability Act [Superfund]	December 17, 1980	—Created a special tax on industrial chemicals that goes to a trust fund, commonly called Superfund —Directed EPA to perform site cleanups or take legal action to force responsible parties to perform the cleanups
NWPA	Nuclear Waste Policy Act	January 7, 1983	—Authorized the design and construction of disposal facilities for spent nuclear fuels and high-level radioactive wastes —Formed the Office of Civilian Radioactive Waste Management
HSWA	Hazardous and Solid Waste Amendments	November 9, 1984	—Established minimum technical requirements for land disposal of wastes —Modified permitting process for treatment, storage, and disposal facilities —Established controls for underground storage tanks
SARA	Superfund Amendments and Reauthorization Act	October 17, 1986	—Increased Superfund financial coverage amounts —Established emergency planning and public right-to-know procedures —Required industry to report toxic release inventories
CAAA-90	Clean Air Act Amendments of 1990	November 15, 1990	—Addressed urban air quality, especially ozone —Dealt with mobile sources comprehensively —Regulated hazardous air pollutants —Established a program for operating permits for air polluters

1.5 Pollution Control, Pollution Prevention, and Sustainable Development

Initially, once the need to protect the environment from human pollution was recognized, the accepted approach to a pollution problem was to design, build, and operate "end-of-the-pipe" treatment processes. In the last twenty years or so, however, people have begun to realize that preventing environmental degradation makes much more sense than cleaning up pollution damage after it occurs. In many cases, it is smarter and more cost effective to *not create* the pollution during the manufacturing process than to put pollution control devices at the tail end of that process. Pollution prevention has become the new emphasis of the EPA, and of many industrial companies.

One good example of the pollution prevention philosophy is the decision by the EPA in the mid-1970s to mandate the removal of lead from gasoline. (Lead is an octane booster in gasoline but is a toxic metal in the environment.) By the early 1990s, motor vehicle emissions of lead to the atmosphere had been reduced dramatically (by more than 98%). Another example deals with chlorofluorocarbons (CFCs). In the mid-1970s it was postulated (and later demonstrated in 1985) that CFCs released into the atmosphere (from leaky automotive air conditioners, styrofoam manufacture, metals degreasing operations, and other industrial processes) were contributing to the destruction of the stratospheric ozone layer that protects life on earth from the sun's ultraviolet rays. Some 36 countries worked together to develop the Montreal Protocol in 1987 to limit and reduce the manufacture and use of CFCs. Industry has since diligently sought ways to reduce its dependence on CFCs. For example, local environmental agencies implemented regulations to control how automotive air conditioners could be recharged. Stores stopped selling individual cans of freon. Many industrial cleaning processes were changed (for example, solvent degreasing of parts in the metals-plating industry was replaced with alkaline aqueous detergent washing), and many product substitutions occurred. Today, there is much less dependence on CFCs than in the 1980s.

The U.S. EPA has endorsed a hierarchical approach to solving pollution problems (see Figure 1.5). At the base is pollution prevention/waste minimization, which is the most preferred approach. Next comes recycling and reuse of waste materials. Third comes treatment, and the fourth and least preferred approach is disposal. These principles are demonstrated with the following example.

Example Problem 1.5

A company is discharging 100,000 gallons per day of a wastewater that contains small amounts of dissolved organic compounds as well as dissolved chromium metal ions. The organics come from a solvent-cleaning process, and the chromium originates from a 2000 gallon-per-day chrome-plating operation. Because of the chromium content, the entire wastewater stream has been declared a hazardous waste, treatment and disposal of which will be extremely costly. You as an engineer are asked to recommend an approach to solving this problem. What are your thoughts?

Solution

First, investigate whether the solvent-cleaning process can be changed to clean the parts to be plated with a simple jet wash using water only, or perhaps soap and water. If we can eliminate all use of the organic solvent, we will be well ahead of the game.

Second, analyze the chrome-plating process to maximize the placement of the chrome onto the product and to reduce the chrome losses to the greatest extent possible. Third, try to reduce the volume of chrome wastewater and separate it from the ordinary wash waters that come from other parts of the plant. Perhaps we can isolate the chromium wastewater to its original 2000 gallons per day.

Fourth, after the volume of the wastewater containing chromium has been reduced as much as possible and separated from the rest of the wastewater, try to use ion exchange or some other process to further separate and recover chromic acid to the extent possible. Finally, chemically treat the remaining small volume of chrome wastewater to remove the chrome as a solid precipitate, and dispose of only the chrome precipitate as hazardous waste. The other 98,000 gallons per day of "ordinary" wastewater now can be treated by conventional means, saving a considerable amount of money for the company, while simultaneously reducing the threat of chrome contamination of the environment.

Figure 1.5
Hierachy of solving pollution problems.

The term sustainable development was introduced in the last section, and refers to development that meets present needs while preserving environmental resources for the future. It recognizes the need for all countries to continue to develop and grow economically, but also emphasizes the needs of the global environment. This term was introduced to the world in 1987 in the United Nations report, "Our Common Future," and was popularized in the 1992 Earth Summit in Rio de Janeiro, Brazil. At the Earth Summit, more than a hundred heads of state and over 1400 other leaders, scientists, and planners from 178 nations met to develop plans and policies for addressing global environmental issues. The major strategies that resulted from that summit are as follows:

- reduce population growth
- reduce poverty
- reduce the wasting of resources—both matter and energy
- emphasize pollution prevention
- make things that last longer and are easier to repair and/or recycle
- protect habitats and preserve biodiversity
- use renewable resources at their natural rates of renewal
- use locally renewable energy resources such as the sun, the wind, flowing water, and biomass

Ways to achieve sustainable development have been given serious attention by planners and leaders from many countries. Sustainable development is an especially important concept with regard to current concerns about global climate change—the so-called "greenhouse effect." For the past three hundred years humans have been clearing forests and burning fossil fuels, both of which contribute carbon dioxide to the atmosphere faster than it is being removed by natural processes. As the CO_2 concentration increases in the atmosphere, more heat is retained and the global climate is affected (more on this in chapter 5).

A good example of sustainable development is the sugarcane industry. Sugarcane is grown in many countries throughout the world, including the United States, where it grows in Florida, Louisiana, Texas, and Hawaii. In south Florida, it is grown in large flat fields located north of the Everglades. As it grows, sugarcane very efficiently absorbs CO_2 from the air. The cane is harvested and brought to a mill. There the cane is cut and crushed to obtain its sugar juice (sucrose in water). Using steam generated in boilers at the mill, the sugar juice is evaporated to produce sugar crystals.

The steam needed for evaporation is produced at the sugarcane factory by burning the crushed cane fiber (called bagasse), thus avoiding the need to burn fossil fuels. When the bagasse is burned CO_2 is generated, of course, but since the carbon in the cane fiber came from CO_2 that was already present in the atmosphere, there is no net increase for the cycle. Bagasse is the primary fuel source for the mills, and supplies over 95% of all fuel needs (fossil fuels are often used for the initial start-up of the boilers). The steam is also used to turn turbines which run the milling machines and generate electrical power needed to run pumps, blowers, and other equipment in the factory. Sometimes excess electricity is sold to the grid. This particular industry is sustainable: it is a fully integrated co-generation operation, fossil fuels are not used, electricity is not purchased, solid wastes are not produced, and the net addition of CO_2 to the atmosphere is zero.

1.6   The History of the Future

Our actions today are making the history of the future! One hundred years from now, when people look back to the beginning of the twenty-first century, will they see continued commitment to environmental protection? Will they see that population control was successful? How will global warming issues be resolved? How will we, as engineers who worked to solve these difficult problems, be judged?

It is the authors' belief that concern for the environment has become an important part of human society. There are still debates about the best ways to implement environmental protection policies, and how to balance human economic interests and environmental interests, but there remains little doubt in the minds of millions of people around the world that protection of our environment is critical to the long-term well being of humanity. This environmental ethic is spreading, and many developing countries are concerned not only about how to develop their economic wealth but how to do so while protecting and preserving their environment. Hopefully, they can learn from some of the mistakes the United States and Europe have made and achieve sustainable development. As environmental engineers we must do everything we can to continue to foster this ethic and to implement sound solutions to environmental problems. In the truest sense of the phrase, we are history makers; let's work to make it a history our grandchildren will study with pride.

End-of-Chapter Problems

1.1   A car manufacturer needs 5 tires to make one car (one tire for the spare). Each tire weighs 32 pounds. How many tires are needed to make 100 cars? How many pounds of tires are needed to make 100 cars?

1.2   How much salt (NaCl) is carried by a river flowing at 30 m^3/s and containing 50 mg/L of salt? Give your answer in kg/day.

1.3   A petroleum refinery loses 0.4% of its input mass of hydrocarbons through leaks and spills. The refinery processes 300,000 barrels per day of crude oil with a specific gravity of 0.95. A U.S. barrel is 42 U.S. gallons. How many tons of hydrocarbons are lost each year from this refinery?

1.4   Consider the combustion of propane: $C_3H_8 + 5\,O_2 \rightarrow 3\,CO_2 + 4\,H_2O$. How many lb-moles of oxygen are required to burn 100 lb-moles of propane?

How many pounds of oxygen are needed?

1.5   A coal-fired power plant burns 8000 metric tons (MT) per day of coal. The coal contains 5% by weight noncombustible ash. Pollution-control devices capture 99% of the ash, and the captured ash is buried. How much ash is buried each year (MT)? How much ash is emitted into the atmosphere each year (MT)?

1.6   A chicken farmer owns a 20-acre farm and started business with 20 chickens. After one year, he has 400 chickens. How many will he have after one more year? Assuming a constant growth rate, how many chickens would he have after a total of 6 years? Does this seem reasonable to you? Why or why not?

1.7   What is the doubling time for the situation described in problem 1.6 above?

1.8   If the farmer in problem 1.6 wants to sell 10,000 chickens per month, what is the chicken population he needs to maintain?

1.9   If the world population is 6.0 billion in 1997, and the growth rate is constant at 1.5%, calculate the population in 2020. If the growth rate is constant for another 30 years, what will be the population in 2050?

1.10 Consider the farmer in problem 1.8 who is selling 10,000 chickens per month. How many chicken wings is he selling? If each wing weighs 32.0 g, how many kg of wings are being sold? If each chicken weighs 1.2 kg, how many kg of chicken does it take to produce 100 kg of wings?

1.11 Find out how much wastewater and solid wastes are generated at your campus. Where are these two wastes sent? What happens to them?

1.12 Assuming a population of 6 billion people, about how much CO_2 is exhaled by humans every year? Assume that the world burns about $300 (10)^{15}$ Btu of fossil fuels (coal, oil, and gas) each year. For the total mix of fossil fuels, the average energy content is 15,000 Btu/pound, and the average carbon fraction is 0.83. How much CO_2 is formed per year from fossil fuel combustion if 3.67 pounds of CO_2 are produced for every pound of carbon burned?

1.13 Discuss in writing the benefits and limitations of solving pollution problems using the hierarchy shown in Figure 1.5.

1.14 Read your local newspaper thoroughly for a few days. Find an article relating to environmental pollution or environmental restoration in your geographical area. Research the issues raised by this article, and write a short paper supporting one side or the other of the controversy.

1.15 Do you think society has been improved by the passage of the laws listed in Table 1.3? Why or why not? Give specific examples to support your argument.

1.16 Use Figure 1.2 to estimate world population in the years 1000, 1500, 1800, and 2000. For each of the three time intervals (1000–1500, 1500–1800, and 1800–2000), use the exponential growth model to estimate the growth rate r. Explain why r is so different for those intervals.

References

Carson, Rachel. 1962. *Silent Spring*. Boston: Houghton Mifflin.

Connolly, John. 1997. "Personal Perspectives: Reflecting on the Past Fifty Years." *Chemical Engineering Progress* 93 (1).

Masters, G. M. 1991. *Introduction to Environmental Engineering and Science*. Englewood Cliffs, NJ: Prentice Hall.

Miller, G. Tyler, Jr. 1996. *Living in the Environment*. 9th ed. Belmont, CA: Wadsworth.

Vesilind, P. Aarne, J. Jeffrey Peirce, and Ruth F. Weiner, 1994. *Environmental Engineering*. 3rd ed. Woburn, MA: Butterworth-Heinemann.

U.S. Environmental Protection Agency. 1980. *Everybody's Problem: Hazardous Waste*. Publication SW-826.

Chapter 2

Chemical Foundations

2.1   Introduction

Chemistry is defined as the science that deals with the composition, structure, and properties of substances and the changes they undergo. Environmental chemistry applies this body of science to understanding and predicting the fate and transport of chemical substances in nature and to the engineering design of systems to reduce or remove pollution. It is absolutely essential that environmental engineers have a good knowledge of environmental chemistry. Unfortunately, too many engineers are inadequately trained in environmental chemistry and often fail to comprehend the complex chemical issues that challenge them in solving environmental problems. This chapter provides a review of certain fundamental information presented in basic chemistry courses and illustrates its application to environmental issues.

2.2   Solutions

When two substances are combined but do not chemically react they may form a suspension, a true solution, or a colloidal dispersion. Figure 2.1 provides an indication of the relative sizes of natural materials. In a **suspension**, matter is suspended in a gas or liquid and will settle out relatively quickly. Usually, suspended matter is greater than 1 micron (μm) in diameter. A **true solution** is composed of matter homogeneously dissolved in a liquid or gas. A true solution cannot be mechanically separated into its component phases without changing the temperature or pressure of the solution. The dissolved matter is referred to as the **solute** while the dissolving material is the **solvent**. Typically, the solute is similar in size to molecules. Somewhere between a suspension and a solution is a **colloidal dispersion**, which is

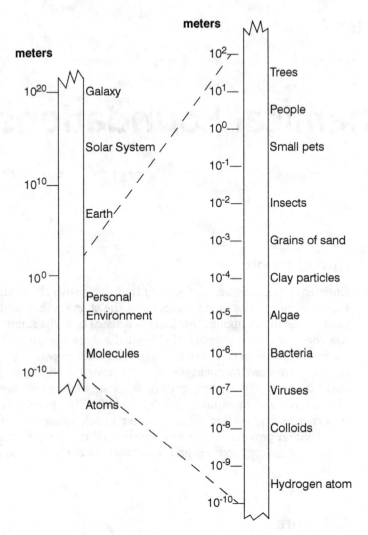

Figure 2.1
Sizes of selected natural objects.

composed of material scattered in a liquid or gas. A colloidal dispersion cannot be separated by gravity; however, it can be mechanically separated. The dispersed material is typically 0.001 to 1 μm in diameter. All natural waters contain complex combinations of suspended, colloidal, and dissolved matter. For example, surface water contains large particles of soil suspended in rivers or lakes (which will settle when the water is allowed to remain quiescent), colloidal algae cells and other small particles that can be removed only by filtration or centrifugation, and salts and gases (solutes) which are dissolved and physically inseparable from the solvent (water).

2.2.1 Solution Concentration Expressions

In order to describe chemical processes, it is necessary to quantitatively define the components of a solution. Solutions are often defined on the basis of mass (or weight—in this text, mass and weight are used interchangeably) or moles of solute per mass or moles of solution. Solutions also can be described based on mass or moles of solute per volume of solution, or volume of solute per volume of solution. Table 2.1 summarizes the conventionally used concentration terms and the types of solutions to which they usually apply.

In this chapter we use customary chemical nomenclature including the term "mole." A **mole** of an element or compound can be defined as the mass, in grams, of that substance that is numerically equal to its atomic or molecular mass. A more precise way to denote a mole, then, would be as a gram-mole or g-mole. Similarly, we can define a lb-mole as the mass, in pounds, of a substance that is equal to its molecular mass, a ton-mole as the mass in tons, and so forth. Obviously, a g-mole and a lb-mole refer to different amounts of a substance. In this chapter, and throughout this book, the term mole means a g-mole; if any other type of mole is desired, it will be clearly noted.

2.2.1.1 Mass and Mole Concentration Expressions. Frequently, the concentration of a solution is based upon the number of moles of solute present. A mole is an amount of matter which contains an Avogadro's Number (6.023×10^{23}) of entities (molecules or atoms in this case). Atomic numbers and weights of all the elements are provided inside the front cover of this text. By definition, one mole of carbon weighs exactly 12.0 grams and contains exactly 6.023×10^{23} atoms.

Table 2.1 Conventionally used concentration expressions.

Concentration Expressions	Solution Type
Mole fraction, n_i/n_T	Gas or liquid mixture, high concentrations
Percent by weight, $M_i/M_T \times 100$	Solid or aqueous mixture, high concentration
Parts per million or billion by weight (ppm or ppb), $M_i/10^6 M_T$ or $M_i/10^9 M_T$	Dilute aqueous solution
Molarity (moles/L solution)	Dilute aqueous solution
g/L or mg/L or μg/L, M_i/V_T	Dilute aqueous solution
μg/m^3, M_i/V_T	Gaseous solution
Normality (equivalents/L of solution)	Dilute aqueous solution
Parts per million or billion by volume (ppm or ppb), V_i/V_T	Gaseous solution

Note: Where n_i = number of moles, i = compound i, M = mass, V = volume, T = solution total

One measure of concentration based on the relative number of moles of each solution component is the **mole fraction**. The mole fraction is the ratio of the number of moles of any one component of a solution to the total number of moles present in the solution. This expression is often used for mixtures in which all components are present at high concentrations. For example, dry air can be approximated by a mixture of gases with a 0.21 oxygen mole fraction and 0.79 fraction for nitrogen. This means there are 21 moles of oxygen and 79 moles of nitrogen for every 100 moles of dry, clean air.

For aqueous solutions, mass per mass expressions can be based on the mass of solute present divided by the mass of solution (solute plus solvent). In this case, concentration is often expressed as a percent by weight. For more dilute solutions, mass per mass concentrations are expressed as parts of solute weight per million parts of solution weight (ppm), parts per billion (ppb), or even parts per trillion (ppt).

2.2.1.2   Mass/Volume Expressions. For dilute solutions, it is common to express concentration as molarity (moles/liter). **Molarity** (M) is defined as the number of moles of solute divided by the total volume of solution in liters. Alternatively, **concentration** can be expressed as the mass of solute per volume of solution in units of g/L, mg/L, or μg/L for aqueous solutions or $\mu g/m^3$ for gaseous solutions. Since in many cases the mass of contaminants is fairly low, one L of aqueous solution weighs approximately 1000 g, so:

$$1 \text{ mg/L} = 1 \text{ mg}/1000 \text{ g} = 1 \text{ g}/10^6 \text{ g} = 1 \text{ ppm (by weight), or}$$

$$1 \text{ } \mu\text{g/L} = 1 \text{ } \mu\text{g}/1000 \text{ g} = 1 \text{ g}/10^9 \text{ g} = 1 \text{ ppb (by weight)}$$

If the specific gravity of the solution is not equal to one, the conversion is as follows:

$$1 \text{ mg/L} = \text{ppm (by weight)} \times \text{specific gravity}$$

Example Problem 2.1

Calculate the NaCl concentration for a 500-mL solution containing 250 mg of NaCl in units of mg/L, M, and ppm by weight. The solution is sufficiently dilute to assume that the specific gravity is 1.0.

Solution

First, calculate the concentration in terms of mg/L:

$$\text{Concentration} = \frac{250 \text{ mg}}{0.5 \text{ L}}$$
$$\text{Concentration} = 500 \text{ mg/L}$$

(Note that 500 mg/L is the upper limit for acceptable salt content of drinking water.)

Next, calculate the number of moles that dissolved:

$$\text{M. W. of } NaCl = 23 + 35.5$$
$$= 58.5 \text{ g/mole}$$
$$\text{Number of Moles} = \frac{0.250 \text{ g}}{58.5 \text{ g/mole}}$$
$$= 4.27 \times 10^{-3} \text{ mole}$$

Next, calculate the molarity of the solution:

$$\text{Molarity} = 4.27 \times 10^{-3} \text{ moles/}0.5 \text{ L}$$
$$= 8.55 \times 10^{-3} \text{ moles/L}$$

Next, calculate the concentration in units of ppm:

$$500 \text{ mg/L} = 500 \text{ ppm by weight } \left(\text{assuming the specific gravity is 1.0}\right)$$

Another expression frequently used is **normality**, defined as the number of equivalents of solute per liter of solution. The number of equivalents is equal to the number of moles present multiplied by a small whole number, n. The value of n is determined by the charge of an ion, the number of available protons for acids, the number of available hydroxyl groups for bases, the number of electrons transferred in an oxidation-reduction reaction, or the charge of the positively charged ion (cation) in a molecule. The equivalent weight (EW) of a compound is its molecular weight (M.W.) divided by n.

The advantage of using normality is that it allows one to equate different masses of substances that have the same reacting capacity. In order for a solution to be electrically neutral, the number of positive equivalents present in a solution must be equal to the number of negative equivalents.

Example Problem 2.2

Drinking water supplies are often obtained from underground porous limestone aquifers. Limestone is primarily calcium carbonate $CaCO_3$, so the water contains calcium ions (which contributes to "hardness" in water—see chapter 6).

Determine the normality and the molarity of water containing 30 mg/L of Ca^{+2} (at this level the water is relatively "soft").

Solution

$$\text{The M. W. of } Ca^{+2} = 40 \text{ g/mole}$$

The Ca^{+2} ion has a charge of 2, therefore n = 2 equivalents/mole

$$\text{The EW of Ca}^{+2} = \text{M.W.}/n$$
$$= 40/2$$
$$= 20 \text{ g/equivalent}$$
$$\text{Concentration of equivalents} = (0.030 \text{ g/L}) / (20 \text{ g/equivalent})$$
$$= 0.0015 \text{ eq/L}$$
$$\text{Normality} = 0.0015 \text{ N}$$
$$\text{Molarity} = 0.030 \text{ g/L} / (40 \text{ g/mole})$$
$$= 0.00075 \text{ M}$$

(Note that the normality equals n times the molarity.)

2.2.1.3 *Volume Per Volume Expressions.*

Gaseous solutions are frequently expressed on the basis of the volume of the gas solute present divided by total gas solution volume. Volumetric concentration can be calculated by determining the volume that the gas solute would occupy if present by itself at the same temperature and pressure as the total solution. This expression is also frequently called ppm by volume, and describes the gas solute volume per million volumetric units of total gas solution. The term ppm by weight discussed in section 2.2.1.1 refers to mass of solute per million mass units of aqueous solution. The use of this concentration expression will be discussed further in section 2.7.1.

2.2.2 Mass Concentration Expressed as $CaCO_3$

Traditionally, the concentration of certain chemical constituents in water and wastewater have been expressed on the basis of calcium carbonate ($CaCO_3$). Expressing different chemicals as one common material makes it easy to add their concentrations to get their total effect. As an analogy, consider a wealthy person with bank accounts in several countries. In order to calculate his or her total wealth, you must express each currency in terms of a common unit (such as ounces of gold). Examples of constituents that are often quantified as $CaCO_3$ are "hardness" species (divalent metal cations such as Ca^{+2} and Mg^{+2}, discussed further in chapter 6) and alkalinity species (hydroxide (OH^-), carbonate (CO_3^{-2}), and bicarbonate (HCO_3^-), discussed further in section 2.5.5). Expressing these species in terms of a single component allows the individual species to be summed, indicating their equivalent total reacting capacity. Therefore, this expression is based on the number of equivalents of each species present. For any species X, to determine its concentration as $CaCO_3$, use the following equation:

$$\text{mg/L as } CaCO_3 = \frac{\text{mg/L as X}}{\text{EW of X}} \times \text{EW of } CaCO_3 \qquad (2.1)$$

Example Problem 2.3

Determine the total hardness of water containing 40 mg/L of calcium and 20 mg/L of magnesium. Report the answer in terms of mg/L as $CaCO_3$.

Solution

$$EW \text{ of } Ca^{+2} = M.W./2$$
$$= 40/2$$
$$= 20$$
$$EW \text{ of } Mg^{+2} = M.W./2$$
$$= 24.3/2$$
$$= 12.2$$
$$EW \text{ of } CaCO_3 = 50$$
$$Hardness = \left(40/20 + 20/12.2\right) \times 50$$
$$= 180 \text{ mg/L as } CaCO_3$$

(Note that this is a relatively hard water.)

2.3   Stoichiometry

Atoms tend to combine during chemical reactions in ratios of small integers (i.e., 2 moles of H_2 react with 1 mole of O_2 to yield 2 moles of H_2O). A chemical reaction always follows the law of conservation of mass which states that matter can neither be created nor destroyed (i.e., 40 g of H_2 plus 320 g of O_2 produce 360 g of H_2O). The elements or molecules that are reacting are called **reactants**; the elements or molecules formed are called **products**.

There are basically four types of reactions. **Metathesis** involves the rearranging of atoms into new molecules and often leads to the separation of one or more products due to a phase change (gas liberation or solid precipitation). An example of this type of reaction is:

$$H_2CO_3 \rightarrow CO_2\left(gas\right) + H_2O$$

Acid-base reactions involve the combination of protons and hydroxide ions to form water. An example of an acid-base reaction is:

$$NaOH + HCl \rightarrow NaCl + H_2O$$

Reduction-oxidation reactions involve the transfer of electrons from a reduced compound to an oxidizing agent. An example of this reaction is:

$$15\,O_2 + 2\,C_6H_6 \rightarrow 12\,CO_2 + 6\,H_2O$$

Finally, some reactions lead to **electron sharing** between two or more reactants. An example of electron sharing is:

$$NH_3 + H^+ \rightarrow NH_4^+$$

The quantitative relationship among the reactants or products is called the **reaction stoichiometry**. An understanding of the stoichiometric relationships among reacting atoms or molecules is essential to the control and design of environmental chemical processes.

2.3.1 Chemical Equations

Just as a sentence expresses ideas and thoughts in written form, the chemical equation describes the behavior of reactants and products. Conventionally, the equation is written with the reactants on the left and the products on the right. A chemical equation is of little use unless it is balanced. The following example illustrates the process required to balance an oxidation-reduction equation.

Example Problem 2.4

Balance the following oxidation-reduction equations:
(a) A simple equation:

$$H_2 + O_2 \rightarrow H_2O$$

(b) A more complicated equation:

$$Cl_2 + NH_3 \rightarrow N_2 + HCl$$

Solution

(a) The first equation can be balanced by inspection.

$$2\,H_2 + O_2 \rightarrow 2\,H_2O$$

(b) The second equation requires a step-wise approach for balancing. First, identify the half-reactions describing the oxidation reaction (reaction involving the loss of electrons) and reduction reaction (reaction involving the gain of electrons).

$Cl_2 \rightarrow Cl^-$ (reduction of zero-valent chlorine to chloride with an oxidation number of –1)

$NH_3 \rightarrow N_2$ (oxidation of –3 nitrogen to produce zero-valent nitrogen)

Next, balance these two reactions using water, hydrogen atoms (H^+), and electrons (e^-).

$$Cl_2 + 2\,e^- \rightarrow 2\,Cl^-$$

$$2\,NH_3 \rightarrow N_2 + 6\,H^+ + 6\,e^-$$

Next, multiply the equations by appropriate common factors so that when the half-reactions are combined the electrons cancel.

$$(Cl_2 + 2e^- \rightarrow 2Cl^-) \times 3$$

$$(2NH_3 \rightarrow N_2 + 6H^+ + 6e^-) \times 1$$

Combine the half-reactions and simplify.

$$3\,Cl_2 + 2\,NH_3 \rightarrow N_2 + 6\,HCl \text{ (final molar stoichiometry)}$$

3(71) 2(17)   2(14) 6(36.5)

213g 34g   28 g 219 g (final mass stoichiometry)

2.3.2 Molar and Mass Ratios

The balanced chemical equation provides the basis of a quantitative relationship among the reactants and products and allows one to calculate masses of reactants consumed and products formed. For example, in the balanced reaction of example problem 2.4, three moles of chlorine react with two moles of ammonia to produce one mole of nitrogen and six moles of hydrochloric acid. On a mass basis, stoichiometry tells us that 213 g of chlorine react with 34 g of ammonia to produce 28 g of nitrogen and 219 g of hydrochloric acid.

Example Problem 2.5

Using stoichiometric relationships developed in example problem 2.4, calculate the amount of ammonia in grams that will react with 100 g of chlorine to produce nitrogen and hydrochloric acid. Also calculate the amount, in grams, of nitrogen and hydrochloric acid produced.

Solution

The solution to the first question will be demonstrated using molar stoichiometry. First, calculate the number of moles of chlorine that are reacting:

$$\text{Number of moles of } Cl_2 \text{ reacting} = 100 \text{ g}/71 \text{ g/mole}$$

$$= 1.41 \text{ moles}$$

From the balanced chemical equation above, two moles of ammonia are consumed per three moles of chlorine reacting. Therefore,

$$\text{Number of moles of } NH_3 \text{ reacting} = 1.41 \text{ moles } Cl_2 \times 2 \text{ moles } NH_3/3 \text{ moles } Cl_2$$

$$= 0.94 \text{ moles}$$

Finally, calculate the weight in g of 0.94 moles of NH_3.

$$\text{Mass of } NH_3 = 0.94 \text{ moles} \times 17 \text{ g/mole}$$

$$= 16.0 \text{ g}$$

The second question will be answered using mass stoichiometry. To calculate the amount of N_2 and HCl produced, note that 28 g of N_2 and 219 g of HCl are produced for every 213 g of chlorine consumed. Therefore,

$$\text{Mass of } N_2 \text{ produced} = 100 \text{ g Cl}_2 \times 28 \text{ g N}_2/213 \text{ g Cl}_2$$

$$= 13.2 \text{ g N}_2$$

$$\text{Mass of HCl produced} = 100 \text{ g Cl}_2 \times 219 \text{ g HCl}/213 \text{ g Cl}_2$$

$$= 103 \text{ g HCl}$$

2.4   Equilibrium Concepts

From chemical experiments, it can be proved that many reactions are reversible; that is, under certain conditions the reactants will produce products, and under other conditions the products will produce reactants. When the reactants are producing products at the same rate as the products are producing reactants, **equilibrium** has been achieved. Equilibrium occurs when the total free energy of the system is at a minimum. The approach to equilibrium is a dynamic process during which small changes in temperature, pressure, or concentrations can cause a shift in the direction of the reaction. The change in free energy for any reaction (which can be calculated knowing the conditions of the reaction) indicates whether the reaction can be expected to proceed in the direction written. Reaction equilibrium occurs when there is no further change in free energy. A reversible reaction can be portrayed by the following general equation:

$$aA + bB \leftrightarrow cC + dD \tag{2.2}$$

2.4.1 Equilibrium Constants

Chemical equilibrium can be quantitatively described by an **equilibrium constant**. The constant is determined from the ratio of the equilibrium molar concentrations (for dilute solutions) of the products and reactants at the reaction temperature and pressure, and is specific to a particular reaction. The equilibrium constant can be calculated from thermodynamic properties of the reactants and products in their standard states. The value of the equilibrium constant is the ratio of the concentrations of products to reactants each raised to their respective stoichiometric number when the reaction is at equilibrium. For the chemical reaction of equation (2.2), the equilibrium constant is written as follows:

$$K = \frac{[C]^c [D]^d}{[A]^a [B]^b} \tag{2.3}$$

The equilibrium constant will have a different value if the temperature or pressure changes. A small K value suggests that only a small fraction of reactants has been converted to products at equilibrium, while a large K value indicates that the bulk of the reactants has been converted. The value of K, however, does not provide information regarding the rate at which the reaction approaches equilibrium. For example, the K for the reaction between hydrogen and oxygen to produce water suggests that water production is highly favored. However, hydrogen and oxygen can exist together in a reaction vessel for a very long time without producing water, but if a catalyst, such as finely divided platinum, is added, or if a spark is provided, the reaction will rapidly proceed to equilibrium. The presence of the catalyst or spark does not affect the value of K, only the rate at which equilibrium was reached.

The units of K are specific to the reaction. The concentration of dissolved solutes is expressed in units of moles/L, or mole fractions (for liquids), while for gases, units are expressed in moles/m^3 or partial pressures such as cm of mercury or atm. The concentration of a pure solid taking part in a reaction is assumed to remain constant, and is defined to be 1. A pure liquid taking part in the reaction is assumed to have a mole fraction of 1.

2.5 Acid-Base Chemistry

Acid-base chemistry is very important in environmental engineering. Acid rain (the presence of nitric and sulfuric acids dissolved in rainwater due to air pollution) lowers the pH of natural waters, and thus has wreaked havoc on aquatic populations throughout the world. The acid gases that cause acid rain often are removed at the industrial sources by caustic scrubbers, described in chapter 5.

Acid-base chemistry is an important application of equilibrium principles. An **acid** is defined simplistically as a compound that is capable of donating a hydrogen ion or proton (H^+) in an aqueous solution. A **base** is a compound that produces hydroxide ions (OH^-) or is capable of accepting a proton in an aqueous solution. Water is a unique compound in that it produces both a proton and hydroxide ion when it dissociates:

$$H_2O \leftrightarrow H^+ + OH^- \tag{2.4}$$

The equilibrium constant describing the dissociation of water can be defined as follows:

$$K = \frac{[H^+][OH^-]}{[H_2O]} \tag{2.5}$$

The concentration of water can be considered to remain constant at 55.5 M and can be incorporated into the equilibrium constant. The equilibrium constant then can be written as a water ionization constant:

$$K_w = [H^+][OH^-] \tag{2.6}$$

In this form, the water ionization constant, K_W, has a value of 1×10^{-14} at 25 °C.

2.5.1 The pH System

Because only a small amount of water dissociates, as indicated by the low value of K_W, the concentrations of the hydrogen and hydroxide ions are very small numbers. In fact, in neutral water, the $[H^+]$ and $[OH^-]$ concentrations are each equal to 10^{-7} moles/liter (at 25 °C). To simplify computations dealing with these ions, the pH system was developed. The pH of an aqueous solution is defined as follows:

$$pH = -\log [H^+] \tag{2.7}$$

Similarly,

$$pOH = -\log [OH^-] \tag{2.8}$$

From the definition of K_W, it can be seen that:

$$[H^+][OH^-] = 10^{-14} \text{ (at 25 °C), and thus}$$

$$pH + pOH = 14$$

This relationship holds no matter what else is in the solution. When the concentrations of hydrogen and hydroxide ions are equal, the pH and pOH are both 7 (at 25 °C). When the solution is acidic, the concentration of hydrogen ions is greater than the concentration of hydroxide ions and the pH is less than 7 (but pH + pOH still equals 14).When the solution is basic, the concentration of hydrogen ions is less than the concentration of hydroxide ions and the pH is greater than 7. The pH values of some common substances are illustrated in Figure 2.2. The importance of pH in controlling certain metal concentrations in water will be illustrated in a subsequent example problem.

2.5.2 Weak and Strong Acids and Bases

When placed in aqueous solution, acids and bases tend to dissociate to different degrees, depending on their chemical composition. Typical reactions of an acid and base are shown below:

$$HA \leftrightarrow H^+ + A^- \tag{2.9}$$

$$BOH \leftrightarrow B^+ + OH^- \tag{2.10}$$

The tendency to dissociate is referred to as the strength of the acid or base. A strong acid dissociates nearly completely, and the equilibrium position is far to the right in equation (2.9). In other words, the concentration of the ion A^- is much greater than the concentration of the un-ionized acid, HA. In equation (2.10), the equilibrium position is far to the right for a strong base, which tells us that the concentration of the ion B^+ is much greater than BOH. A weak acid tends to remain largely un-ionized, with the equilibrium position

pH Scale

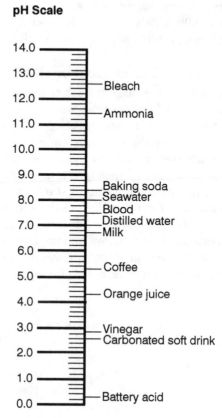

Figure 2.2
Values of pH of some common liquids.

far to the left in equation (2.9); that is, the concentration of HA exceeds that of A⁻. In equation (2.10) for a weak base, the equilibrium position is far to the left—the concentration of BOH far exceeds that of B⁺. The strength of the acid or base is determined by the magnitude of its equilibrium constant, defined in equation (2.11) for an acid.

$$K_A = \frac{[H^+][A^-]}{[HA]} \tag{2.11}$$

For a weak acid the value of K_A is quite small, so for convenience the p-operator system is often used for equilibrium constants as follows:

$$pK_A = -\log K_A \tag{2.12}$$

Typically, the pK for a weak acid or base is a positive number; for example, the K_A for acetic acid is 2.5×10^{-5} (at 25 °C) and the pK_A is 4.6. Although rarely used, a pK can be determined for a strong acid or base; in this case K will be a large number and pK will be negative. For example, the K_A for hydrochloric acid is 1000 at 25 °C and the pK_A is −3. Appendix B provides examples of commonly encountered acid and base equilibrium constants.

2.5.3 The pH of an Acid or Base

When an acid is placed in a solution of pure water, the pH will decline. Conversely, the pH will increase when a base is placed in solution. The pH is a function of both the concentration of the acid or base after placing it in solution and the equilibrium constant for the weak acid or base. At equilibrium, the concentrations of the various species involved (for example, H^+, OH^-, A^-, and HA, for a typical monoprotic acid solution) are controlled by a set of four equations: the equilibrium constant for acid or base dissociation, the ionization constant for water, the mole balance for the chemical species of interest, and the electroneutrality balance. An electroneutrality balance sums the equivalents of charged ions and requires that the equivalents of positive ions be equal to the sum of the equivalents of the negative ions. For example, for a typical monoprotic acid, HA, the equations can be written as follows.

$$K_A = \frac{[H^+][A^-]}{[HA]} \qquad (2.11)$$

$$K_w = [H^+][OH^-] \qquad (2.6)$$

$$C_{T,A} = [HA] + [A^-] \qquad (2.13)$$

$$1[H^+] = 1[OH^-] + 1[A^-] \qquad (2.14)$$

These four equations can be manipulated to solve problems related to acid and base equilibrium. Often, however, simplifying assumptions can be made, as illustrated in the following example.

Example Problem 2.6

What is the pH of a 0.01 M solution of hydrochloric acid (HCl)? HCl is a strong acid.

Solution

From equation (2.13), a mole balance for chloride can be written:

$$C_{T,Cl} = [HCl] + [Cl^-]$$

Since HCl is a strong acid, $[Cl^-] >>>> [HCl]$; $[HCl]$ can be neglected, and:

$$C_{T,Cl} \cong [Cl^-]$$

$$C_{T,Cl} \cong 0.01\,M$$

From the charge balance in equation (2.14):

$$1[H^+] = 1[OH^-] + 1[Cl^-]$$

Combining equations (2.6) and (2.14), $[OH^-]$ can be eliminated:

$$[H^+] = K_w/[H^+] + [Cl^-]$$

Rearranging and substituting values for K_w and chloride produces the following quadratic equation (assuming 25 °C) :

$$[H^+]^2 - 0.01[H^+] - 10^{-14} = 0$$

Solving:

$$[H^+] = 0.01\,M$$

Therefore,

$$pH = ^-\log[H^+]$$

$$pH = ^-\log(0.01)$$

$$pH = 2$$

(Note that for a concentrated solution of a strong acid, the concentration of H^+ is equal to the initial concentration of HCl, but this is not true for a weak acid, as shown in the following example.)

Example Problem 2.7

What is the pH of a 0.01 M solution of acetic acid (HAc)? HAc is a weak acid with a pK_A of 4.6.

Solution

From equation (2.13), a mole balance for acetate can be written:

$$C_{T,Ac} = [HAc] + [Ac^-]$$

$$C_T - [Ac^-] = [HAc]$$

From the charge balance:

$$1[H^+] = 1[OH^-] + 1[Ac^-]$$

and,

$$[Ac^-] = [H^+] - [OH^-]$$

Substituting for [Ac-] and [HAc] in equation (2.11):

$$K_A = \frac{[H^+][Ac^-]}{[HAc]}$$

$$K_A = \frac{[H^+]\left([H^+] - [OH^-]\right)}{[C_T] - \left([H^+] - [OH^-]\right)}$$

Substituting $K_W/[H^+]$ for [OH-]:

$$K_A = \frac{[H^+]\left([H^+] - \dfrac{K_w}{[H^+]}\right)}{C_T - \left([H^+] - \dfrac{K_w}{[H^+]}\right)}$$

This expression expands to the following equation:

$$[H^+]^3 + \left(K_A - K_w\right)[H^+]^2 - K_A\, C_T[H^+] + K_A\, K_w = 0$$

which is tedious to solve without using numerical analysis techniques described in chapter 3. This problem can be solved in a more simplistic manner by assuming that the hydroxide ion concentration is negligible in comparison to the hydrogen ion concentration. Let X represent the number of moles/L of the acetic acid which ionize at equilibrium. Stoichiometrically, it can be seen that for every mole of HAc that ionizes, one mole of H^+ and one mole of Ac- are produced. Then:

$$[HAc] = C_T - X$$
$$[H^+] = X$$
$$[Ac^-] = X$$

Substituting into equation (2.11):

$$K_A = \frac{X^2}{C_T - X}$$

X can be solved using the quadratic equation:

$$X^2 + K_A X - C_T K_A = 0$$

$$X = \frac{-K_A \pm \sqrt{K_a^2 + 4C_T K_A}}{2}$$

Substituting:

$$X = 4.89 \times 10^{-4}$$

or since $[H^+] = x$

$$pH = 3.3$$

Frequently, X is far less than C_T and can be ignored in the denominator, then the expression simplifies to:

$$K_A = \frac{X^2}{C_T}$$

Then:

$$X = \sqrt{K_A C_T}$$

Since $[H^+] = X$, converting to the logarithmic system:

$$pH = 1/2(pK_A + pC_T) \qquad (2.15)$$

Substituting into equation (2.15), pH is determined to be 3.3 as above. The simplifying assumptions must be checked. C_T of 0.01 M is much greater than the concentration of H^+ (4.89 x 10^{-4} M), and at a pH of 3.3, the hydroxide ion concentration is negligible compared to the hydrogen ion concentration.

Frequently, an industrial waste will be encountered that has an extremely high or low pH and must be neutralized to a more acceptable pH. The equivalence point is reached when an equal number of acid and base equivalents are present. The amount of acid or base required to reach the equivalence point is determined through a procedure called a **titration** (illustrated in Figure 2.3 on the following page). When combining strong acids and bases, the pH at the equivalence point is equal to that of a neutral solution, or 7.0. The pH of a solution at the equivalence point for a strong acid or base added to a weak acid or base is generally not 7.0, as illustrated in example problem 2.8.

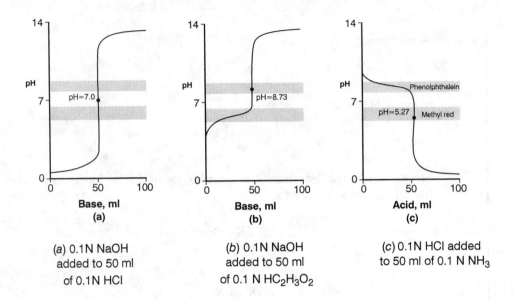

(a) 0.1N NaOH added to 50 ml of 0.1N HCl

(b) 0.1N NaOH added to 50 ml of 0.1 N HC$_2$H$_3$O$_2$

(c) 0.1N HCl added to 50 ml of 0.1 N NH$_3$

Figure 2.3
Examples of titration curves.

Example Problem 2.8

What volume of 0.1 N sodium hydroxide (NaOH, a strong base) is required to reach the equivalence point with 1000 L of 0.005 N acetic acid? What will the pH be after adding the acid?

Solution

By definition, the equivalence point is reached when the number of acid and base equivalents is equal:

$$V_A N_A = V_B N_B \tag{2.16}$$

where:

V_A = volume of acid, L
N_A = normality of acid, eq/L
V_B = volume of base, L
N_B = normality of base, eq/L

Therefore, the volume of base required to neutralize the acid is:

$V_B = V_A N_A / N_B$, and
$V_B = 1000(0.005)/(0.1)$
$V_B = 50$ L

To calculate the pH at neutrality, let's begin by looking at the mole balance on acetate.

$$C_{T,Ac} = [HAc] + [Ac^-]$$

and

$$C_{T,Ac} - [Ac^-] = [HAc]$$

A charge balance can be written for this solution as well:

$$1[Na^+] + 1[H^+] = 1[Ac^-] + 1[OH^-]$$

Since the number of equivalents of the base present in solution is equal to the number of equivalents of acid, the normality of the base (N_B) and acid (N_A) are equal at the equivalence point.

$$N_B = N_A$$
$$N_B = n_B C_B$$
$$N_A = n_A C_A$$
$$n_B = n_A = 1$$

Therefore:

$$C_B = C_A = C_{T_{Ac}}$$

Since NaOH is a strong base which completely ionizes ($[Na^+] >>>> [NaOH]$), the following can be stated:

$$C_B = [Na^+], \text{ and}$$
$$C_B = C_T, \text{ therefore}$$
$$C_T = [Na^+]$$

After adding 50 L to 1000 L, C_T has changed:

$$C_{T_{Ac}} = \frac{V_A C_{T_i}}{V_A + V_B}$$

where:

C_{T_i} = initial concentration of acid

$C_{T_{Ac}} = 4.8 \times 10^{-3} \text{ M}$

Substituting into the charge balance:

$$C_{T_{Ac}} + [H^+] = [Ac^-] + [OH^-]$$
$$0.0048 + [H^+] = [Ac^-] + [OH^-]$$

From equations (2.6), (2.11), (2.13), and (2.14) we find the following:

$$10^{-4.6} = \frac{[H^+][Ac^-]}{[HAc]}$$

$$10^{-14} = [H^+][OH^-]$$

$$0.0048 = [HAc] + [Ac^-]$$

We now have a system of four equations and four unknowns that can be solved using the numerical analysis techniques described in chapter 3 to yield a pH of 8.14.

A more convenient expression for the pH at the equivalence point can be developed by making several simplifying assumptions: (1) upon neutralizing a weak acid, the pH is usually sufficiently greater than 7 (usually pH > 8 satisfies this) to assume that [OH-] >>> [H$^+$] and (2) the hydrogen ion concentration is sufficiently small to be neglected relative to [Na$^+$] (shown previously to be equal to C_T) in the charge balance equation. The following can be written:

$$C_{T_{Ac}} = [Ac^-] + [OH^-], \text{ and rearranging}$$

$$C_{T_{Ac}} - [Ac^-] = [OH^-]$$

From the mass balance:

$$C_{T_{Ac}} = [Ac^-] + [HAc], \text{ and rearranging}$$

$$C_{T_{Ac}} - [Ac^-] = [HAc], \text{ therefore:}$$

$$[HAc] \approx [OH^-]$$

Inserting into equation (2.11):

$$K_A = \frac{[H^+][Ac^-]}{[HAc]}$$

$$K_A = \frac{[H^+][Ac^-]}{[OH^-]}$$

At the equivalence point, practically all of the HAc has been converted to acetate (that is, the concentration of HAc can be assumed to be a very low number), therefore [HAc] can be neglected relative to [Ac-], and from the mole balance:

$$C_T \approx [Ac^-]$$

and

$$[OH^-] = K_w / [H^+]$$

therefore substituting into equation (2.11):

$$K_A = \frac{[H^+]^2 C_T}{K_w}$$

Solving for [H$^+$] and transforming to pH:

$$pH = 1/2 \left(\log C_T + pK_A + pK_w \right) \tag{2.17}$$

$$pH = 1/2 \left(\log(0.0048) + 4.6 + 14 \right)$$

$$pH = 8.14$$

This pH is quite close (within 0.01 pH units) to the more exact solution obtained using numerical techniques (see chapter 3). Assumptions must always be checked to ensure that appropriate conditions are met. At a pH of 8.14, the concentration of Ac^- is 720,000 times the concentration of HAc, therefore HAc can be neglected. Also, the concentration of OH^- is 190 times that of H^+, therefore hydrogen can be neglected, and the concentration of Na^+ is 660,000 times that of H^+, therefore hydrogen can be neglected relative to sodium as well.

2.5.4 Buffers

Many solutions exist in nature that are capable of withstanding the addition of strong acids and bases with little change in pH. These solutions are called **buffers**. They are generally combinations of weak acids and their salts. For example, a combination of sodium bicarbonate ($NaHCO_3$) and sodium carbonate (Na_2CO_3) will form a buffered solution with a pH near the pK_A of bicarbonate, 6.35. A buffer functions because the salt (carbonate) is able to absorb any new acids (hydrogen ions) that may be added, and the acid (bicarbonate) is able to absorb any new bases (hydroxide ions) that may be added so that the resulting equilibrium concentrations of H^+ and OH^- do not change dramatically. The ability of a buffered solution is not infinite, however, and works best within $\pm$ 1 pH unit of the system pK_A.

2.5.5 The Carbonate System

One of the most important buffering systems in nature is the carbonate system, composed of carbon dioxide (CO_2), carbonic acid (H_2CO_3), and bicarbonate (HCO_3^-) and carbonate (CO_3^{2-}) ions. The aqueous carbonate system develops from both atmospheric carbon dioxide and the many solid carbonate species on the earth. Carbon dioxide is involved in both biological respiration (where it is produced) and photosynthesis (where it is consumed). The precipitation of calcium carbonate ($CaCO_3$) in pipes and process tanks can lead to clogging and loss of reactor volume. The carbonate system helps stabilize pH in many lakes and rivers to provide a constant pH for aquatic life. Natural waters are in equilibrium with these carbonate sinks (see Figure 2.4) as indicated by the following equilibrium equations (all values are for 25 °C).

The dissolution of carbon dioxide in water:

$$CO_2(gas) \leftrightarrow CO_2(aq) \tag{2.18}$$

$$K = \frac{[CO_2]_{(aq)}}{[CO_2]_{(gas)}} = 10^{-1.47} \tag{2.19}$$

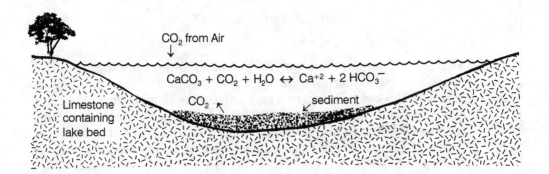

Figure 2.4
Calcium carbonate/carbon dioxide equilibria in a lake.

The hydrolysis of carbon dioxide to form carbonic acid:

$$CO_2(aq) + H_2O \leftrightarrow H_2CO_3 \tag{2.20}$$

$$K = \frac{[H_2CO_3]}{[CO_2]_{(aq)}[H_2O]} = 10^{-2.80} \tag{2.21}$$

The first dissociation step of carbonic acid to form bicarbonate:

$$H_2CO_3 \leftrightarrow H^+ + HCO_3^- \tag{2.22}$$

$$K_1 = \frac{[H^+][HCO_3^-]}{[H_2CO_3]} = 10^{-6.35} \tag{2.23}$$

The second dissociation step to form carbonate:

$$HCO_3^- \leftrightarrow H^+ + CO_3^{2-} \tag{2.24}$$

$$K_2 = \frac{[H^+][CO_3^-]}{[HCO_3^-]} = 10^{-10.33} \tag{2.25}$$

The reaction between calcium and carbonate to form sparingly soluble calcite (calcium carbonate):

$$CaCO_3 \,(precipitated) \leftrightarrow Ca^{2+} + CO_3^{2-} \tag{2.26}$$

$$K = \frac{[Ca^{2+}][CO_3^{2-}]}{[CaCO_3]_s} \tag{2.27a}$$

Since $[CaCO_3] = 1$:

$$K_{SP} = [Ca^{2+}][CO_3^{2-}] = 10^{-8.34} \tag{2.27b}$$

where: K_{SP} is the solubility product, discussed in section 2.6.

The ability of a natural water to withstand pH changes is measured by its acidity or alkalinity. **Acidity** is defined as the capacity to neutralize a strong base and is calculated as follows:

$$\text{Acidity (eq/L)} = 2[H_2CO_3] + [HCO_3^-] + [H^+] - [OH^-] \tag{2.28}$$

Recall that the brackets [ ] symbolize the units of moles/L.

Alkalinity is defined as the capacity to neutralize a strong acid and is calculated as follows:

$$\text{Alkalinity (eq/L)} = [HCO_3^-] + 2[CO_3^{2-}] + [OH^-] - [H^+] \tag{2.29}$$

In the field of environmental chemistry, acidity and alkalinity values are conventionally expressed in units of mg/L as $CaCO_3$ using the following formula:

$$\text{mg/L as } CaCO_3 = \left(\text{eq/L of acidity or alkalinity} \times 50\,g\,CaCO_3\,/\,eq\right) \times \left(1000\,mg/g\right) \tag{2.30}$$

The alkalinity of most natural waters is associated primarily with bicarbonate. The presence of alkalinity in natural waters allows the pH to remain fairly constant and near neutral. A neutral pH is important because many aquatic species have stringent pH requirements for survival. Acid rain depresses the pH of surface waters. Figure 2.5 provides the pH of precipitation in the continental United States. The presence of the natural carbonate buffer system in the form of carbonic acid and bicarbonate (a weak acid and its salt) maintains the pH above 6.0 unless the acid dose exceeds the buffering capacity. Clearly, in many cases, the alkalinity of lakes and streams has been insufficient to withstand the addition of acid rain because pH has been measured well below 5.0 in many surface waters.

Equations (2.18) through (2.27b) demonstrate the interrelationship among the carbonate species and pH. Given the pH and the alkalinity, the species concentrations can be calculated, as shown in example problem 2.9.

Example Problem 2.9

A 1-m deep lake with a surface area of 4,000 m² has an initial pH of 6.5 buffered by the carbonate system (alkalinity 20 mg/L as $CaCO_3$). Calculate the pH of the lake after receiving 5 cm of acid rain with a nitric acid (a strong acid) concentration of 2×10^{-4} M.

Solution

First calculate the increase in the concentration of H^+ in the lake due to the addition of the acid rain:

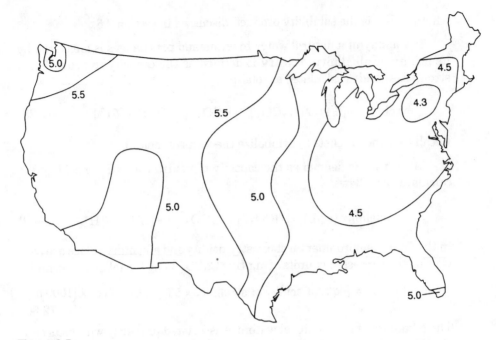

Figure 2.5
Acidity (as pH) of rainfall in the United States in 1996. (For more information, see web site at: http://nadp.nrel.colostate.edu/nadp/isopeths.maps)

$$0.05 \text{ m rain x } 4000 \text{ m}^2 \text{ surface area} = 200 \text{ m}^3$$

$$= 200,000 \text{ L of acid rain}$$

$$200,000 \text{ L x } 0.0002 \text{ moles/L HNO}_3 = 40 \text{ moles H}^+ \text{ added}$$

$$\text{The final lake volume} = 4000 \text{ m}^2 \text{ x } 1\text{m} + 200,000\text{L} = 4200 \text{ m}^3$$

$$= 4,200,000 \text{ L}$$

Thus the concentration of H^+ *added* is:

$$[H^+] = 40 \text{ moles}/4,200,000 \text{ L} = 9.5 \text{ x } 10^{-6} \text{ M}$$

Because the pH is at 6.5, the concentration of carbonate is negligible and the buffering is provided by bicarbonate and carbonic acid. Now, calculate the concentration of bicarbonate and carbonic acid before the addition of the acid rain. Using equation (2.23) and substituting for $[H^+] = 10^{-6.5}$ M:

$$K_1 = \frac{(10^{-6.5})[HCO_3^-]}{[H_2CO_3]} = 10^{-6.35}$$

$$[H_2CO_3] = 0.71[HCO_3^-]$$

$$\text{Alkalinity} = 20 \text{ mg/L as CaCO}_3$$

Converting to eq/L (using equation 2.30):

$$(0.020 \text{ g/L}) / (50 \text{ g/eq}) = 4 \times 10^{-4} \text{ eq/L}$$

And, from equation (2.29):

$$4 \times 10^{-4} \text{ eq/L} = [HCO_3^-] + 2[CO_3^{2-}] + [OH^-] - [H^+]$$

Because the pH is low (6.5), we can neglect both carbonate and hydroxide concentrations.

$$[HCO_3^-] = 4 \times 10^{-4} \text{ M}$$

and

$$[H_2CO_3] = 0.71 \times 4 \times 10^{-4}$$
$$= 2.8 \times 10^{-4} \text{ M}$$

With the addition of 9.5×10^{-6} M H^+ in the acid rain, 9.5×10^{-6} M of bicarbonate is converted to carbonic acid with the following resulting concentrations:

$$[HCO_3^-] = 4 \times 10^{-4} - 9.5 \times 10^{-6}$$

and

$$= 3.9 \times 10^{-4}$$

$$[H_2CO_3] = 2.8 \times 10^{-4} + 9.5 \times 10^{-6}$$
$$= 2.9 \times 10^{-4}$$

From equation (2.23) the pH can be determined:

$$K_1 = \frac{[H^+](3.9 \times 10^{-4})}{(2.9 \times 10^{-4})} = 10^{-6.35}$$

$$[H^+] = 3.32 \times 10^{-7} \text{ M}$$

$$\text{pH} = 6.48$$

(Note that because of the effective buffer system present, the pH depression was limited to 0.02 units.)

2.6   Solubility Product

Another example of the application of the equilibrium concept is the solubility of solids. The **solubility product** constant is an equilibrium constant that describes the dissolution of a solid into ions in aqueous solutions, and is widely used in designing methods to remove toxic metal ions dissolved in water. For the general case:

$$A_a B_b \text{ (s)} \leftrightarrow aA^{b+} + bB^{a-} \qquad (2.31)$$

the equilibrium constant is:

$$K = \frac{[A^{b+}]^a\,[B^{a-}]^b}{[A_aB_b]_s}$$ (2.32)

Since the concentration of $A_aB_b(s)$ is 1 M, by definition, the solubility product constant can be written as follows (for a system at equilibrium):

$$K_{SP} = [A^{b+}]^a\,[B^{a-}]^b$$ (2.33)

Solubility product constants are presented in appendix B for commonly encountered compounds.

A solution can be described as being saturated, unsaturated, or supersaturated. A **supersaturated** solution can be created by dissolving a solid at an elevated temperature and then allowing it to cool. Once cooled, precipitation may not occur, although the solution is not at equilibrium. Precipitation will occur if the reaction vessel is shaken or otherwise disturbed. An **unsaturated** solution is not at equilibrium and can dissolve more solid. Only a **saturated** solution, which cannot dissolve more solid unless the temperature or pressure is changed, is at equilibrium. The equilibrium state of any solution can be determined by comparing the product of the ion concentrations raised to their appropriate stoichiometric coefficients (or ion product, IP) to the solubility product constant. If the IP is greater than the K_{SP} then the solution is supersaturated, if the IP is less than the K_{SP} then the solution is unsaturated, and if the IP is equal to the K_{SP} the solution is at equilibrium and is saturated.

The term "solubility" is not the same as the solubility product, although the two are related. **Solubility** is the amount of solid that will dissolve. In most cases the solubility will increase with increasing temperature but this is not always the case. For example, calcium carbonate solubility decreases as temperature increases, which can create problems in water heaters and boilers. The solubility also is affected by the presence of other ions in solution which may react with one of the dissolving products. The solubility is particularly affected by the addition of an ion common to one of the product ions.

Example Problem 2.10

Calculate the solubility of barium sulfate, $BaSO_4$, in pure water at 25 °C in mole/L. What is the equilibrium concentration of barium (Ba^{+2}) in water that initially contains 10^{-3} M sulfate, SO_4^{2-}? At 25 °C, the K_{SP} for $BaSO_4$ is 1.0×10^{-10}.

Solution

Let x equal the number of moles/L of $BaSO_4$ which will dissolve in pure water as follows:

$$BaSO_4(s) \leftrightarrow Ba^{2+} + SO_4{}^{2-}$$

If x moles of $BaSO_4$ dissolve, then x moles of Ba^{2+} and x moles of $SO_4{}^{2-}$ will enter the solution.

$$K_{SP} = 1.0 \times 10^{-10}$$
$$= \left[Ba^{2+}\right]\left[SO_4{}^{2-}\right]$$
$$= x^2$$

and

$$x = 1.0 \times 10^{-5} \text{ moles/L}$$

If the water initially contains 10^{-3} M $SO_4{}^{2-}$, after x moles/L of $BaSO_4$ dissolve, the solution will contain $(x + 10^{-3})$ M $SO_4{}^{2-}$ and x moles/L of Ba^{2+}. Then:

$$K_{SP} = 1.0 \times 10^{-10}$$
$$= \left[Ba^{2+}\right]\left[SO_4{}^{2-}\right]$$
$$= x(x + 10^{-3})$$

and using the quadratic formula:

$$x = \frac{-10^{-3} \pm \sqrt{10^{-6} + 4\left(1.0 \times 10^{-10}\right)}}{2}$$

and

$$x = 1.0 \times 10^{-7} \text{ M}$$

In the presence of 0.001 M of the common ion, sulfate, the concentration of the heavy metal, barium, is two orders of magnitude lower than in pure water.

For many compounds the K_{SP} is a very small number, indicating that the solid is only sparingly soluble. The concentration of many heavy metals can be controlled by precipitation with ions such as sulfide, hydroxide, or carbonate because their precipitates (solid forms) have extremely small K_{SP} values (see Figure 2.6). Note, however, in Figure 2.6, that at extremely high pH, solubility of these metals can actually increase due to the fact that soluble charged metal hydroxides complexes can form such as $Fe(OH)_4{}^-$. The addition of lime, dolomite ($CaCO_3$), or iron sulfide can reduce the concentration of many heavy metals to environmentally acceptable ranges as shown in example problem 2.11.

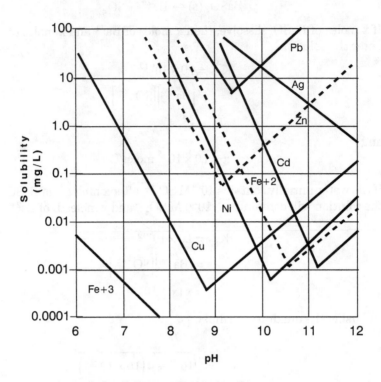

Figure 2.6
Solubility of some heavy metal hydroxides as a function of pH. (*U.S. EPA, 1983*)

Example Problem 2.11

(a) Calculate the concentration of cadmium (a toxic heavy metal) as the pH of a solution decreases from 11 to 10 to 9. Assume that the solubility of cadmium is controlled only by hydroxide. The K_{SP} for cadmium hydroxide, $Cd(OH)_2$, at 25 °C is 2.0×10^{-14}. (b) The national groundwater drinking water standard for cadmium is 0.005 mg/L. Calculate the minimum hydroxide concentration required to meet the groundwater standard.

Solution

(a) From the K_{SP} and the concentration of OH^-, the concentration of Cd^{2+} can be calculated.

$$Cd(OH)_2 \ (s) \leftrightarrow Cd^{2+} + 2\,OH^-$$

$$K_{SP} = [Cd^{2+}][OH^-]^2$$

At pH = 11, the concentration of OH^- is 10^{-3} M, so the concentration of Cd^{2+}

is:

$$[Cd^{2+}] = 2.0 \times 10^{-14} / (10^{-3})^2 = 2.0 \times 10^{-8}\,M\ (pH = 11)$$

At pH = 10, the concentration of OH^- is 10^{-4} M, so the concentration of Cd^{2+} is:

$$[Cd^{2+}] = 2.0 \times 10^{-14} / (10^{-4})^2 = 2.0 \times 10^{-6}\,M\ (pH = 10)$$

At pH = 9, the concentration of OH^- is 10^{-5} M, so the concentration of Cd^{2+} is:

$$[Cd^{2+}] = 2.0 \times 10^{-14} / (10^{-5})^2 = 2.0 \times 10^{-4}\,M\ (pH = 9)$$

As you can see, as the pH decreases (as the hydroxide concentration gets lower) the cadmium concentration gets much higher.

(b) In order to calculate the concentration of hydroxide when cadmium is 0.005 mg/L, first the concentration must be expressed in units of moles/L.

$$[Cd^{2+}] = (5 \times 10^{-6}\ g/L) / (112.4\ g/mole)$$

$$= 4.45 \times 10^{-8}\ M$$

From the K_{SP}, the minimum hydroxide concentration can be calculated:

$$[OH^-] = \left[\left(2.0 \times 10^{-14} \right) / \left(4.45 \times 10^{-8} \right) \right]^{1/2}$$

$$= 6.71 \times 10^{-4}$$

(Note that this corresponds to a pH of 10.8. So, to control Cd in water, we would need to add a base, like NaOH, to raise the pH to at least 10.8.)

2.7   Gas Laws

Gases take part in many environmental processes. For example, the transfer of oxygen into water is essential to support aquatic life and the biological degradation of organic compounds. Stripping of nuisance gases such as ammonia, carbon dioxide, or hydrogen sulfide involves the transfer of these gases from water to the atmosphere. The design and operation of processes involving gas transfer require knowledge about gas behavior in mixtures and solutions.

2.7.1 Ideal Gas Law

The Ideal Gas Law states that the product of the pressure, P, and the volume, V, of a given quantity of an ideal gas is directly proportional to the temperature, T. The Ideal Gas Law can be expressed more universally as shown below:

$$PV = nRT \tag{2.34}$$

where:

n = number of moles of gas present

R = the universal gas constant

The units of the universal gas constant, R, are a function of the units of the pressure, volume, and temperature terms. Common values of R are provided in appendix B. Although the Ideal Gas Law applies only to ideal gases, most gases behave ideally at low pressures and ambient temperatures.

The Ideal Gas Law can be used to determine the volume of one mole of gas at standard temperature and pressure (STP: 25 °C, 1 atm) as follows:

$V/n = RT/P$

$V/n = (0.08205 \text{ L-atm/mol-K}) (25 + 273 \text{ °K})/(1 \text{ atm})$

$= 24.45 \text{ L/mole}$

It must be kept in mind that the volume of one mole of gas will vary with the temperature and pressure of the gas. The Ideal Gas Law also can be used to convert between two popularly used expressions of gas concentrations, ppm (by volume) and $\mu g/m^3$. An example of this conversion using the previous calculation is shown in equation (2.35) for a gas at 25 °C, 1 atm.

$$\mu g / m^3 = \frac{\text{ppm} \times \text{M.W.} \times 1000}{24.45} \qquad (2.35)$$

For other temperatures and pressures, equation (2.35) must be recalculated.

2.7.2 Dalton's Law

Dalton's Law of Partial Pressure can be expressed as follows:

In a mixture of gases, each gas exerts pressure independently of other gases present. The partial pressure exerted by each gas is proportional to the amount (by volume) of gas present.

Dalton's Law can be expressed mathematically by the following equations.

$$P_T = P_1 + P_2 + P_3 + \cdots P_i = \sum P_i \quad \text{or} \quad V_T = \sum V_i \qquad (2.36)$$

where:

P_i = the partial pressure exerted by gas i (if gas i were the only gas present in the total volume V_T)

V_i = the partial volume occupied by gas i (at the total pressure P_T)

And, from the Ideal Gas Law:

$$P_i = n_i RT / V_T \quad \text{or} \quad V_i = n_i RT / P_T$$

$$P_T = n_T RT / V_T \quad \text{or} \quad V_T = n_T RT / P_T$$

$$\frac{P_i}{P_T} = \frac{n_i}{n_T} \quad \text{or} \quad \frac{V_i}{V_T} = \frac{n_i}{n_T} \tag{2.37}$$

where:

n_i = number of moles of gas i

n_T = total number of moles of gas

2.7.3 Raoult's Law

A volatile liquid or mixture of liquids may be in equilibrium with the gas phase above it. The partial pressure of the component in the gas phase will be directly proportional to the mole fraction of the component in the liquid mixture and to the volatility of the component as measured by its vapor pressure. Raoult's Law may be expressed by the following equation:

$$P_i = X_i P_V \tag{2.38}$$

where:

P_i = partial pressure of component i in the gas phase

X_i = mole fraction of i in the liquid mixture

P_V = vapor pressure of component i

The vapor pressures of many liquids are commonly reported in chemistry and physics handbooks, and are strongly dependent on temperature. Raoult's Law has many applications in environmental chemistry. For example, Raoult's Law can be used to predict the vapor phase concentration of components of gasoline spilled into the subsurface when the gasoline is present as a floating pool (nonaqueous phase liquid) on the water table.

2.7.4 Henry's Law

Often, environmental water pollutants are present in very dilute solutions and Raoult's Law is not applicable. For dilute solutions, a variation on Raoult's Law called Henry's Law can be applied. Henry's Law states that, under equilibrium conditions, the concentration of a gaseous solute dissolved in a liquid is proportional to the solute's concentration in the gas phase that is in contact with the liquid. The proportionality constant is called Henry's Constant and takes on many different units, depending on the units of the gas and liquid concentration terms. For example, in the following equation Henry's Constant has units of moles/L-atm because the concentration of the component of interest in the liquid is expressed in moles/L and in the gas, in atm.

$$C_{aq} = K_H P_g \tag{2.39}$$

Alternatively, if the concentration of the gas in both phases is expressed in units of mole/L or mg/L, then Henry's Constant is unitless. Appendix B provides Henry's Constants for several environmentally important gases expressed in different units. Keep in mind that Henry's Constant is also dependent on temperature since it is actually an equilibrium constant. Henry's Law has important environmental application to situations such as determining the solubility of oxygen in water, the partitioning of hydrogen sulfide between wastewater and the atmosphere, the stripping of volatile organic compounds from groundwater using a stream of air, and others.

Example Problem 2.12

A drinking water must be treated to control taste and odor due to the presence of 6.4 mg/L of H_2S. It is proposed to remove the H_2S from the water by transferring it to an air stream in a stripping tower. Within the stripping tower, water flows downward at 40 million gallons/day (MGD) and air flows upward at 120,000 standard cubic feet per min (scf/min). The temperature is 25 °C and the pressure is 1 atm. First, calculate the air concentration of H_2S in $\mu g/m^3$ and ppm assuming that the H_2S is completely removed from the water. Next, calculate the equilibrium concentration of H_2S in the air assuming the water concentration remains at 6.4 mg/L.

Solution

First, calculate the H_2S mass flow rate in g/sec.

Mass Flowrate $= C$, g/L x Q, L/sec

$\qquad Q = (40 \text{ MGD})(10^6 \text{ gal/MG})(3.785 \text{ L/gal})(86,400 \text{ sec/d})$

$\qquad Q = 43.8 \text{ L/sec}$

Mass Flowrate $= (0.0064 \text{ g/L})(43.8 \text{ L/sec}) = 11.17 \text{ g/sec}$

Now calculate the concentration of the H_2S in the air, assuming that all H_2S is transferred to gas.

$\qquad C_{gas}, \mu g/m^3 = (\text{Mass Flowrate}, \mu g/sec)/(Q_{Air}\ m^3/sec)$

$\qquad Q_{Air}, m^3/sec = (120,000 \text{ scf/min})/(60 \text{ sec/min})(35.31 \text{ ft}^3/m^3)]$

$\qquad\qquad\qquad = 56.6 \text{ m}^3/sec$

$\qquad\qquad C = (11.17 \times 10^6 \mu g/sec)/ (56.6 \text{ m}^3/sec)$

$\qquad\qquad\quad = 197,200 \ \mu g/m^3$

Rearranging equation (2.34) and substituting:

$$C_{ppm} = \frac{(197,200 \,\mu g/m^3)(24.45)}{(34 \text{ g/mole})(1000)}$$

$$C_{ppm} = 142 \text{ ppm}$$

Calculate the concentration of H_2S in the air if it were in equilibrium with the incoming water using Henry's Law. Henry's Constant for H_2S is 0.1022 mole/L-atm.

$$P_g K_H = C_{aq}$$
$$P = (0.0064 \text{ g/L})/[(34 \text{ g/mole})(0.1022 \text{ mole/L-atm})]$$
$$= 0.00184 \text{ atm}$$

From the Ideal Gas Law:

$$C_g, \text{ mole/L} = n/V = P_g/RT$$
$$= (0.00184 \text{ atm})/[(0.08206)(298)]$$
$$= 7.52 \times 10^{-5} \text{ moles/L}$$

$$C_g = (7.52 \times 10^{-5} \text{ moles/L})(R, \text{ L-atm/mole-K})(T, \text{ K}) \times \frac{10^6}{P, \text{ atm}}$$

$$C_g = [(7.52 \times 10^{-5})(0.08206)(298)]10^6/1$$
$$C_g = 1840 \text{ ppm}$$

The equilibrium gas concentration of H_2S (1840 ppm) is much higher than the actual exit gas concentration (142 ppm) achieved by completely stripping the H_2S from the water. Since the actual exit gas concentration is much lower than the equilibrium concentration, the air flow is probably sufficiently high to achieve the desired removal of H_2S.

2.8   Chemical Reaction Kinetics

2.8.1 Overview

Characterization of chemical or biological reaction rates lies in the realm of kinetics, and involves fundamentally different concepts than the study of chemical equilibrium (thermodynamics) discussed previously. Equilibrium considerations are useful for determining whether an intended reaction is spontaneous (possible) for a defined system; however, equilibrium does not provide useful information to quantify how fast that reaction will proceed. For this kind of information, knowledge of reaction kinetics is required, and thus kinetics is a major field of study within chemistry.

Complete definition of reaction rates requires knowing the reaction stoichiometry, the reaction order, and having numerical values for all rate constants. Evaluation of reaction order and estimation of rate constants must be based on experimental data; rate relationships cannot be deduced from a balanced chemical reaction. Laboratory—or pilot—scale testing is often conducted to define reaction kinetics in conjunction with the design of pollution control facilities. Thus, kinetic relationships are useful in the design (sizing) of chemical or biological reactors. The sizing of reactors is explored in more detail in chapter 3.

In order to quantify reaction rates and apply kinetic relationships to natural and engineered systems, certain conventions have been widely adopted in environmental engineering practice. The key nomenclature is reviewed in the context of the hypothetical reaction noted below:

$$a\,A + b\,B \rightarrow c\,C + d\,D \tag{2.40}$$

Based on the reaction stoichiometry, the rates of destruction of reactants A and B are related to the rates of production of products C and D. In terms of molar quantities, this relationship is stated as follows:

$$\frac{-R_A}{a} = \frac{-R_B}{b} = \frac{R_C}{c} = \frac{R_D}{d} \tag{2.41}$$

$$R_X = \frac{d\left(\dfrac{N_X}{V}\right)}{dt} \tag{2.42}$$

where:

R_X = intrinsic molar production rate per unit volume
N_X = number of moles of compound X
V = reactor volume
t = time

For a constant volume reaction:

$$R_X = \frac{d[X]}{dt} \tag{2.43}$$

where: $[X]$ = molar concentration of compound X

It should be noted that a negative sign must be included in equation (2.41) for the reactant species (A and B) because the number of moles of the reactant species decreases as a result of reaction, and because R is defined as a **production** rate.

In environmental engineering practice, we often employ mass units rather than molar units. Equations (2.41) and (2.43) may be rewritten as follows for a constant volume reaction in terms of intrinsic reaction rates in mass concentrations:

$$\frac{-r_A}{a\,M.W._A} = \frac{-r_B}{b\,M.W._B} = \frac{r_C}{c\,M.W._C} = \frac{r_D}{d\,M.W._D} \tag{2.44}$$

$$r_X = \frac{d\,C_X}{dt} \left(= M.W._X\,R_X\right) \tag{2.45}$$

where:

r_X = intrinsic mass production rate per unit volume
$M.W._X$ = molecular weight of compound X
C_X = mass concentration of compound X

Example Problem 2.13

A hard-chrome-plating operation generates wastewater with a chromate concentration of 0.010 mole/L. The chromate is reduced (using SO_2) to trivalent chromium at low pH, and then precipitated upon adding a base:

$$2\,H_2CrO_4 + 3\,SO_2 \rightarrow 2\,Cr^{3+} + 3\,SO_4^{2-} + 2\,H_2O$$

$$2\,Cr^{3+} + 6\,NaOH \rightarrow 2\,Cr(OH)_3 + 6\,Na^+$$

Determine the (average) intrinsic reaction rates in molar and mass units of chromate (CrO_4^{2-}) and sulfur dioxide (SO_2) for a short time interval, assuming that the reaction achieves 5% completion in the first minute.

Solution

After 1 minute and at 5% completion, the "final" chromate concentration is 0.0095 mole/L. The molar intrinsic reaction rate for chromate is determined first:

$$-R_{chromate} = -\frac{d[C]}{dt} = -\frac{[C_{final}] - [C_{initial}]}{t_{final} - t_{initial}} = 0.0005\ \text{mole/L-min}$$

The mass intrinsic reaction rate for chromate is then obtained by multiplying by the molecular weight of chromate (116):

$$-r_{chromate} = (0.0005\ \text{mole/L-min})(116\ \text{g/mole}) = 0.058\ \text{g/L-min}$$

The molar and mass intrinsic reaction rates for sulfur dioxide are obtained by applying algebraic modifications of equations (2.41) and (2.44), respectively:

$$-R_{SO_2} = (0.0005\ \text{mole/L-min})(3\ \text{mole}\ SO_2/2\ \text{mole chromate})$$

$$= 0.00075\ \text{mole}\ SO_2/\text{L-min}$$

$$-r_{SO_2} = (0.00075\ \text{mole}\ SO_2/\text{L-min})(64\ \text{g/mole}) = 0.048\ \text{g}\ SO_2/\text{L-min}$$

2.8.2 Elementary Reactions

A chemical or biological process may actually consist of many individual steps in series and/or parallel. These individual reactions, which together comprise the overall mechanism of reaction, are referred to as elementary steps or elementary reactions (Smith, 1970). The reaction mechanism is the compilation of these elementary reactions. In most cases in environmental engineering practice, it is not feasible or necessary to fully characterize all elementary reactions for a particular process. It is more common to rely on empirical kinetic models that adequately describe an overall reaction. For example, the thermal oxidation of hydrocarbons to carbon dioxide and water involves hundreds of elementary reactions, but waste hydrocarbon incinerators are designed using simple empirical models based on the rate of destruction of one or two specific compounds.

A limited discussion of selected elementary reactions is presented to introduce concepts and terminology that are critical to applying the more widely used empirical kinetic models. Incorporating kinetic relationships in the solution of problems in subsequent chapters does not require a determination of elementary processes. In general, evaluation of environmental processes is based almost exclusively on the limited empirical models reviewed in section 2.8.3. However, the kinetic concepts are best learned by discussion of elementary reactions.

For the general reaction presented as equation (2.40), the intrinsic reaction rate is often, but not always, described by an expression of the following form:

$$-r_A = kC_A^n C_B^m \tag{2.46}$$

where k, n, and m are constants that are specific to the reaction. The reaction rate constant is designated as k. The constants n and m (which need not be integers) equal the reaction order with respect to compound A and compound B, respectively. The overall reaction order equals the sum of the orders with respect to each reactant (n + m). The order of the reaction (n or m) does not necessarily equal the stoichiometric coefficient (a or b), even for elementary reactions. However, for *many* elementary reactions, collision theory has been proposed to justify equality of reaction order and stoichiometric coefficient (Smith, 1970).

2.8.2.1 First-Order Elementary Reactions.
One of the simplest forms of elementary reaction is described by a first-order relationship between the intrinsic rate and the reactant concentration:

$$A \rightarrow C \tag{2.47}$$

$$-r_A = -\frac{dC_A}{dt} = kC_A \tag{2.48}$$

In spite of the simplicity of this relationship, first-order kinetics have been used with success to describe many processes of interest in environmental engineering, including radioactive decay.

The differential equation in equation (2.48) can be solved by separation of variables and integration, as outlined below. Note that the integration step is completed with definite integrals. The lower integration limits correspond with the initial conditions of $C = C_{A_0}$ and $t = 0$.

$$\frac{dC_A}{C_A} = -k\,dt \tag{2.49}$$

$$\int_{C_{A_0}}^{C_A} \frac{dC_A}{C_A} = \int_0^t -k\,dt \tag{2.50}$$

$$\ln\left(\frac{C_A}{C_{A_0}}\right) = -k\,t \tag{2.51}$$

$$C_A = C_{A_0}\exp(-k\,t) \tag{2.52}$$

First-order reactions may be characterized by the rate constant (k) or by specification of the half-life ($t_{1/2}$). The latter approach has been adopted for characterization of radioactive decay. The half-life corresponds with the time duration that is required to achieve a 50% reduction in concentration. The relationship between the half-life and the first-order rate constant can be developed from equation (2.51):

$$t_{1/2} = \frac{\ln\left(\dfrac{0.5\,C_{A_0}}{C_{A_0}}\right)}{-k} = \frac{0.693}{k} \tag{2.53}$$

where:

$t_{1/2}$ = half-life
k = first-order rate constant

It should be noted that for first-order processes, the half-life is independent of the species concentration.

2.8.2.2 Second-Order Elementary Reactions. Two types of second-order processes may be identified, as illustrated in equations (2.54) and (2.55) or equations (2.56) and (2.57).

$$2\,A \rightarrow C \tag{2.54}$$

$$-r_A = kC_A^{\,2} \tag{2.55}$$

$$A + B \rightarrow C \tag{2.56}$$

$$-r_A = kC_A C_B \tag{2.57}$$

The latter form—equations (2.56) and (2.57)—finds greater application in environmental engineering practice.

2.8.2.3 Variable-Order Processes. Variable-order reactions have been widely used for enzyme-catalyzed biological reactions (Lehninger, 1970) in which an enzyme-reactant complex (EA) is formed.

$$E + A \leftrightarrow EA \rightarrow E + P \tag{2.58}$$

The formation of the enzyme-reactant complex is assumed to be reversible, while the dissociation of the complex to form the product is assumed to be irreversible. The overall reaction can be resolved into a series of elementary reactions as described below.

The rate of formation of the enzyme-reactant complex from the reactants (equation 2.59) is defined by equation (2.60):

$$E + A \rightarrow EA \tag{2.59}$$

$$-R_A = k_1 [E] [A] \tag{2.60}$$

The rate of reversion of the enzyme-reactant complex to the reactant species (equation 2.61) is defined by equation (2.62):

$$EA \rightarrow E + A \tag{2.61}$$

$$R_A = -R_{EA} = k_2 [EA] \tag{2.62}$$

The rate of product formation from the enzyme-reactant complex (equation 2.63) is defined by equation (2.64):

$$EA \rightarrow E + P \tag{2.63}$$

$$-R_{EA} = k_3 [EA] \tag{2.64}$$

If it is assumed that the system has achieved steady state, then accumulation of the enzyme-reactant complex does not occur and the net molar rate of consumption of reactant A must equal the molar rate of production of product P:

$$-R_A = k_1 [A] [E] - k_2 [EA] = Rp = k_3 [EA] \tag{2.65}$$

The total molar concentration of enzyme (ET) is defined as the sum of the free enzyme concentration (E) and the enzyme-reactant complex (EA):

$$[ET] = [E] + [EA] \tag{2.66}$$

Equations (2.65) and (2.66) can be solved simultaneously to determine the concentration of the enzyme-reactant complex:

$$[EA] = \frac{[ET] [A]}{[A] + \dfrac{k_2 + k_3}{k_1}} \tag{2.67}$$

The overall rate of reaction can then be expressed as follows:

$$-R_A = R_P = k_3 [EA] = \frac{k[A]}{k_A + [A]} \tag{2.68}$$

where:

$$k = k_3 \left[ET \right] \tag{2.69}$$

$$k_A = \frac{k_2 + k_3}{k_1} \tag{2.70}$$

The relationship between the intrinsic reaction rate and the reactant concentration is illustrated in Figure 2.7. The rate approaches a maximum (or saturated) value equal to k at elevated reactant concentration. The rate constant k is therefore referred to as the maximum rate. When the reactant concentration is equal to k_A, the rate is equal to one-half of the maximum rate. The rate constant k_A is commonly referred to as a **half-saturation constant**.

Examination of Figure 2.7 and equation (2.68) is useful to understand the rate dependence on reactant availability at the extremes of high and low reactant concentration.

For very large reactant concentrations, the rate approaches the maximum value:

$$[A] >> k_A \tag{2.71}$$

$$-R_A = \frac{k[A]}{k_A + [A]} \cong \frac{k[A]}{[A]} = k \tag{2.72}$$

For these conditions, the reaction rate is nearly constant and does not depend on the reactant concentration. This situation corresponds with zero-order kinetics (section 2.8.3).

If the concentration is very small, the reaction rate is approximately proportional to the reactant concentration. This situation was defined previously as a first-order kinetic relation:

$$[A] << k_A \tag{2.73}$$

$$-R_A = \frac{k[A]}{k_A + [A]} \cong \frac{k[A]}{k_A} = k^*[A] \tag{2.74}$$

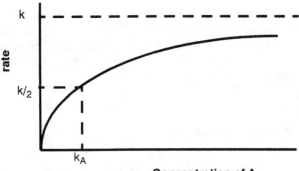

Figure 2.7
Variable-order kinetics.

$$k^* = \frac{k}{k_A} \tag{2.75}$$

where: k^* = first-order rate constant

As shown in the preceding discussion, the process will follow first-order kinetics at limiting (low) reactant concentrations and zero-order kinetics under saturation (high) reactant conditions. This characteristic is the basis of the variable-order designation. Many biological processes exhibit this type of kinetic behavior.

2.8.3 Empirical Models

As noted previously, full definition of elementary reactions is rarely pursued in environmental engineering practice. Design is normally achieved using empirical kinetic models that are selected by comparing experimental data with several common candidate models to identify an appropriate reaction order and to estimate rate constants.

2.8.3.1 n^{th} Order Models. Examples of nth-order kinetic models were discussed earlier in conjunction with first- and second-order elementary reactions. The general form of the rate expression is stated in equation (2.76):

$$-r_A = k C_A{}^n \tag{2.76}$$

In environmental engineering practice, many reactions are described adequately by zero-order (equation 2.77) or first-order (equation 2.78) kinetics. Accordingly, emphasis is placed on these kinetic relationships in this text.

$$-r_A = k \tag{2.77}$$

$$-r_A = kC_A \tag{2.78}$$

Example Problem 2.14

A constant volume, batch chemical reactor achieves a reduction in concentration of compound A from 100 mg/L to 5 mg/L in one hour. If the reaction is known to follow zero-order kinetics, determine the value of the rate constant with appropriate units. Repeat the analysis if the reaction is known to follow first-order kinetics.

Solution

Assuming zero-order kinetics,

$$-r_A = -\frac{dC_A}{dt} = k C_A{}^0 = k$$

$$\int_{C_{A_0}}^{C_A} d\, C_A = \int_0^t - k\, dt$$

$$C_A - C_{A_0} = -k\,t$$

$$k = \frac{C_{A_0} - C_A}{t} = \frac{100\ mg/L - 5\ mg/L}{1\ hr} = 95\ \frac{mg}{L\text{-}hr}$$

Assuming first-order kinetics,

$$-r_A = -\frac{d\, C_A}{dt} = k\, C_A^{\ 1} = k\, C_A$$

$$\int_{C_{A_0}}^{C_A} \frac{d\, C_A}{C_A} = \int_0^t -k\, dt$$

$$\ln\!\left(\frac{C_A}{C_{A_0}}\right) = -k\,t$$

$$k = \frac{\ln\!\left(\dfrac{C_A}{C_{A_0}}\right)}{-t} = \frac{\ln\!\left(\dfrac{5\ mg/L}{100\ mg/L}\right)}{-1\ hr} = 3.0\ hr^{-1}$$

It is important to recognize that the units associated with the rate constant are specific for the reaction order. Careful attention to units is critical to proper evaluation and application of kinetic relationships.

Example Problem 2.15

A first-order reaction occurring in a batch reactor is 40% complete after 30 minutes. How long is needed to achieve 95% completion? 99% completion?

Solution

The rate constant is determined with the equations developed in example problem 2.14 for first-order kinetics:

$$k = \frac{\ln\!\left(\dfrac{C_A}{C_{A_0}}\right)}{-t} = \frac{\ln\!\left[\dfrac{(1-0.40)\,C_{A_0}}{C_{A_0}}\right]}{-30\ min} = 0.0170\ min^{-1}$$

The reaction duration to achieve 95% conversion is determined with an algebraic manipulation of the same equation:

$$t = \frac{\ln\left(\dfrac{C_A}{C_{A_0}}\right)}{-k} = \frac{\ln\left[\dfrac{(1-0.95)C_{A_0}}{C_{A_0}}\right]}{-0.0170 \text{ min}^{-1}} = 176 \text{ min}$$

For 99% conversion:

$$t = \frac{\ln\left(\dfrac{C_A}{C_{A_0}}\right)}{-k} = \frac{\ln\left[\dfrac{(1-0.99)C_{A_0}}{C_{A_0}}\right]}{-0.0170 \text{ min}^{-1}} = 270 \text{ min}$$

Example Problem 2.16

A radioactive waste decays with a half-life of 10 days. Determine the percentage removal that is achieved in one month (30 days). Repeat the calculation for 2, 3, and 4 months.

Solution

Radioactive decay follows first-order kinetics. Since the half-life is known, the rate constant may be determined using equation (2.53):

$$k = \frac{0.693}{t_{1/2}} = \frac{0.693}{10 \text{ days}} = 0.0693 \text{ day}^{-1}$$

The percent removal is determined with equation (2.52):

$$\text{Pct Removal} = \left(1 - \frac{C_A}{C_{A_0}}\right)100\% = \left[1 - \exp\left(-k\,t\right)\right]100\%$$

The removal percentages are summarized below for various durations.

Time (months)	Time (days)	Removal %
1	30	87.50
2	60	98.44
3	90	99.80
4	120	99.98

The results from example problems 2.15 and 2.16 illustrate the "law of diminishing returns," a common dilemma in environmental engineering practice in which the rate of removal decreases as the target efficiency of removal increases. Treatment costs are often inversely related to the removal rate, with the result that it is usually much more difficult and expensive to achieve additional removal of contaminants as the efficiency of removal is increased.

2.8.3.2 Pseudo First-Order Models. Consider a reaction with the following stoichiometry which is assumed to follow a first-order kinetic relationship with respect to each reactant:

$$A + B + C \rightarrow D \tag{2.79}$$

$$-r_A = k\,C_A\,C_B\,C_C \tag{2.80}$$

If compounds B and C are present in excess relative to the concentration of reactant A, then the concentration of reactants B and C may not change significantly during the reaction even though compound A is exhausted. For these conditions, compound A would be the limiting reactant, and the rate expression could be approximated by a pseudo first-order expression:

$$-r_A \simeq k^* \, C_A \tag{2.81}$$

$$k^* = k\,C_B\,C_C \tag{2.82}$$

where: k^* = pseudo first-order rate constant

In many applications where removal of a target compound must achieve high efficiency, great care (and expense) is taken to provide an excess supply of all other reactants to guarantee that the target compound is the limiting reactant. Examples of this strategy include biological wastewater treatment for removal of organic compounds and incineration of volatile organic compounds. In both cases, oxygen is supplied in excess of stoichiometric requirements to achieve the desired organic removal efficiency.

2.8.3.3 Variable-Order Models. A variable-order kinetic model was reviewed previously in the context of a specific reaction mechanism for enzyme-catalyzed reactions:

$$-r_A = \frac{k\,C_A}{k_A + C_A} \tag{2.83}$$

Monod (1949) reported empirical evidence to support use of a similar rate expression to describe bacterial growth and decomposition of wastes. Variable-order kinetic expressions are commonly identified as Monod kinetics in environmental engineering practice. This model has great flexibility to describe many applications, including zero- and first-order situations, and is widely used for characterization of biological wastewater treatment processes.

2.8.4 Temperature Effects
The rates of chemical and biological reactions are strongly influenced by temperature, and we often heat reactants to achieve faster reaction rates. A practical example using extreme heating is the incineration of hazardous waste. Another common example is the mild heating of the liquid during the anaerobic digestion of sludge, a process in which the microorganisms are particu-

larly sensitive to low temperatures. In other cases (such as municipal wastewater treatment), adding heat to raise the temperature is not practical due to the large volume of wastewater. Nevertheless, quantification of the effect of temperature on reaction kinetics is important to account for seasonal or regional variation in performance. Two models are presented for adjustment of reaction rate constants for temperature variation:

1. Arrhenius Equation model:

$$k = A \exp\left(-\frac{E_a}{R\,T}\right) \qquad (2.84)$$

where:

A = frequency factor
E_a = activation energy
R = universal gas law constant (in energy units)
T = temperature (absolute scale)

2. Temperature Correction Factor model:

$$k_T = k_{20}\,\Theta^{(T-20)} \qquad (2.85)$$

where:

k_T = rate constant at temperature = T
k_{20} = rate constant at temperature = 20 °C
Θ = temperature correction factor
T = temperature in °C

The Arrhenius relation is particularly applicable to chemical reactions that potentially span a very large temperature range. Use of the temperature correction factor is common for biological processes that do not normally experience large-scale temperature variations.

Example Problem 2.17

A chemical reaction is reported to have an activation energy of 12,000 cal per mole. Determine the ratio of the rate constant at operating temperatures of 30 °C and 20 °C.

Solution

The rate constant is determined with equation (2.84), with conversion of temperature to an absolute temperature scale (Kelvin). The appropriate value for the gas law constant is

$$1.98\,\frac{cal}{°K\text{-mole}}$$

$$k_{30} = A \exp\left[-\frac{12{,}000\,\frac{cal}{mole}}{\left(1.98\,\frac{cal}{°K\text{-}mole}\right)(30+273\,°K)}\right] = 2.06 \times 10^{-9}\,A$$

$$k_{20} = A \exp\left[-\frac{12{,}000\,\frac{cal}{mole}}{\left(1.98\,\frac{cal}{°K\text{-}mole}\right)(20+273\,°K)}\right] = 1.04 \times 10^{-9}\,A$$

$$\frac{k_{30}}{k_{20}} = \frac{2.06 \times 10^{-9}\,A}{1.04 \times 10^{-9}\,A} = 1.98$$

For this specific value of activation energy, the rate of the chemical reaction is observed to double for a 10 °C increase in temperature.

Example Problem 2.18

A biological wastewater treatment process is known to exhibit first-order kinetics with a temperature correction factor equal to 1.02. Laboratory studies at 20 °C established a value for the rate constant of 5.0 day^{-1}. Determine the required reaction time in a constant volume batch reactor to achieve 90% conversion for summer conditions in Florida (assume a wastewater temperature of 30 °C). Repeat for winter conditions in Alaska (assume 5 °C).

Solution

The rate constant is determined with equation (2.85) for the specified temperatures:

$$k_T = k_{20}\,\Theta^{(T-20)} = (5.0\,day^{-1})(1.02)^{(T-20)}$$
$$k_{30} = (5.0\,day^{-1})(1.02)^{(30-20)} = 6.1\,day^{-1}$$
$$k_5 = (5.0\,day^{-1})(1.02)^{(5-20)} = 3.7\,day^{-1}$$

The reaction time is determined for this first-order process with equation (2.51):

$$t = \frac{\ln\left(\frac{C_A}{C_{A_0}}\right)}{-k}$$

$$t_{30} = \frac{\ln\left[\dfrac{(1 - 0.90)\, C_{A_0}}{C_{A_0}}\right]}{-6.1 \text{ day}^{-1}} = 0.38 \text{ days}$$

$$t_5 = \frac{\ln\left[\dfrac{(1 - 0.90)\, C_{A_0}}{C_{A_0}}\right]}{-3.7 \text{ day}^{-1}} = 0.62 \text{ days}$$

The calculated difference in reaction time underscores the need to consider during the design stage any site-specific factors (such as temperature) which may influence reaction kinetics. Empirical design guidelines that have been established for one region may be wrong for other climatic regions; engineers cannot simply use an "off-the-shelf" design in every part of the country.

2.8.5 Catalysis

This introductory discussion of reaction kinetics would be incomplete without some mention of catalytic reactions, many of which are vitally important in environmental engineering practice. Several general characteristics of catalytic reactions are noted below (Smith, 1970):

1. A catalyst accelerates the reaction rate by providing alternate pathways to products with an associated reduction in the activation energy.
2. The active catalyst is combined with at least one reactant during the reaction cycle and is then released with the appearance of product.
3. The catalyst is reused many times so that relatively small quantities are effective for generation of large amounts of product.
4. Equilibrium is not altered by the presence of catalysts.

The discussion of variable-order elementary reactions introduced one class of catalytic reactions. In this case, enzymes served to catalyze biological reactions. In a more general context, the ability of microorganisms to serve as catalysts is utilized in biological wastewater treatment, fermentation of alcoholic beverages, and production of pharmaceuticals. Other important applications of catalysts in environmental engineering practice include automobile catalytic converters, and fixed- or fluidized-bed catalytic oxidizers for control of volatile organic compounds (VOCs) from paint spray booths.

2.9   Summary

Subsequent chapters in this text provide a description of processes that are used for pollution abatement. A high percentage of these applications rely on chemical and/or biological reactions. An in-depth knowledge of chemical reaction stoichiometry is essential in environmental engineering practice, as is knowledge of solubility, acids and bases, and the gas laws. A quantitative description of these processes requires a numerical expression of reaction rates. The most commonly used kinetic expressions in environmental engineering practice are zero-, first-, pseudo first-, and variable-order models. Chemistry is used extensively in the design of engineered systems to control pollution emissions or to remediate contaminated sites. Chemistry is one of the *foundations* of environmental engineering, and it is *crucial* that all environmental engineering students master the fundamentals of chemistry. The applications of chemistry will be seen throughout the remainder of this text.

End-of-Chapter Problems

2.1 Balance the following equations:
   a. $Na_2CO_3 + HCl \rightarrow NaCl + CO_2 + H_2O$
   b. $Cl_2 + KOH \rightarrow KCl + KClO_3 + HCl$
   c. $Fe(OH)_2 + H_2O + O_2 \rightarrow Fe(OH)_3$
   d. $FeSO_4 + K_2Cr_2O_7 + H_2SO_4 \rightarrow Fe_2(SO_4)_3 + Cr_2(SO_4)_3 + K_2SO_4 + H_2O$
   e. $FeS + HCl \rightarrow FeCl_2 + H_2S$

2.2 Thirty mg of calcium (Ca^{2+}) and 50 mg of magnesium (Mg^{2+}) are dissolved in 0.50 L of water. What is the total hardness of the solution expressed in mg/L as $CaCO_3$?

2.3 The compound, $C_7H_4N_3O_6$, burns explosively with oxygen to produce CO_2, water, and N_2. Write a balanced chemical equation for this reaction and calculate the mass of oxygen required (in grams) to burn 100 g of this compound.

2.4 One method of removing phosphate from wastewater effluents is to precipitate it with aluminum sulfate. A plausible stoichiometry (but not exact because aluminum and phosphate can form many different chemical materials) is:

$$2 PO_4^{3-} + Al_2(SO_4)_3 \rightarrow 2 AlPO_4 + 3 SO_4^{2-}$$

If the concentration of phosphate (PO_4^{3-}) is 30 mg/L, how many kg of aluminum sulfate must be purchased annually to treat 40 L/sec of wastewater? How many kg of precipitate will be formed as sludge if all of the phosphate is precipitated as $AlPO_4$?

2.5 If equal moles of magnesium chloride and aluminum chloride are dissolved in a dilute acid solution, will magnesium hydroxide or aluminum hydroxide precipitate first as the pH of the solution is increased? The K_{SP} of $Mg(OH)_2$ is 1×10^{-11}; for $Al(OH)_3$ it is 6×10^{-23} at 25 °C.

2.6 What is the ratio of the undissociated form of acetic acid to acetate at a pH of 5.2? The pK_A for acetic acid is 4.6 at 25 °C.

2.7   What is the pH of the following solutions?
      a. 0.01 M NaOH (strong base)
      b. 0.01 M HNO_3 (strong acid)
      c. 0.01 M $HClO_3$ (weak acid, pK_A = 7.6)
      d. 10^{-6} M HCl (strong acid)

2.8   How many kg/day of NaOH must be added to neutralize a waste stream generated by an industry producing 90.8 million g/day of sulfuric acid, if 0.1% of the sulfuric acid produced is lost to wastewater? The wastewater flow is 750,000 L/day.

2.9   The solubility of carbon dioxide in pure water is given by its Henry's Constant. Calculate the concentration of CO_2 in rainwater in mole/L at 25 °C, given that the atmosphere contains 365 ppm CO_2. Assume that $[CO_2]_{aq}$ = $[H_2CO_3]$.

2.10 Once CO_2 is absorbed in water, it reacts according to equations (2.20) through (2.23). Calculate the pH of natural rainwater described in problem 2.9 above.

2.11 What is the alkalinity (expressed in mg/L as $CaCO_3$) of a solution containing 0.01 M HCO_3^- and 0.02 M CO_3^{2-} at a pH of 10.6?

2.12 Nickel ion (Ni^{2+}) and OH^- are in equilibrium with solid $Ni(OH)_2$ in a 1-L solution formed by dissolving pure $Ni(OH)_2$ in water. What is the equilibrium concentration of Ni^{2+} ions (in mole/L and mg/L)? What is the final concentration of Ni^{2+} ions in solution (in mg/L) after the pH is adjusted to 7 using HCl (ignore the small increase in solution volume)? The K_{SP} of $Ni(OH)_2$ is 6 x 10^{-18} at 25 °C.

2.13 Balance the following equation:

$$C_2H_5OH + O_2 \rightarrow CO_2 + H_2O$$

      Calculate the volume, in L, of CO_2 produced from the oxidation of 100 g of ethanol (C_2H_5OH) with excess O_2 at 25 °C and 1 atm.

2.14 If 100 g of HCl reacts with excess sodium carbonate at 35 °C and 1 atm, how many moles of carbon dioxide gas will be produced and what volume will the CO_2 occupy? (See problem 2.1a.)

2.15 If the Henry's Constant for H_2S is 9.74 atm-L/mole at 25 °C, calculate the mole fraction of H_2S gas present at 1 atm above a solution containing H_2S/HS^- at a concentration of 0.01 M and a pH of 7.2. You may assume that the concentration of S^{2-} is negligible compared to the concentrations of H_2S and HS^-. The pK_1 of H_2S is 7.04.

2.16 How many kilograms of oxygen and nitrogen are contained in a 200-ha lake (1 ha = 10,000 m^2) with an average depth of 5 m, if the water is in equilibrium with the atmosphere which contains 79% N_2 and 21% O_2? Assume P = 1 atm and T = 25 °C.

2.17 Determine the correct units for the rate constant(s) in each of the following kinetic expressions:

a. $-r_A = \dfrac{k\,C_A}{k_A + C_A}$ b. $-r_A = k\,C_A{}^2$ c. $-r_A = k\,C_A$

d. $-r_A = k\,C_A\,C_B$ e. $-r_A = k$

2.18 A chemical reaction occurs in a batch reactor. The initial concentration of compound A is 100 mg/L. The concentration which remains after 10 minutes is 90 mg/L. The concentration which remains after 20 minutes is 80 mg/L. Determine the reaction order and the rate constant (with appropriate units).

2.19 A second-order reaction occurring in a batch reactor of the type described by equation (2.55) is known to be 40% complete after 30 minutes. How long is required to achieve 95% completion? Assume the initial concentration of compound A is 100 mg/L.

2.20 A zero-order reaction occurring in a batch reactor is known to be 40% complete after 30 minutes. How long is required to achieve 95% completion? Assume the initial concentration of the reactant is 100 mg/L.

2.21 A variable-order reaction with a half-saturation constant of 50 mg/L and a maximum rate of 4 mg/L-hr occurs in a batch reactor. If the initial concentration is 500 mg/L, how long is required to achieve 75% completion? Is a zero-order approximation reasonable for this application? Is a first-order approximation reasonable for this application?

2.22 A variable-order reaction with a half-saturation constant of 50 mg/L and a maximum rate of 4 mg/L-hr occurs in a batch reactor. If the initial concentration is 10 mg/L, how long is required to achieve 75% completion? Is a zero-order approximation reasonable for this application? Is a first-order approximation reasonable for this application?

2.23 A zero-order reaction is known to have a temperature correction factor (Θ) of 1.01. A batch reactor was used in the laboratory (25 °C) with a reaction time of 3 hours to achieve a reduction in the concentration of a synthetic organic compound from 3.0 to 0.1 mg/L. How long will the reaction take under field conditions (12 °C) to achieve the same treatment efficiency?

References

Lehninger, Albert L. 1970. *Biochemistry.* New York: Worth Publishers.

Monod, J. 1949. "The Growth of Bacterial Cultures." *Annual Review of Microbiology* 371.

Smith, J. M. 1970. *Chemical Engineering Kinetics.* 2nd ed. New York: McGraw-Hill.

U.S. Environmental Protection Agency. 1983. *Development Document for Effluent Limitation Guidelines and Standards for the Metal Finishing Point Source Category.* EPA 440/1–83/091 (June).

Chapter 3

Mathematical and Physical Foundations

3.1 An Approach to Solving Problems

Engineers are problem solvers—that's our claim to fame. We are trained to identify a problem, to analyze it into "bite-size pieces," and then to synthesize a cost-effective solution. One simple (but effective) way to think about problem solving is as a three-step process:

1. find out where you are (here),
2. decide on where you want to go (there), and
3. figure out the best way to get from here to there.

Step one is the first part of identifying the problem. It may begin as an assignment from a supervisor, by investigating a complaint, or simply by personal observation of an improper or inefficient operation. Step two is the second part of problem identification, and the first part of the analysis phase. It involves researching and defining acceptable goals for improvement; for pollution control projects, these goals may be meeting all applicable emission standards or regulations, or perhaps satisfying internal company policies. Step three includes making the traditional engineering design calculations, the first ones being the material and energy balances, which are the main topics of this chapter.

As an example, consider the situation of an engineer working for a petrochemical company who is called in one day by the plant manager. The manager tells the engineer that she has received a complaint from a local resident about a visible plume of "smoke" from the plant, and asks him to solve the problem. First, the engineer must find out exactly what the problem is: which stack was emitting, what was it emitting, why it was doing so (i.e., normal or upset operations), when it happened, and for how long. Next he must determine if there are any state regulations that apply to these stack emissions, and determine how far the plant must go to clean up the emissions to meet regulations or to satisfy the resident's complaint. Then, and only then, can he begin to formulate the best solution for bringing the operation from where it is to where it should be.

Material and energy balances are key analytical tools for any engineer working with an existing industrial or municipal process. These calculations are invaluable in analyzing the details of an existing process, and supplement the information obtained from instruments that measure various process parameters. Real-time measurements of the important parameters are essential for good operation and control of the process, but we cannot measure every flowing stream in the whole unit—it is simply too expensive, and in the case of fugitive emissions (i.e., leaks) not technically possible. In addition, material and energy balance calculations are the only way that engineers can predict how a new process (one that has not yet been built) will behave.

Material and energy balances are used by engineers in many ways. They are used to check the accuracy (or at least the consistency) of flow meters. They are used to help predict the outcome of a proposed change in the way an existing unit is being operated. But, most importantly, material and energy balances are essential in the design of a new process (we cannot measure flows or other variables in a process that does not yet exist). Whether we are assessing the impact on the local environment of a proposed plant, or calculating the profit or loss expected from a new chemical process, or sizing equipment to be purchased, material and energy balances are essential.

The use of fundamental conservation equations (mass, energy, and momentum) is always an appropriate approach for describing natural or engineered systems. Sufficient experience and existing databases may be available in selected cases to support the development of empirical equations or heuristic design guidelines. But, public health and safety concerns may preclude conducting experiments to generate the data (as in the case of accidental radioactive releases from nuclear power facilities). So, in the absence of data, conservation equations may represent the only method available.

The development of system descriptive equations from the fundamental concepts of conservation of mass and energy is emphasized throughout this text. Such development will reinforce methodology that is basic to other engineering courses. Force and momentum balances are critical in mechanics courses (statics, dynamics, mechanics of materials). Conservation of mass underlies the continuity equation in fluid mechanics and mass transfer equations in chemical engineering. Principles of energy conservation are incorporated into Bernoulli's Equation (fluid mechanics), Kirchoff's Law (electrical engineering circuits), and heat-work relationships (thermodynamics). These basic equations (which are used throughout engineering practice) are all derived from fundamental conservation laws. Mastery of the concepts in chapter 3 is essential to the proper development of the needed equations in this course as well as in many other engineering courses.

3.2   Flow of Materials—Mass Balances

All industries that manufacture products have at least two things in common: energy usage and the flow of raw materials into (and finished products out of) the plant. **Conservation of mass** is a fundamental principle of engineering that states, simply, that matter can neither be created nor destroyed. While not strictly true for nuclear reactions, the principle can be viewed as being exact for ordinary physical and chemical processes. Put another way, this principle tells us that if one or more streams of material are flowing into a defined region of space (such as a tank, a reactor, or other process unit) then material must be either flowing out of that region at the same total mass flow rate, or else material will accumulate in, or be withdrawn from, the region.

Whether we are measuring volumetric flow rate of a gas, or the linear velocity of a liquid flowing in a pipe, or simply counting the number of fifty-pound bags of a dry chemical received per hour, we must convert our measured flow rates to proper units to make a material balance. The only units that are *always* correct are mass/time or moles/time. In certain special cases, it is permissible (and more convenient) to make a volumetric flow balance, but the assumptions implicit in this approach must be understood. Specifically, for volume balances all streams must have the same density—a condition often approximated closely by streams of water (whether slightly polluted or not). These conservation equations require dimensional consistency. Careful attention to units is therefore essential. It is worth emphasizing that identification and diagnosis of many calculational errors begins with the examination of units for dimensional consistency. Calculations performed in contemporary American practice often switch between English and metric units. Familiarity with the different units systems continues to be necessary for most engineering applications.

The **continuity equation** allows one to convert from a linear flow velocity to a mass flow rate provided the density of the fluid and the area normal to flow are known:

$$\dot{m} = \rho u A \qquad (3.1)$$

or

$$\dot{m} = \rho Q \qquad (3.2)$$

where:

$\dot{m}$ = mass flow rate, kg/s

ρ = fluid density, kg/m^3

u = fluid linear velocity, m/s

A = area normal to the flow, m^2

Q = volumetric flow rate, m^3/s

The continuity equation is useful for calculating fluid linear velocities or volumetric flow rates at different points along a conduit since the mass flow rate does not change under steady conditions.

Example Problem 3.1

Water is flowing in a 4-inch, Schedule 40 pipe with a velocity of 3.00 m/s. The pipe splits into two 2-inch, Schedule 40 pipes with an equal volumetric flow in each. Calculate the linear velocity and mass flow rate in one of the 2-inch pipes. The density of water is 1000 kg/m³ and the inside diameters of the 4-inch pipe and the 2-inch pipe are 10.23 cm and 5.250 cm, respectively.

Solution

First calculate the mass flow rate in the 4-inch pipe:

$$\dot{m}_4 = 1000 \frac{kg}{m^3} \, 3.00 \frac{m}{s} \times \frac{\pi}{4} \left(\frac{10.23 \text{ cm}}{100 \text{ cm/m}} \right)^2 m^2 = 24.66 \text{ kg/s}$$

Exactly half of this flow goes into each 2-inch pipe. The volumetric flow in one 2-inch pipe is:

$$Q_2 = 12.33 \frac{kg}{s} \times \frac{1}{1000 \text{ kg/m}^3} = 0.01233 \text{ m}^3/s$$

and the linear velocity in that pipe is

$$u_2 = 0.01233 \text{ m}^3/s \, \frac{1}{\frac{\pi}{4}(0.0525 m)^2} m^2 = 5.70 \text{ m/s}$$

Note the increase in *velocity* of the fluid even though the mass flow rate is constant.

3.2.1 The Basic Balance Equation

A material balance is simply an accounting of all materials going into and out of an identifiable process area. Consider first an overall mass balance. We want to account for the total mass flow rate regardless of chemical type. Since mass can neither be created nor destroyed, we can say with certainty that *whatever mass flows into the process area must either flow out of the process area or accumulate within it.* In other words, the rate of accumulation of mass within the designated process area is equal to the inflow rate minus the outflow rate. Mathematically, we write

$$\frac{dM}{dt} = \dot{m}_{in} - \dot{m}_{out} \tag{3.3}$$

where: M = total mass within the boundaries of our system.

Usually there are several different chemical components flowing through a process. Quite often, a chemical reaction is employed to change a less valuable raw material into a valuable product. Other times, the process is designed to use chemical or biological reactions to change harmful waste products into less harmful ones. An independent material balance equation can be written for each chemical component (or for all components but one, if an overall balance is also made). In the general case, a material balance equation has the following form for component i:

(Accumulation Rate)$_i$ = (Input Rate)$_i$ – (Output Rate)$_i$ + (Generation Rate)$_i$

or

$$AR_i = IR_i - OR_i + GR_i \tag{3.4}$$

Usually, we are interested in steady-state operations, which means no accumulation. And if there are no chemical reactions generating (or destroying) component i, the balance equation becomes very simple:

$$(IR)_i = (OR)_i \tag{3.5}$$

In most cases, the mass flow of a particular substance will be the product of the volume flow of the stream carrying that substance times the concentration of that substance in the stream.

3.2.2 Choosing the Basis and the Boundaries

A key step in making a material balance is clearly defining the region in space that is to be material balanced, including the boundaries across which mass is flowing. To aid in doing this, process engineers draw simplified flow diagrams. Blocks represent reactors, tanks, separation columns, and so forth, and lines with arrows indicate material flow. For example, Figure 3.1 could be used to represent a process in which a wastewater stream is aerated, and the pollutants converted in the reactor (aeration basin) to sludge, which is then discharged separately along with the treated water.

In Figure 3.1, the dotted line represents the boundaries for material balance. In this case, it represents a processing unit material balance, rather than a reactor balance or a clarifier balance. Often we make several material balances on a single unit and its individual equipment to better understand the process.

Since the material is flowing, we may choose as a basis either a unit of time (an hour, a day, etc.) or a unit of material (1,000,000 gallons of wastewater, one metric ton of sludge produced, etc.). The most general approach is to use the equations as written and solve in terms of flow rates. However, often it is more convenient to pick a specified time interval or amount of material as the basis and solve in terms of mass of material only.

After drawing a diagram (with boundaries) and deciding on a basis, the next step is to write down the general unsteady-state equation represented by equation (3.4). Next, label the known streams and write down given data. Once everything known is identified on the diagram, you are ready to solve the problem. Begin with the general equation, simplify as appropriate, substitute for terms in the equation, and solve.

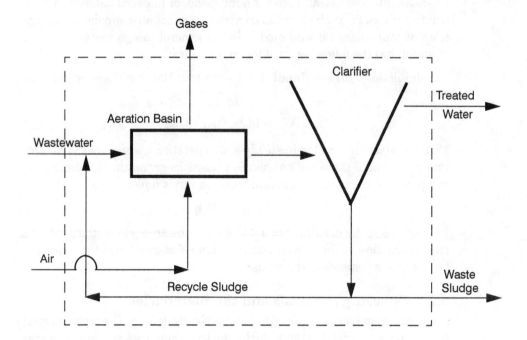

Figure 3.1
A simplified process flow diagram.

3.2.3 Steady-State Flow Processes

Steady state is an important condition for design and analysis. By definition, when a process is operating at steady state, *nothing changes with time*. This means all flow rates, temperatures, pressures, liquid levels, etc. are constant. While true steady state is rarely achieved for extended periods of time, it is often approximated to a reasonable degree. Design is usually based on steady-state conditions, and achieving steady state is a goal of most process operators. Thus, processes can be analyzed neglecting the accumulation rate term (at steady state there can be no accumulation—either positive or negative). This is a big help, because the accumulation rate term usually leads to differential equations. Without this term, often only algebraic equations remain.

Example Problem 3.2

A treated wastewater stream flowing at 5000 L/s contains 75 mg/L of solid particles. It flows into a river that was flowing at 35,000 L/s and carrying 5 mg/L of solids. What is the concentration of particles in the combined stream (river plus wastewater)?

Solution

Step 1—Draw the diagram. A diagram with boundaries is presented below.

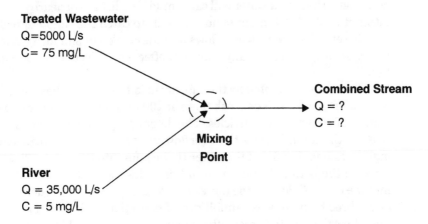

Treated Wastewater
Q=5000 L/s
C= 75 mg/L

Combined Stream
Q = ?
C = ?

**Mixing
Point**

River
Q = 35,000 L/s
C = 5 mg/L

Step 2—Choose a basis. The basis we pick is the continuous flow rate through the "mixing point," the point where the two streams join and mix (assumed to be complete and instantaneous mixing).

Step 3—Write the general equation.

$$AR = IR - OR + GR$$

Step 4—Simplify the general equation.

In this problem we have no accumulation in the mixing point, and no generation of particles or water by reaction. We can write:

$$IR = OR$$

Note that two mass balances are necessary; the first is on the water, the second is on the particles.

Step 5—Substitute known terms and solve the simplified equations.

Water balance: $5000 \text{ L/s} \times \rho + 35,000 \text{ L/s} \times \rho = Q_m \times \rho$

Since the density of each stream is almost identical, we can divide through by density and simply add the volumetric flow rates:

$$Q_m = 40,000 \text{ L/s}$$

Solids balance: (for each stream, mass flow of solids = volume flow x concentration)

$$5000 \text{ L/s} \times 75 \text{ mg/L} + 35,000 \text{ L/s} \times 5 \text{ mg/L} = Q_m \times C_m$$
$$C_m = 13.75 \text{ mg/L}$$

Example Problem 3.3

Consider a hilly area in which new homes are being constructed. As rain falls on the site, the runoff water will carry mud and dirt into a nearby creek—polluting it badly. A detention pond was built to catch the runoff and allow the mud to settle before the water flows into the creek. For simplicity, assume the rain is steady, and ignore any transient effects on the pond water concentration.

The runoff generated by the rainstorm is 10 m³/s, and has a particle concentration of 5.0 g/L. The creek flows at 200 m³/s and normally has a concentration of 15.0 mg/L of particles. The detention pond will catch and remove most of the mud, leaving the overflow water with a concentration of 400 mg/L. Assume that 5% of the water that comes into the pond will seep down through the bottom (filtering out all particles), while the rest will overflow into the creek. Calculate the mass of mud accumulated in the detention pond after three hours of this rainfall/runoff operation. Calculate the concentration of particles in the creek after mixing with the overflow from the pond.

Solution

First, solve the mass balance for the pond.

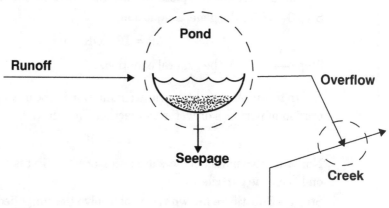

$AR = IR - OR$

$AR = 10\,\mathrm{m^3/s} \times 1000\,\mathrm{L/m^3} \times 5\,\mathrm{g/L} \times 1000\,\mathrm{mg/g} - 9500\,\mathrm{L/s} \times 400\,\mathrm{mg/L}$
$\quad - 500\,\mathrm{L/s} \times 0\,\mathrm{mg/L}$

$\quad = 5.0\,(10)^7\,\mathrm{mg/s} - 3.8\,(10)^6\,\mathrm{mg/s}$

$\quad = 4.62\,(10)^7\,\mathrm{mg/s}$

(mud is accumulating at the bottom of the pond at the rate of 46.2 kg/s)

After three hours like this, the total amount of mud in the pond is:

46.2 kg/s × 3600 s/hr × 3 hr = 499,000 kg

The concentration of particles in the creek after mixing with the overflow is:

$$9500 \text{ L/s} \times 400 \text{ mg/L} + 200{,}000 \text{ L/s} \times 15 \text{ mg/L} = Q_{total} \times C_{mix}$$

$$C_{mix} = \frac{9500 \text{ L/s} \times 400 \text{ mg/L} + 200{,}000 \text{ L/s} \times 15 \text{ mg/L}}{209{,}500 \text{ L/s}}$$

$$C_{mix} = 32.5 \text{ mg/L}$$

In the next example, the material balance approach is applied to a steady-state flow process in a system that has a recycle loop. If the boundaries are drawn such that the recycle loop is totally enclosed within boundaries, it is *as if the recycle stream does not exist.* We simply ignore the recycle stream because it does not cross the boundaries.

Example Problem 3.4

Consider the diagram and the data given below representing some sort of reaction between two liquids (F_1 and F_2) to produce a third liquid (P) and a gas (G). The gas is produced at a mass ratio of 1 unit of gas for every 1000 units of liquid product.

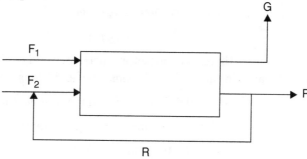

Data

Stream	Flow Rate	Density
Feed F_1	1000 L/min	1.5 kg/L
Feed F_2	6000 kg/min	1.2 kg/L
Product P	Not given	1.3 kg/L
Recycle R	0.5 times P	same as P
Gas G	Not given	1.0 kg/m³

Calculate P, G, and R in L/min. What is the total feed rate into the reactor in kg/min?

Solution

First draw the boundaries around the whole system.

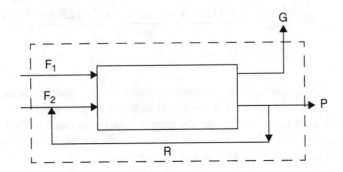

The general balance for total mass simplifies to Input Rate = Output Rate or $F_1 + F_2 = G + P$ (note that the recycle stream R does not appear in this equation).

$$1000 \text{ L/min} \times 1.5 \text{ kg/L} + 6000 \text{ kg/min} = 0.001 \text{ P} + \text{P}$$

$$P = 7492.5 \text{ kg/min}$$

Converting to volumetric flow rate, we get

$$P = 5763 \text{ L/min}$$

Note: Even if there were no gas products, we could not have made a direct volume balance in this problem because the liquid densities are different. Recall:

G = 0.001 times P on a mass basis, therefore

G = 7.49 kg/min or using the density of the gas,

G = 7490 L/min

Finally,

$$R = 0.5P = 3746 \text{ kg/min} = 2882 \text{ L/min}$$

The fresh feed rate to the reactor is $F_1 + F_2$; the total feed rate is $F_1 + F_2 +$ R which equals 11,247 kg/min.

In most processes with a reaction step, a total mass balance by itself does not answer all our questions. Indeed, a reactor is built specifically to convert at least one reactant into at least one product. Hence, individual mole balances for each of the components of interest are usually required. In the general balance equation, the term "generation rate" refers to an overall rate of generation by chemical reaction within the system boundaries. It is usually represented as an intrinsic reaction rate times the system volume. Note that if a component is being used up in the reaction, then its rate of generation is negative. An intrinsic reaction rate was defined mathematically in the previous chapter (equation 2.42) as the time rate of change in the number of moles of a given component per unit volume (due to reaction only). In the laboratory we can most easily obtain intrinsic reaction rates from constant-volume, well-stirred batch reactors. In practice, we use intrinsic reaction rates to design continuous flow reactors.

Recall that for a simple A→B reaction, a kinetic model to predict R_A and R_B is:

$$R_A = -kC_A^n \tag{3.6}$$

$$R_B = kC_A^n \tag{3.7}$$

where:

R_i = production rate of species i, moles/L-sec

k = a temperature-dependent rate constant

n = an empirically determined order

Note that the production or generation rate of A is negative because A is being used up. Note also that the molar generation rate of the product B is equal to the molar destruction rate of A (and is expressed as a function of the concentration of *reactant* A).

Two common models of ideal chemical reactors are the completely mixed flow reactor, better known as the continuous flow stirred tank reactor (CSTR) and the plug flow reactor (PFR). We must approach the material balances for these two types of reactors differently.

The CSTR model (see Figure 3.2) is that of an overflowing tank in which the contents are rapidly and continuously mixed. There is no difference in concentrations of any species anywhere in the tank. Since the outlet stream is continuously withdrawn from the tank and the contents of the tank have the same composition everywhere, the concentrations in the outlet stream are identical to those in the tank. *These are the concentrations at which all reactions occur.* A steady-state material balance on a reacting component (A→B, first order) in a CSTR results in a simple algebraic expression as shown below:

$$0 = Q_0 C_{A_0} - Q_e C_{A_e} - kC_{A_e} V \tag{3.8}$$

Again, we emphasize that the concentration to be used in the generation rate term is the in-tank concentration, which for a CSTR equals C_{A_e}, the exit concentration.

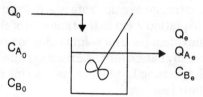

Figure 3.2
Schematic diagram of a continuous flow stirred tank reactor (CSTR).

Example Problem 3.5

Calculate the CSTR volume required for 98% conversion of component A. The kinetics are:

$R_A = -kC_A$ with $k = 0.10$ s^{-1}. The inflow rate is 75 L/s.

Solution

Solving equation (3.8) for the volume of the reactor and noting that $Q_e = Q_0 = Q$, we get:

$$V_R = \frac{Q_0(C_{A_0} - C_{A_e})}{kC_{A_e}} = \left(\frac{Q}{k}\right)\left(\frac{1 - C_{A_e}/C_{A_0}}{C_{A_e}/C_{A_0}}\right)$$

Substituting and solving:

$$V_R = \left(\frac{75 \text{ L/s}}{0.1 \text{ s}^{-1}}\right)\left(\frac{0.98}{0.02}\right) = 36{,}750 \text{ liters}$$

The other widely used ideal reactor model is that of a plug flow reactor (PFR). In this model, the reactor is pictured as a long, narrow tube, through which fluid is flowing. Flow is assumed to be one-dimensional; velocity, concentration, and temperature are all constant with radial position, and longitudinal mixing is assumed to be negligible. Often, temperature does not vary much with axial position, and the analysis is very much simplified if the reactor can be assumed to be isothermal. We shall use the material balance approach to develop the basic steady-state design equation for an isothermal plug flow reactor. Refer to Figure 3.3.

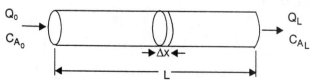

Figure 3.3
Schematic diagram of a plug flow reactor (PFR).

Consider the simple reaction A→B with kinetics as described by equations (3.6) and (3.7). Start by making a material balance on component A in a small volume increment of the reactor ($\Delta V = S\Delta x$, where S is the cross-sectional area) located at an arbitrary position x along the length of the reactor. The mole balance equation (subject to our assumptions) becomes:

$$S\Delta x \frac{\partial C_A}{\partial t} = Q_x C_{A_x} - Q_{x+\Delta x} C_{A_{x+\Delta x}} + R_A S\Delta x \qquad (3.9)$$

Assume that S remains constant throughout the reactor and that Q, the volumetric flow rate, does not change with x. Dividing through by $S\Delta x$ and noting that $Q/S = U$, the linear velocity, we obtain

$$\frac{\partial C_A}{\partial t} = \frac{U\left(C_{A_x} - C_{A_{x+\Delta x}}\right)}{\Delta x} + R_A \qquad (3.10)$$

Now let Δx approach zero and take the limit. Substituting equation (3.6) for R_A, equation (3.10) becomes

$$\frac{\partial C_A}{\partial t} = -U\frac{\partial C_A}{\partial x} - kC_A^n \qquad (3.11)$$

At steady state the accumulation term is zero, so this partial differential equation can be written as an ordinary differential equation

$$U\frac{dC_A}{dx} = -kC_A^n \qquad (3.12)$$

which we can solve by integration. For a first-order reaction (n = 1), we write

$$\int_{C_{A_0}}^{C_{A_L}} \frac{dC_A}{C_A} = \frac{-k}{U}\int_0^L dx \qquad (3.13)$$

which upon integration is

$$\ln\frac{C_{A_L}}{C_{A_0}} = -k\frac{L}{U} \qquad (3.14)$$

or

$$C_{A_L} = C_{A_0} e^{-k\tau} \qquad (3.15)$$

where: τ = reactor residence time (= L/U or V/Q).

Example Problem 3.6

Rework the problem of example 3.5, except calculate the volume of a PFR required for the same 98% conversion.

Solution

From equation (3.14),

$$\ln\left\{\frac{.02\, C_{A_0}}{C_{A_0}}\right\} = 0.1\,s^{-1}\tau$$

Solving for τ we get $\tau = 39.1$ seconds. But,

$$\tau = V/Q$$

so

$$V = Q\tau = (75 \text{ L/s})(39.1 \text{ s})$$

$$V = 2930 \text{ L}$$

Note that the PFR volume is much less than the CSTR volume.

3.2.4 Transient Processes

Up to now, we have always dropped the rate of accumulation term in our general material balance equation by assuming steady state. Although steady-state operations are desirable, and knowledge of the steady-state condition is useful for design, the transient (nonsteady-state) response of a process is also important. An in-depth study of the dynamic behavior of systems is a complete course in itself. However, we need to be aware of transient processes; therefore, some examples of responses of simple systems to simple forcing functions are presented in Table 3.1. Finally, the use of the unsteady-state material balance equation is illustrated with the following example problem.

Example Problem 3.7

A company has been discharging its nonreactive waste into a holding pit for a long time. For the last several years the pit has been full and has overflowed into a local river. The waste concentration has been nearly constant at 10 mg/L of the pollutant of interest (as has the pit overflow stream). At this concentration, there have been no adverse effects on the stream. Suddenly there is a process change and the company's waste stream concentration goes up to 100 mg/L. Given that the waste stream flow rate is 100,000 L/day and the holding pit volume is one million liters, calculate the concentration of the pit's overflow stream after 10 days. Assume the pit is well mixed.

Solution

$$AR = IR - OR + GR$$

The pollutant balance on the pit is:

$$\frac{d(C_e V)}{dt} = QC_i - QC_e + 0$$

where C_e is the concentration in the exit stream at any time. Rearranging,

$$\frac{dC_e}{dt} + \frac{Q}{V}C_e = \frac{QC_i}{V}$$

The hydraulic residence time of the pit, τ, is defined as $\dfrac{V}{Q}$, so the above equation becomes

$$\frac{dC_e}{dt} + \frac{1}{\tau}C_e = \frac{1}{\tau}C_i$$

Since both τ and C_i remain constant with time after the concentration jumps to 100 mg/L, the solution to this linear, first-order differential equation is

$$C_e(t) = C_{e_0} + (C_i - C_{e_0})(1 - e^{-t/\tau})$$

where C_{e_0} is the concentration in the overflow stream when the inlet stream first jumps to 100 mg/L.

Note that this solution checks with our intuitive understanding of the situation: At $t = 0$, $C_e = C_{e_0}$, and at $t = \infty$, C_e will equal C_i.

Substituting for this particular problem

$$C_e = 10 \text{ mg/L} + (100 - 10) \text{ mg/L} (1 - e^{-10/10})$$
$$C_e = 10 + 90 (0.632)$$
$$C_e = 67 \text{ mg/L after 10 days}$$

Table 3.1 Transient responses of some simple systems to some simple forcing functions.

System	Forcing Function	Forcing Graph	Response Graph
CSTR (A→B)	Step increase in feed flow rate		
	Step increase in concentration of reactant A		
	Spike increase in concentration of reactant A		
(PFR) (A→B)	Step increase in feed flow rate		
	Step increase in concentration of reactant A		
	Spike increase in concentration of reactant A		

3.3   Flow of Energy—Energy Balances

3.3.1 Various Forms of Energy and Power

Energy is often defined as work or the capacity to do useful work. (However, we also classify useless heat as energy.) **Power**, on the other hand, is the rate of doing work or the rate of expending energy. On a global scale, we have a continuous flow of energy into and through the environment. Unlike matter, the earth is not a closed system with respect to energy. We depend on a continuous flow of high quality energy (in the form of solar radiation) into the biosphere, just as we must have the flow of low quality thermal radiation away from the earth. Not only does the sun drive our hydrologic cycle, power our winds, and drive other physical processes, but through green plants it is the basis for energy flow up the food chain.

To do work requires energy; but no process is 100% efficient in converting energy into useful work. There will always be waste heat produced in any natural or human-constructed energy conversion process. However, energy *units* can be mathematically converted among the many forms of energy without incurring "losses," just as mass units can be converted from pounds to grams. Also, energy and power can be related mathematically by dividing or multiplying by a unit of time.

Energy has many forms (thermal, electrical, chemical potential, etc.), and many units of measure. Power may also take on many different units, although the most common deal with electrical power (kilowatts, megawatts) or with mechanical power (horsepower). Various units of measure and conversion factors for energy and power are presented in appendix A. Because fuel combustion is so important in today's way of life, Table 3.2 presents some nominal values of the energy content of various fuels.

Table 3.2 Nominal energy content of various fuels.

Form	Btu	kwh	kJ
Higher heating values of fuel			
1000 std cubic feet (MSCF) of natural gas	$1.03(10)^6$	302	$1.08(10)^6$
1 barrel (B) of crude oil (42 gallons)	$6.0(10)^6$	1,760	$6.33(10)^6$
1 gallon of gasoline	$1.26(10)^5$	36.9	$1.33(10)^5$
1 gallon of ethanol (anhydrous)	83,800	24.6	88,500
1 pound of coal			
—Pennsylvania anthracite	13,500	3.96	14,200
—Illinois bituminous	11,500	3.37	12,100
—N. Dakota lignite	7,200	2.11	7,600

Example Problem 3.8

Let us suppose that a family of four uses 1250 kwh of electrical energy on average every month. (a) How many gallons of gasoline is this equivalent to? (b) Where does this family use more energy in a year, operating their house or operating two cars (say about 20,000 miles at 20 miles per gallon)? (c) If gasoline is $1.25/gallon and electricity is 8.33 cents/kwh, where does the family spend more money?

Solution

(a) 1250 kwh/mo. x 1 gal. gasoline/36.9 kwh = 33.87 gal./mo. equivalent

(b) 33.87 gal. gasoline/mo. x 12 mo./yr = 406.4 gal./yr (electricity)
    20,000 miles/yr x 1 gal/20 miles = 1000 gal/yr (cars)

Therefore, the family expends more energy operating the cars each year.

(c) 1250 kwh/mo. x 12 mo./yr x $ .0833/kwh = $1250/yr (for electricity) vs.
    1000 gal/yr x $1.25/gal = $1250/yr (for gasoline)

Thus the family spends the same amount of money on both forms of energy.

Example Problem 3.9

(a) What is the annual average electrical *power* delivery to the family in example problem 3.8? Actually, most of the electricity used is delivered to homes in relatively short periods of time—the "peak" hours are 6–9 A.M. and 4–8 P.M. on week days. Assume that peak power delivery to a home is 15 kw. Calculate the electrical energy used if the 15 kw load is sustained (b) for one hour, or (c) for one year.

Solution

(a) 15,000 kwh/1 yr x 1 yr/365 days x 1 day/24 hours = 1.71 kw
(b) 15 kw x 1 hour = 15 kwh
(c) 15 kw x 1 yr x 365 day/yr x 24 hr/day = 131,000 kwh

3.3.2 Energy Balances

Just like mass, energy is conserved. In the study of thermodynamics, this simple statement is called the **first law of thermodynamics**. Regardless of what it is called, in the analysis of any process all energy inputs and outputs as well as energy accumulation must be taken into account. The form of energy can change through an energy conversion process (i.e., when natural gas is burned, its chemical potential energy is converted into thermal energy), but basically, the energy that goes into a process unit must come out or accumulate. Thus, an energy balance is very similar to a mass balance.

However, it must be recognized that, unlike mass, energy can be transferred by radiation and/or by conduction through the walls of containers and pipelines in addition to being transferred along with material by bulk flow. We will make use of mass and energy balances throughout this course, just as engineers make use of them throughout their careers.

A flowing stream of material carries with it an associated energy flow. If we consider only thermal energy for the moment, a more useful property to engineers than energy is enthalpy. **Enthalpy** is a thermodynamic property of material which depends on temperature, pressure, and composition. Enthalpy has a precise mathematical thermodynamic definition, but here we will use the definition "heat content." The change in enthalpy that occurs when the material passes from one set of conditions to another often is more useful than the absolute value of enthalpy. In many real-world situations, the change in enthalpy of a fixed amount of material can be approximated as:

$$\Delta H = mC_p\Delta T \qquad (3.16)$$

where:

ΔH = enthalpy change

m   = mass

C_p = specific heat at constant pressure

ΔT = temperature difference

In equation (3.16), any dimensionally consistent set of units may be used for ΔH, m, C_p, and ΔT. Two sets frequently encountered are (Btu, lb, Btu/lb-°F, and °F) and (cal, g, cal/g-°C, and °C), respectively.

Equation (3.16) can easily be extended to matter that is flowing from one place to another as well as matter that is stationary, as long as a constant flowing mass is considered. The assumptions inherent in equation (3.16) are that C_p is constant over the range of temperatures, that the effect of pressure is negligible (or that pressure is constant), that any change in composition has negligible effect on enthalpy, and that there has been no change of phase during the process. Phase changes of a pure compound can be accounted for easily by the following equation:

$$\Delta H = m\lambda \qquad (3.17)$$

where:

λ = the heat of phase change (i.e., vaporization, liquefaction, or sublimation).

Figure 3.4 traces the change in temperature and enthalpy as heat is added to a pure solid (ice) until it is converted to pure vapor, and example problem 3.10 illustrates some typical calculations with enthalpy.

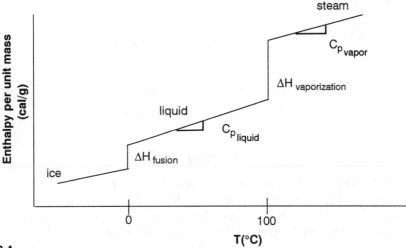

Figure 3.4
Temperature-enthalpy diagram for water.

Example Problem 3.10

A 30,000 L/day stream of wastewater must be heated from 15 °C to 40 °C for a treatment process to work properly. At what rate must we put heat into the stream? Assume the stream has thermal properties similar to water.

Solution

The heating process can be pictured as shown below.

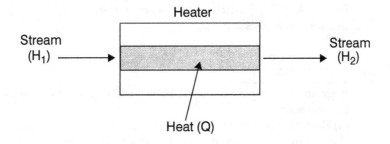

The general form of the energy balance is the same as that for a material balance.

$$AR = IR - OR + GR$$

which for this problem simplifies to:

$$0 = H_1 + Q - H_2 + 0$$

(Note: Q is often used as the symbol for heat flow as well as for volumetric flow rate.)
Thus,

$$\Delta H = Q \text{ but } \Delta H = \dot{m}C_p \Delta T,$$

so

$$\Delta H = 30{,}000 \text{ L/day} \times 1 \text{ kg/L} \times (4.18 \text{ kJ/kg-°C}) \times (25 \text{ °C})$$

$$\Delta H = 3.14 \ (10)^6 \text{ kJ/day}$$

Thus,

$$Q = 3.14 \ (10)^6 \text{ kJ/day}$$

Note: The way this problem was worded (i.e., calculate the heat put into the stream), there are no heat losses (heat transfer inefficiencies) to consider. In a real situation, in order to transfer that much heat into the stream, a device called a heat exchanger is used. In this device, as in all real-world devices, there are always inefficiencies—this is the topic of the next section!

3.3.3 Real-World Energy Conversion Processes

Based on common sense, we can accept the idea that there is no real-world process that is 100% efficient in using energy to do useful work, or in transferring energy from one useful form to another useful form. As an example, if we use electrical energy to lift an elevator full of people, not all of the electrical energy will be used to do the lifting. Some portion of the electrical energy will be "lost" as heat in the motor wiring, as friction in the gears, etc. In the study of thermodynamics, this common-sense idea is known as the **second law of thermodynamics**. Note that "losing energy" is not a violation of the first law: the total amount of energy is conserved. But the second law requires that some of the energy input to this process must be dissipated in a useless form (such as low-temperature heat). The following example demonstrates this concept.

Example Problem 3.11

Regarding example problem 3.10, assume that a steam heater is used to heat the wastewater.
(a) Calculate the heat that must be contained in the steam in order to transfer the required amount of heat into the water stream. Assume that this old heater has heat losses to the surroundings of 15%. (b) Next, assume that the steam is produced in a furnace burning a fuel oil that is similar in heat content to crude oil. Assume that the production of steam from the chemical potential energy in the oil has a thermal efficiency of only 60%; the remaining heat goes out the stack with the hot gases, or is lost through the hot walls of the pipes or through the furnace walls, or other places. Calculate the hourly feed rate of oil, in gal/hour.

Solution

(a) From the previous example, the heat required by the wastewater was 3.14 $(10)^6$ kJ/day. This is one "output" of the heater. The other output is the "heat losses." At steady state, the input energy to the heater (the heat contained in the steam) must be the sum of the two outputs, or:

$$\text{Input} = 3.14\ (10)^6 + \text{losses} = 3.14\ (10)^6 + 0.15 \times \text{Input}$$

Therefore, the Input (heat contained in the steam) is:

$$\frac{3.14(10)^6\,\text{kJ/day}}{0.85} = 3.69(10)^6\,\text{kJ/day}$$

(b) The heat contained in the steam comes from burning oil, but only 60% of the heat released by burning the oil actually ends up in the steam. Therefore,

$$\text{Heat released by oil} = \frac{3.69\ (10)^6\ \text{kJ/day}}{0.6} = 6.15\ (10)^6\ \text{kJ/day}$$

The oil feed rate is

$$\frac{6.15\ (10)^6\ \text{kJ/day}}{6.33\ (10)^6\ \text{kJ/B}} \times \frac{42\ \text{gal}}{1\ \text{B}} \times \frac{1\ \text{day}}{24\ \text{hr}} = 1.70\ \text{gal/hr}$$

3.4 Combined Material and Energy Balances

We do nothing different when we combine material and energy balances. In fact the two complement each other and assist the engineer in truly understanding the process. The previous example problem showed how doing an energy balance can help determine a required firing rate of a fuel. In addition, combined energy and material balances are quite useful in gaining a preliminary understanding of the major environmental effects of large electricity-generating plants. The use of combined energy and materials balances are demonstrated in the following example problem.

Example Problem 3.12

A 1000 MW coal-burning power plant is burning anthracite coal from Pennsylvania, which has 6% ash and 2.5% sulfur. The plant has a thermal efficiency of 40%. Assume that the overall removal efficiencies for ash and SO_2 are 99.5% and 88%, respectively. Calculate: (a) the rate of heat emitted to the environment (kJ/s), (b) the rate of coal input to the furnace (kg/day), (c) the rate of ash emissions to the atmosphere (kg/day), and (d) the rate of SO_2 emissions to the atmosphere (kg/day).

Solution

First, do an energy balance on the plant (recall that power is a rate of energy).

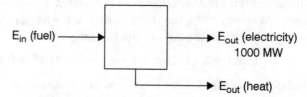

With a thermal efficiency of 40%, only 40% of the input energy is converted to useful electricity, so

$$E_{in} = \frac{1000 \text{ MW}}{0.40} = 2500 \text{ MW}$$

(a) The heat emitted to the environment is:

$$\text{Heat} = (1 - 0.40)\ 2500 \text{ MW} = 1500 \text{ MW, or}$$

$$\text{Heat} = 1500 \text{ MW} \times \frac{1000 \text{ kw}}{1 \text{ MW}} \times \frac{1 \text{ kJ/s}}{1 \text{ kw}} = 1.5(10)^6 \text{ kJ/s}$$

(b) To calculate the coal input rate, we need the energy content. From Table 3.2, 1 pound of this coal contains 14,200 kJ or 3.96 kwh.

$$\text{Coal input} = 2500 \text{ MW} \times \frac{1000 \text{ kw}}{1 \text{ MW}} \times \frac{24 \text{ hr}}{1 \text{ day}} \times \frac{1 \text{ lb}}{3.96 \text{ kwh}} \times \frac{1 \text{ kg}}{2.2 \text{ lb}}$$

$$= 6.89\ (10)^6 \text{ kg/day}$$

(c) To get ash emissions, first calculate ash input,

$$\text{Ash input} = 6.89(10)^6 \text{ kg/day} \times \frac{0.06 \text{ kg ash}}{1 \text{ kg coal}} = 4.13(10)^5\ \frac{\text{kg ash}}{\text{day}}$$

Then multiply by (1.0 – removal efficiency)

$$\text{Ash emissions} = 4.13\ (10)^5 \text{ kg/day} \times (1.00 - 0.995) = 2065 \text{ kg/day}$$

(d) Assume that all the incoming sulfur is oxidized to SO_2, i.e.,

$$S + O_2 \rightarrow SO_2$$

The incoming sulfur is just:

$$S_{in} = 6.89\ (10)^6 \times 0.025 = 1.72\ (10)^5 \text{ kg/day}$$

Since the mass ratio of SO_2 to S is 64 to 32, or 2 to 1, and since the problem specifies 88% removal efficiency, SO_2 emissions are:

$$SO_2 = 1.72(10)^5 \text{ kg/day} \times \frac{2 \text{ kg } SO_2}{1 \text{ kg S}} \times (1 - 0.88) = 4.13(10)^4 \text{ kg/day}$$

3.5   Computer Applications

Personal computer developments over the past two decades have greatly expanded access to computing capability. Prior to the mid-1970s, the high costs of mainframe computers effectively limited availability to highly capitalized organizations such as corporations, universities, and government agencies. In particular, during the period from 1990 to the present, powerful personal computers have become affordable for individual use. The personal computing market (hardware and software) now targets the needs of individual users, including undergraduate engineering students.

The problems presented in this text are not so advanced that computers are necessary to obtain solutions. In most cases, solutions are obtained using algebra, differential and integral calculus, graphical methods, and/or trial and error calculations. In some cases, numerical methods are also provided for selected example problems which involve solution of unsteady-state processes, or which require iterative solution (simultaneous linear or nonlinear equations). The reason we illustrate elementary numerical techniques is so the student may use them when numerical methods may be the only way to solve the more complex problems that engineers sometimes face.

The next few example problems are fully developed to include an analytic solution and the pseudocode for numerical solution. The pseudocode shows the logical thought processes involved in developing a numerical solution. For some of the example problems, a spreadsheet solution or a BASIC program is shown in appendix C to demonstrate the ease with which these techniques can be applied. The authors recognize the difficulty of "teaching" spreadsheets in a textbook. Some basic knowledge of computers and spreadsheets on the part of the student is assumed. It is hoped and recommended that students experiment on their own with spreadsheets to improve their skills.

3.5.1 Simulation of Unsteady-State Processes

Nonsteady-state equations may often be solved more easily using numerical methods than by analytic approaches. A general solution method is presented for a first-order differential equation using Euler's Method (also known as the point-slope method). Higher order differential equations may be converted to first-order cases by creation of state variables (Smith and Corripio, 1985). Euler's Method is valid for linear or nonlinear equations. Nonlinearities are present when a variable is exponentiated, squared, or raised to a power, for example, or when two variables are multiplied or divided. Other more sophisticated methods are available that would achieve greater computational efficiency, however the simplicity of Euler's Method makes it attractive.

In order to use Euler's Method to numerically integrate an equation, it is necessary to define the initial state of the system to permit determination of a time derivative $\left(\dfrac{dy_i}{dt} \right)$ for all response variables (y_i). The new value of the response variable (y_{new}) is calculated by stepping through calculations using the old value (y_{old}), the time derivative, and the time step size (deltime) as shown below:

initial state definition

$$t = 0 \tag{3.18}$$

$$y_{old} = y(0) \tag{3.19}$$

iteration

$$\frac{dy}{dt} = f(y_{old}, t) \tag{3.20}$$

$$y_{new} = y_{old} + \frac{dy}{dt} \times deltime \tag{3.21}$$

$$y_{old} = y_{new} \tag{3.22}$$

$$t = t + deltime \tag{3.23}$$

The calculations in equations (3.20) to (3.23) are repeated for the desired duration of the simulation.

The accuracy of the simulation can be improved (that is, the truncation error can be reduced) by using smaller increments for the time step size. Reducing the step size increases the number of computations, which would certainly be a prime consideration if the calculations were performed manually. With modern personal computers, decreases in step size may not noticeably increase the length of time necessary for completion of the simulation, however the use of extremely small time increments may introduce significant round-off errors. Thus their use is not desirable due to numerical (round-off) errors and excessive computer time. The trade-off between round-off and truncation error is reviewed by Chapra and Canale (1988).

Example Problem 3.13

A first-order reaction ($A \rightarrow B$) is conducted in a constant volume batch reactor. The rate constant is 0.1 min^{-1} and the initial concentration is 100 mg/L. Determine the concentration of A which remains after 30 minutes using both analytical and numerical solution methods.

Solution

Develop the governing differential equation from a system mass balance for conservation of the mass of the reactant:

$$AR_A = IR_A - OR_A + GR_A$$

$$\frac{d(V\,C_A)}{dt} = 0 - 0 + Vr_A$$

$$V \frac{d\,C_A}{dt} = -V\,k\,C_A$$

$$\frac{d\,C_A}{C_A} = -k\,dt$$

The analytic solution for this case was developed in chapter 2 by separation of variables and integration:

$$C_A = C_{A_0} \exp(-k\,t)$$

After 30 minutes of reaction:

$$C_A = (100\ \text{mg/L})\ \exp[-(0.1\ \text{min}^{-1})(30\ \text{min})] = 4.9787\ \text{mg/L}$$

Even though not needed for this simple case, a numerical solution was developed to demonstrate Euler's Method using different values of the time step size. The numerical procedures are presented below in a pseudocode form. A spreadsheet for this example is contained in appendix C. The simulation results for various time step sizes are compared with the analytical (true) solution result. As can be seen, reducing the time step size from 10 minutes to 0.01 minute greatly improves the accuracy. However, at very small time steps, overflow occurs in a spreadsheet, and round-off errors increase in coded programs.

Pseudocode

Begin   Define System Parameters—input data
        k = 0.1 per minute
End   Define System Parameters

Begin   Define Initial Conditions—input more data
        C = 100 mg/L
        t = 0 minutes
End   Define Initial Conditions

Begin   Define Simulation Parameters—input desired values to test
        Deltime = 10, 1, 0.1, or 0.01 minutes
        Endtime = 30 minutes
End   Define Simulation Parameters

Begin   Time Domain Simulation—start calculations
DO   For t = 0 to Endtime
        dCdt = − k x C
        C = C + dCdt x deltime

$$t_{new} = t_{old} + deltime$$

ENDDO

End   Time Domain Simulation

Simulation Results

Deltime min	C @ 30 min mg/L	Error mg/L	Pct Error %
10	0.0000	4.9787	100.00
1	4.2391	0.7396	14.86
0.1	4.9041	0.0746	1.50
0.01	4.9712	0.0075	0.1

3.5.2 Iterative Solution of Systems of Equations

The complete description of a system may require many mass and/or energy balances. Independent equations may be obtained by making balances using different boundaries or different conservative properties. The resulting system of equations must be solved simultaneously. Obtaining an analytic solution may be very difficult if there are many unknowns and/or if nonlinearities exist in the equations. Systems of linear algebraic equations can be solved using matrix methods. However, nonlinear equations often require numerical solution methods.

The solution methodology presented in this section relies exclusively on iterative calculations. This unsophisticated, brute-force approach requires very minimal knowledge of numerical methods or computer programming. The methods are readily adapted to spreadsheets or simple programming. Although the lack of sophistication associated with this approach is viewed as an advantage for this course, it is conceded that the methods do not always converge readily and may require accurate initial estimates of the actual solution to achieve convergence efficiently (or to converge at all). Study of other more versatile and powerful numerical methods may be warranted (Chapra and Canale, 1988).

The solution approach involves the following steps:

1. Make initial estimates for all unknown variables (y_{old}).
2. Algebraically manipulate each system equation to obtain a set of equations which are used for direct (iterative) calculation of each unknown variable.
3. By sequential calculation, determine values for each unknown (y_{temp}).
4. Calculate a new estimate of each variable (y_{new}) as a weighted average using y_{old} and y_{temp}.
5. Replace all y_{old} values with y_{new} values.
6. Repeat steps 3 through 5 until satisfactory convergence is achieved.

The weighted average is calculated as follows:

$$y_{new} = y_{temp} \, x + y_{old} \, (1-x) \tag{3.24}$$

where:

y_{new} = value of unknown variable from the current iteration

y_{old} = value of unknown variable from the previous iteration

y_{temp} = value of unknown variable obtained in step 3 above

x   = weighting factor (varies between 0 and 1)

Values of the weighting factor close to 1 tend to achieve convergence more rapidly for "well-behaved" functions; however, use of small values for x may be necessary to achieve convergence for those equations that are not well-behaved. Application of these calculation methods is illustrated with a series of examples which include linear and nonlinear systems of equations.

Example Problem 3.14

A complete-mix reactor (CSTR) is used to treat an industrial wastewater to remove a toxic organic compound. The reaction follows first-order kinetics with a rate constant of 0.4 per minute. The flow rate is 100 L/min, the reactor volume is 1000 L, and the initial concentration is 100 mg/L. Determine the steady-state effluent concentration using algebraic and numerical techniques.

Solution

Develop a mass balance for the reactor for the mass of the toxic compound:

$$AR = IR - OR + GR$$

$$V \frac{dC}{dt} = Q\, C_0 - Q\, C + rV$$

At steady state:

$$0 = Q\, C_0 - Q\, C - k\, C\, V$$

This single linear equation may be solved readily using algebraic methods:

$$C = \frac{C_0}{1 + k\dfrac{V}{Q}} = \frac{100\ \text{mg/L}}{1 + \left(0.4\ \text{min}^{-1}\right)\dfrac{1000\ \text{L}}{100\ \text{L/min}}} = 20\ \text{mg/L}$$

A numerical solution is also presented (the logic and the results are presented here; see the spreadsheet in appendix C for details of the implementation). The mass balance equation is rearranged to solve for the unknown concentration:

$$C_{temp} = C_0 - \frac{k\, V\, C_{old}}{Q}$$

Note that this is an implicit solution; that is, the new calculated (temporary) value of C depends on the old value of C. The preceding equation is solved iteratively with a weighting factor (x) = 0.1 and an initial concentration estimate of 100 mg/L.

Pseudocode

Begin	Define System Parameters
	$k = 0.4$ min^{-1}
	$Q = 100$ Liters per min
	$V = 1000$ Liters
	$C_0 = 100$ mg/L
End	Define System Parameters
Begin	Define Simulation Parameters
	$x = 0.1$
End	Define Simulation Parameters
Begin	Define Initial Estimate of Unknown
	$C = 100$ mg/L
End	Define Initial Estimate of Unknown
Begin	Iterative Calculations
DO	For I = 1 to Convergence

$$C_{temp} = C_0 - \frac{V\,k\,C}{Q}$$

$$C = C_{temp}\,x + C\,(1-x)$$

ENDDO

End   Iterative Calculations

(Note that with a spreadsheet or BASIC program where the student interacts directly with the computer, the iterations can be repeated as many times as deemed necessary without fancy logic to tell the computer when convergence has been obtained.)

Results of Iterative Calculation of C
Weighting Factor = 0.1

I	C
0	100.00000
1	60.00000
2	40.00000
3	30.00000
4	25.00000
5	22.50000
6	21.25000
7	20.62500
8	20.31250

9	20.15625
10	20.07813
15	20.00244
20	20.00008
25	20.00000

Note that convergence to within 0.1% of the correct value of 20 mg/L is obtained after only 15 iterations in spite of the highly inaccurate initial estimate of 100 mg/L. It is emphasized that successful numerical results were obtained because of the relatively small weighting factor ($x = 0.1$).

For comparison, a second series of calculations was performed with a weighting factor of one-half and a comparatively accurate initial estimate of the unknown concentration ($C = 21$ mg/L). As shown below, the calculations diverged for these specifications, with failure to obtain a solution in spite of the accuracy of the initial estimate for the unknown. This failure exposes certain shortcomings of this simplistic numerical solution method. It is recommended that final values for the unknown variable(s) obtained by iteration be checked by substitution into the operative equation(s) to validate the solution.

Iterative Calculation of C
Weighting Factor = 0.5

I	C
0	21.00000
1	18.50000
2	22.25000
3	16.62500
4	25.06250
5	12.40625
6	31.39063
7	2.91406
8	45.62891
9	- 18.44336
10	77.66504

Example Problem 3.15

Repeat example problem 3.14 for variable-order kinetics:

$$-r = \frac{k\,C}{k_C + C} \qquad k = \frac{45\,\text{mg}}{\text{L - min}} \qquad k_C = 40 \text{ mg/L}$$

Solution

The mass balance equation may be manipulated to yield a quadratic equation which is solved explicitly:

$$AR = IR - OR + GR$$

$$V \frac{dC}{dt} = Q\,C_0 - Q\,C + rV$$

At steady state:

$$0 = Q\,C_0 - Q\,C - V\frac{k\,C}{k_C + C}$$

$$Q\,C^2 + (Q\,k_C + V\,k - Q\,C_0)\,C - Q\,C_0\,k_C = 0$$

$$100\,C^2 + 39{,}000\,C - 400{,}000 = 0$$

$$C = \frac{-39{,}000 \pm \sqrt{(39{,}000)^2 - 4(100)(-400{,}000)}}{2(100)} = -195 \pm 205$$

The positive root is selected; C = 10 mg/L.

A numerical solution approach is developed which is virtually identical to the procedure presented previously for first-order kinetics. The correct result of 10 mg/L (±0.1%) is obtained after eight iterations (see appendix C for the spreadsheet solution).

Pseudocode

Begin   Define System Parameters

$$k = 45\ \frac{mg}{L\text{-}min}$$
$$k_C = 40\ mg/L$$
Q = 100 Liters per min
V = 1000 Liters
$$C_0 = 100\ mg/L$$

End   Define System Parameters

Begin   Define Simulation Parameters
x = 0.1
End   Define Simulation Parameters

Begin   Define Initial Estimate of Unknown
C = 100 mg/L
End   Define Initial Estimate of Unknown

Begin   Iterative Calculations

DO   For I = 1 to Convergence

$$C_{temp} = C_0 - \frac{V\,k\,C}{Q\,(k_C + C)}$$

$$C = C_{temp}\,x + C\,(1-x)$$

ENDDO

End   Iterative Calculations

Iterative Calculation of C

I	C
0	100.00000
1	67.85715
2	42.76017
3	25.23375
4	15.30345
5	11.32081
6	10.26222
7	10.04818
8	10.00871
9	10.00157
10	10.00028

The preceding examples demonstrate the application of numerical solution techniques to solving a single linear or nonlinear equation with one unknown. The numerical approach requires virtually identical programming and computational effort regardless of whether the equation involves a nonlinearity. The algebraic approach cannot be depended upon to yield an explicit solution for all nonlinearities. In general, numerical methods become more attractive as the complexity of the mathematics increases, either by inclusion of nonlinear terms or by increasing the number of independent equations and unknowns. Application of iterative solution methods to a system of equations with three unknowns is illustrated in example problem 3.16. A final, complex example is presented for simultaneous solution of four highly nonlinear equations in example problem 3.17.

Example Problem 3.16

In example problem 3.14, it was shown that a single CSTR with a volume of 1000 Liters could achieve an effluent concentration of 20 mg/L for a particular wastewater. For the same wastewater (Q = 100 L/min, first-order kinetics with k = 0.4/minute, and concentration = 100 mg/L), determine the total volume (V) of reactor required to achieve the same effluent concentration of 20 mg/L using a three-compartment reactor modeled as three equal-volume

CSTRs in series.

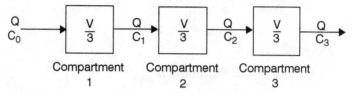

Solution

Three independent steady-state mass balance equations may be generated representing one equation for each reactor compartment:

$$0 = Q\,C_0 - Q\,C_1 - \frac{V}{3}\,k\,C_1$$

$$0 = Q\,C_1 - Q\,C_2 - \frac{V}{3}\,k\,C_2$$

$$0 = Q\,C_2 - Q\,C_3 - \frac{V}{3}\,k\,C_3$$

The equations may be solved simultaneously to obtain an analytical solution as follows:

$$C_1 = \frac{C_0}{1 + \dfrac{k\,V}{3\,Q}} \qquad C_2 = \frac{C_1}{1 + \dfrac{k\,V}{3\,Q}} \qquad C_3 = \frac{C_2}{1 + \dfrac{k\,V}{3\,Q}}$$

$$C_3 = \frac{C_0}{\left[1 + \dfrac{k\,V}{3Q}\right]^3}$$

$$V = \left[\left(\frac{C_0}{C_3}\right)^{1/3} - 1\right]\frac{3\,Q}{k} = \left[\left(\frac{100 \text{ mg/L}}{20 \text{ mg/L}}\right)^{1/3} - 1\right]\frac{3(100 \text{ L/min})}{0.4 \text{ min}^{-1}} = 532 \text{ Liters}$$

A substantial reduction in reactor volume is achieved by dividing the reactor into multiple cells in series (532 Liters instead of 1000 Liters). The reduction in volume would be even more dramatic for applications which operate at a greater removal efficiency.

Numerical solution methods may be employed to achieve the same result. In this example, a large number of iterations (100+) is necessary to achieve convergence (due in part to the use of a small weighting factor). The increase in computational effort is barely noticeable using a personal computer. In appendix C, this problem solution is presented using a simple BASIC program.

Pseudocode

Begin   Define System Parameters

$k = 0.4 \ \text{min}^{-1}$

$Q = 100 \ \text{Liters per min}$

$C_3 = 20 \ \text{mg/L}$

$C_0 = 100 \ \text{mg/L}$

End   Define System Parameters

Begin   Define Simulation Parameters

$x = 0.1$

End   Define Simulation Parameters

Begin   Define Initial Estimates of Unknowns

$C_1 = 100 \ \text{mg/L}$

$C_2 = 100 \ \text{mg/L}$

$V = 100 \ \text{Liters}$

End   Define Initial Estimates of Unknowns

Begin   Iterative Calculations

DO   For I = 1 to Convergence

$$C_{1temp} = C_0 - \frac{V \, k \, C_1}{3 \, Q}$$

$$C_1 = C_{1temp} \, x + C_1 \, (1-x)$$

$$C_{2temp} = C_1 - \frac{V \, k \, C_2}{3 \, Q}$$

$$C_2 = C_{2temp} \, x + C_2 \, (1-x)$$

$$V_{temp} = \frac{3 \, (Q \, C_2 - Q \, C_3)}{k \, C_3}$$

$$V = V_{temp} \, x + V \, (1-x)$$

ENDDO

End   Iterative Calculations

Iterative Calculations of C_1, C_2, and V

I	C_1	C_2	V
1	98.82353	98.71972	385.1990
10	63.97634	58.31544	1366.6450
20	47.45539	33.24564	977.9205
30	49.87777	27.63188	549.6943
40	57.67387	31.80998	435.8147
50	60.29835	35.12552	499.6213
60	59.20722	35.04047	546.0599
70	58.28276	34.25233	544.8803

80	58.23968	34.01500	533.3420
90	58.44880	34.12603	529.7913
100	58.52795	34.21996	531.3990
110	58.50344	34.22382	532.7756
120	58.47646	34.20255	532.8380
130	58.47353	34.19466	532.5274
140	58.47906	34.19721	532.4109
150	58.48157	34.19993	532.4476
200	58.48039	34.19952	532.4805
250	58.48035	34.19953	532.4821

Example Problem 3.17

Determine the pH of the neutralized acid solution described in example problem 2.8 using iterative calculations for simultaneous solution of the system of equations.

Solution

The system of equations is repeated below:
1. Chemical Equilibria

$$K_W = [H^+][OH^-] = 10^{-14}$$

$$K_A = \frac{[H^+][Ac^-]}{[HAc]} = 10^{-4.6}$$

2. Electroneutrality (Charge Balance)

$$1\,[H^+] + 1\,[Na^+] = 1\,[OH^-] + 1\,[Ac^-]$$

3. Conservation of Mass (Moles)

$$C_T = [HAc] + [Ac^-]$$

From the problem statement, 1000 L of 0.005 N acetic acid are neutralized by adding 50 L of 0.10 N sodium hydroxide. Values for C_T and $[Na^+]$ are calculated below:

$$C_T = \frac{(1000\ L)(0.005\ eq/L)(1\ mole/eq)}{(1000\ L + 50\ L)} = 0.004762\ M$$

$$[Na^+] = \frac{(50\ L)(0.10\ eq/L)(1\ mole/eq)}{(1000\ L + 50\ L)} = 0.004762\ M$$

The system of four equations is determinant for the four unknowns: $[H^+]$, $[OH^-]$, $[HAc]$, and $[Ac^-]$. Pseudocode is provided on the next page. Two logic statements are included in the iterative calculations to prevent the occurrence of negative values for $[H^+]$ or $[Ac^-]$. This procedure imposes appropriate constraints on the calculated values to facilitate convergence.

This system of equations does *not* converge readily due to the *highly* non-linear nature of the equations. Application of the electroneutrality condition involves subtraction of numbers which differ by *six orders of magnitude*; very small relative changes in acetate ion concentration exert a pronounced effect on the calculated hydrogen ion concentration. Consequently, use of a small weighting factor, coupled with many iterations, is required to achieve convergence. The final values obtained after 2000 iterations are summarized below:

$$[H^+] = 7.25 \times 10^{-9} \quad [OH^-] = 1.38 \times 10^{-6}$$

$$[HAc] = 1.37 \times 10^{-6} \quad [Ac^-] = 4.76 \times 10^{-3}$$

Pseudocode

Begin   Define System Parameters

$K_W = 10^{-14}$

$K_A = 10^{-4.6}$

$C_T = 0.004762$

$[Na^+] = 0.004762$

End   Define System Parameters

Begin   Define Simulation Parameters

$x = 0.01$

End   Define Simulation Parameters

Begin   Define Initial Estimates of Unknowns

$[H^+] = 10^{-7}$

$[OH^-] = 10^{-7}$

$[Ac^-] = C_T$

$[HAc] = 0$

End   Define Initial Estimates of Unknowns

Begin   Iterative Calculations

DO   For I = 1 to Convergence

$$[OH^-]_{temp} = \frac{K_W}{[H^+]}$$

$$[OH^-] = [OH^-]_{temp} \, x + [OH^-] \, (1 - x)$$

$$[HAc]_{temp} = \frac{[H^+][Ac^-]}{K_A}$$

If $[HAc]_{temp} > C_T$ Then $[HAc]_{temp} = C_T$

$$[HAc] = [HAc]_{temp} \, x + [HAc] \, (1 - x)$$

$$[Ac^-] = C_T - [HAc]$$

$$[H^+]_{temp} = [Ac^-] + [OH^-] - [Na^+]$$

$$\text{If } [H^+]_{temp} < 0 \text{ Then } [H^+]_{temp} = 0$$

$$[H^+] = [H^+]_{temp} \, x + [H^+] \, (1 - x)$$

ENDDO

End   Iterative Calculations

The resulting answer (pH = 8.14) from the simultaneous numerical solution of the exact system of equations is virtually identical to the approximate solution obtained in example problem 2.8 (pH = 8.14). The approximate method provides a satisfactory result in this case because the two simplifying assumptions are very accurate:

$$C_T >> [H^+]$$

$$[Ac^-] >> [HAc]$$

For more dilute solutions of acid or for a lower experienced pH (near the pK_A), these assumptions may introduce significant inaccuracies, in which case the numerical solution would be expected to yield results which are superior to the approximate method used in chapter 2.

End-of-Chapter Problems

3.1   Five million kilograms per day of coal are burned in an electric power plant. The coal has an ash content of 12% by mass. Forty percent of the ash falls out the bottom of the furnace. The rest of the ash is carried out of the furnace with the hot gases into an electrostatic precipitator (ESP). The ESP is 99.5% efficient in removing the ash that comes into it. Draw a diagram representing this process and calculate the mass emissions rate of ash into the atmosphere from this plant.

3.2   Given the following diagram and data, calculate the quantity of steam required to regenerate the carbon in step 2.

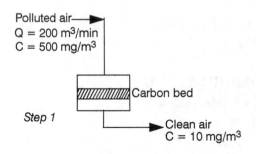

Polluted air⟶
Q = 200 m³/min
C = 500 mg/m³

Carbon bed

Step 1

Clean air
C = 10 mg/m³

In step 1, we allow a polluted air stream to flow through a bed of activated carbon. Most of the pollutant molecules adsorb onto the carbon and are thus removed from the air. This step lasts for 6 hours, then the pollutant-laden bed is replaced with a new bed.

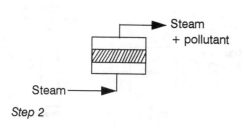

Steam + pollutant

Steam

Step 2

After a bed gets loaded with pollutant, it is cleaned with steam. The steam removes all of the pollutant from the carbon. It has been found that a ratio of steam to pollutant of 7 kg steam/1 kg pollutant is necessary for complete cleaning. Calculate the quantity of steam required to do this job, kg/day. Assume the process runs 24 hr/day.

3.3   The following is a clarifier-thickener system to separate solids and liquids. Ignore the effect of solids content on density of each stream (i.e., assume density of solids = density water).

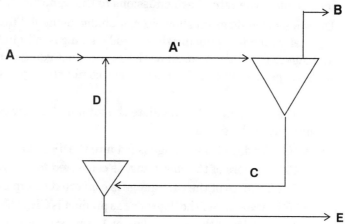

Complete the following material balance table.

Stream	Flow rate, L/s	Solids, mg/L
A	100	3000
B	95	15
C		6000
D	50	
E		
A'		

3.4   A 600 MW coal-burning power plant is burning Illinois bituminous coal with 8% ash content. The plant is 38% efficient, 35% of the ash drops out in the furnace, and the electrostatic precipitator is 99.0% efficient. A simplified sketch appears on the following page.

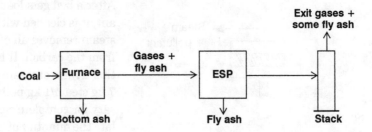

(a) Draw an energy balance diagram for the plant and calculate the rate of heat emitted to the environment, in J/s.

(b) Calculate the rate of coal input to the furnace, in kg/day.

(c) Calculate the rate of ash emissions to the atmosphere, in kg/s.

3.5   Draw a labeled diagram showing raw sludge being fed to an anaerobic digester which is producing three products: a gas, a liquid supernatant, and digested sludge. Also, show the gas product going to a furnace along with a separate stream of air, and show combustion gases leaving the furnace.

Given the following data, calculate the total molar flow rate of CO_2 gas coming out of the furnace.

- 1000 kg/day of raw sludge is fed into the digester.
- 40% by mass of the raw sludge is converted to digested sludge.
- 58% by mass of the raw sludge is converted to supernatant.
- 70% by volume of the digester gas product is CH_4, the other 30% is CO_2.
- The density of the gas product at 1 atm. pressure and 25 °C is 1 kg/m³.
- Enough air is used to burn all of the methane (CH_4) to CO_2 and H_2O.

3.6   *Given*: A wastewater treatment plant as shown below. Influent total suspended solids (TSS) = 260 mg/L. Alum is added at a rate of 30 lb/million gallons as a settling aid. Effluent requirements call for 90% reduction in TSS.

Find: If the primary settler removes 45% of initial TSS, find the percent removal required in the secondary settler in order to achieve the 90% removal requirement for the plant as a whole. Assume all alum settles out.

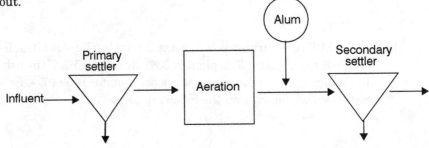

3.7 The stack gas from a process contains 4.0 g of particulate matter (PM) per m^3 of gas. The gas flow rate is 5 m^3/s. If an electrostatic precipitator removes 1000 kg of PM per day, what is the emission rate of PM? Give answer in kg/day.

3.8 A 1000 MW plant operates at 40% efficiency. It consumes 10,000 tons of coal/day. The plant uses coal that is 4% sulfur and 10% ash. The ash leaves the furnace as bottom or fly ash. Virtually all the sulfur is converted to sulfur dioxide, and the plant is equipped with a 90% efficient SO_2 scrubber.

(a) Calculate the amount of sulfur dioxide emitted in gaseous form.

Assume that one-third of the ash in the coal forms bottom ash, and the remainder forms fly ash. Pollution control devices capture 98% of the fly ash.

(b) Determine the amount of bottom ash and fly ash collected in one day.

(c) Determine the daily emissions of fly ash to the atmosphere.

3.9 A mining process discharges a silty water into a clean brook at the rate of 5 cubic meters per minute. The silt content is 200 mg/L. What is the concentration in the brook during the rainy season (the brook flows at 50 cubic meters/min), and during a severe dry spell (the brook flows at 5 cubic meters/minute)?

3.10 Calculate the minimum rate at which cooling water must be pumped through the condensers of a 800 MW nuclear power plant if the maximum cooling water temperature rise is 15 °F. The efficiency of the plant is 32%. Assume no heat is lost to the atmosphere directly from the system. Give your answer in pounds/hr, and ft^3/sec. Assuming the plant is designed with five identical pipes in parallel to carry this cooling water, calculate the diameter of each pipe if the water velocity is 6.0 ft/sec.

3.11 A new state regulation requires that the wastewater (WW) from a tannery be routed to a municipal wastewater treatment plant (WWTP) for final treatment. However, the tannery WW must be pretreated to remove certain organic pollutants before it goes to the WWTP. The pretreatment process is a biological anaerobic process that has a limit on total dissolved solids (TDS) of no more than 20,000 mg/L. But the tannery WW contains 95,000 mg/L of TDS, so it must be diluted somehow. A proposed treatment process is shown in the accompanying figure. Calculate the flow rate of the Recycle that is required in order to keep the TDS in the mixed input stream going into the pretreatment process at 20,000 mg/L.

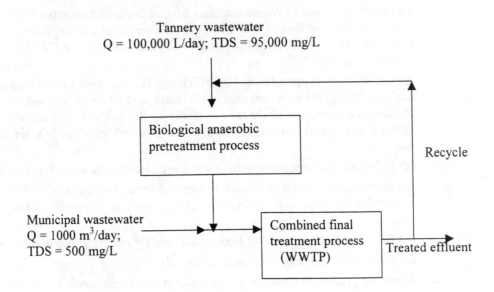

3.12 A gas containing equal parts (on a molar basis) of H_2, N_2, and H_2O is passed through a column of silica gel which absorbs 96% of the water and none of the other gases. The column packing was initially dry, and had a mass of 2 kg. Following five hours of continuous operation, the pellets are reweighed and are found to have a mass of 2.18 kg. Calculate the molar flow rate (in mols/hr) of the feed gas and the mole fraction of water vapor in the product gas.

3.13 In order to meet a certain octane number specification, it is necessary to produce a gasoline containing 83% (by weight) isooctane and 17% n-heptane. How many gallons of a high-octane gasoline containing 92% isooctane and 8% n-heptane must be blended with a straight-run gasoline containing 63% isooctane and 37% n-heptane to obtain 10,000 gal of the desired gasoline? The density of each of the liquids is 6.7 lb/gal.

3.14 The analysis of the waste gas from a burner fueled with natural gas (essentially pure CH_4) is as follows:

    N_2 72.88 mol percent
    O_2 3.87 mol percent
    CO_2 7.75 mol percent
    H_2O 15.5 mol percent

What is the ratio of moles of air (assume air is 79% N_2 and 21% O_2) per mole of natural gas fed to the burner?

3.15 A CSTR is being fed a 100 L/minute stream containing 0.10 mole/L of reactant A. The reaction proceeds according to:

$$2\,A \rightarrow B \qquad \text{with} - R_A = k\,[A]^3$$

(a) Calculate the volume of the reator if k has a value of:

$$3.0\,\frac{(\text{liters})^2}{(\text{mole})^2\,\text{min}} \quad \text{and the outlet concentration of A is 0.05 mole/L.}$$

(b) Calculate the production rate of B in moles/day.

3.16 Consider a second-order reaction (2 A→B) occurring in a CSTR with a volume of 250 L. Given an inlet flow rate of 50 L/s, an inlet concentration of A of 0.20 mole/L, and an outlet concentration of A of 0.08 mole/L, calculate the rate constant (with appropriate units).

3.17 For the same reaction and inlet conditions as problem 3.16 above, but given a PFR with a volume of 250 L, calculate the expected outlet concentration of A.

3.18 A PFR can be simulated as many small CSTRs in series. Develop a spreadsheet or simple BASIC program to solve problem 3.17 above, modeling the PFR as (a) 5 CSTRs of 50 L each, and (b) 25 CSTRs of 10 L each.

3.19 A coal slurry is a mixture of crushed coal and water which can be pumped via pipeline across the country. A power plant needs 20 million kg/day of coal (dry basis). A 50% coal/50% water slurry is pumped in the pipeline to the power plant. When the slurry is received at the power plant, the coal is separated from the water. However, the separation process is inefficient. The separated water stream contains 2% (by weight) coal, and the separated coal contains 20% (by weight) of water. Fill in the following material balances table for the flow rates received at the power plant. (Hint: First draw a diagram of the separation process).

Stream	Mass Flow Rate (millions of kg/day)		
	Coal	Water	Total
Pipeline Slurry			
Separated Water			
Separated Coal	20		

3.20 For the wastewater described in this chapter in example problem 3.14 (flow = 100 L/min, C = 100 mg/L, and k = 0.4 per minute), determine the volume of a single CSTR which is required to achieve 95% removal (an effluent concentration of 5 mg/L). Compare this volume with that required for 80% removal (example problem 3.14).

3.21 For the wastewater described in example problem 3.14 (flow = 100 L/min, C = 100 mg/L, and k = 0.4 per minute), determine the total volume of three, equal-volume CSTRs in series which is required to achieve an effluent concentration of 5 mg/L. Compare this total volume with the volume of a single CSTR (problem 3.20).

3.22 For the wastewater described in example problem 3.14 (flow = 100 L/min, C = 100 mg/L, and k = 0.4 per minute), determine the volume of an ideal plug flow reactor which is required to achieve an effluent concentration of 20 mg/L. Compare this volume with the requirements for a single CSTR (example problem 3.14) and three, equal-volume CSTRs in series (example problem 3.16).

3.23 Develop a numerical simulation using Euler's Method for the batch reaction described in chapter 2, end-of-chapter problem 2.19.

3.24 Develop a numerical simulation using Euler's Method for the batch reaction described in chapter 2, end-of-chapter problem 2.21.

3.25 Develop a pseudocode solution using Euler's Method for the nonsteady-state situation described in example problem 3.7.

3.26 Develop a spreadsheet-based numerical simulation for example problem 3.7.

3.27 Develop a spreadsheet-based numerical simulation for example problem 3.14 with the following nonlinear kinetics:

$$-r = k\,C^{1.5} \quad k = 0.060 \; \frac{L^{\frac{1}{2}}}{mg^{\frac{1}{2}} - min}$$

References

Chapra, Steven C., and Raymond P. Canale. 1988. *Numerical Methods for Engineers*. 2nd ed. New York: McGraw-Hill.

Smith, Carlos A., and Armando B. Corripio. 1985. *Principles and Practice of Automatic Process Control*. New York: John Wiley & Sons.

Chapter 4

Biological Foundations

4.1   Application of Biological Processes in Environmental Engineering

An understanding of biological processes is important in describing natural systems and in designing engineered systems to control the adverse effects of waste discharges. In natural systems, biological processes affect the fate and transport of pollutants in surface waters, groundwaters, vegetation, and soil. In addition, biological processes are important in the stimulation of plant and algal growth, processes that can be accelerated by nutrients (phosphorus and nitrogen) in wastewater effluents, agricultural runoff, and urban runoff. In this chapter, we focus on details of the impact of organic waste materials on the dissolved oxygen content in surface waters. In engineered systems, biological reactions are routinely used to remove undesirable constituents in municipal wastewaters, industrial wastewaters, solid wastes, and contaminated groundwaters. Details of these wastewater treatment applications are presented in chapter 6.

In all of these applications, a quantitative description of the biological reactions is needed in order to predict the detailed impacts of pollutants introduced into natural systems, or to support the design and operation of waste management facilities. Models of natural systems are used by regulatory agencies to determine maximum allowable discharge rates in order to prevent unacceptable degradation of the environment. Other models are used by engineers to design facilities. In all cases, a complete description of the biological process includes definition of the reaction stoichiometry and kinetics.

4.2   Quantification of Substrate

All organisms require a source of energy to support biological activity. Energy may be obtained from solar radiation or by oxidation of chemical species. Photosynthetic organisms (plants, algae, and certain bacteria) capture radiant energy and are able to synthesize cellular material using carbon dioxide as a carbon source. The majority of organisms of interest to the study of environmental engineering are not photosynthetic and require a supply of organic or inorganic chemicals that are oxidized to provide energy to support life processes. For the nonphotosynthetic reactions, it is appropriate to define the **substrate**, or food source, for the organism as the compound that is oxidized to supply energy. Remembering that oxidation is defined as loss of electrons, the substrate may therefore be referred to as the electron donor.

For many organisms, including humans, various organic compounds are used as substrate. Other organisms, principally certain bacterial species, are able to oxidize inorganic compounds (for example, ammonia in nitrification processes) to supply energy. For this latter case, the substrate would be ammonia rather than organics. However, in most biological processes of interest in environmental engineering, the electron donor is an organic compound(s). Therefore, it is customary to define substrate as equivalent to the organic content. Various analytical methods are used to measure the quantity of organics present, including the biochemical oxygen demand (BOD) and chemical oxidation demand (COD). **BOD** can be defined loosely as the oxygen needed by bacteria to decompose (oxidize) a substrate; **COD** is defined similarly except that the oxidation occurs by a chemical (rather than biological) reaction.

4.3   Biochemical Oxygen Demand Test

The biochemical oxygen demand is a surrogate parameter that is used to measure the quantity of organic material present in a sample of water or wastewater. The BOD test provides an indirect measure of the organic material by measuring the oxygen consumed during the biological oxidation of that organic material. This process may be understood by considering the half-reactions associated with biological utilization of substrate (organics). Recall from previous chemistry courses that an oxidation/reduction process may be resolved into two half-reactions: an oxidation (loss of electrons) and a reduction (gain of electrons). Substrate has been defined generally as an electron donor. For organic electron donors, the product of the oxidation is carbon dioxide:

$$\text{organics} \rightarrow CO_2 + e^- \tag{4.1}$$

In an aerobic environment, oxygen is present to serve as the electron acceptor. The corresponding reduction half-reaction yields water as a product:

$$O_2 + 4\,H^+ + 4\,e^- \rightarrow 2\,H_2O \tag{4.2}$$

It is possible to quantify the substrate in a sample by measuring the quantity of oxygen that is consumed during the biological oxidation of the organics to carbon dioxide. It is apparent from the half-reactions that the two parameters (substrate and oxygen consumed) are related, as both are proportional to the number of electrons transferred in the reaction. A single measurement of oxygen consumption is much easier than measuring the concentrations of the numerous organic compounds that are present in a water or wastewater sample.

The BOD test is a batch bioassay analysis. The test is conducted in a closed reactor to prevent oxygen transfer from the atmosphere. Exposure to light is prohibited to exclude photosynthesis. The temperature is maintained at 20 °C to ensure consistency in the results. The sample is diluted with seeded dilution water, which contains a viable bacterial seed and all necessary inorganic nutrients to support the desired biological reaction. The dilution water does not contain any organic materials, so that the only source of substrate in the diluted sample is from the environmental sample. The dilution water must be aerated prior to use to ensure an adequate supply of oxygen to maintain aerobic conditions throughout the BOD test.

At the beginning of the batch test, a measured volume of the environmental sample is placed in a BOD bottle which is then filled with seeded dilution water and sealed to prevent oxygen transfer from the atmosphere. The dissolved oxygen is measured at the beginning and end of the test, which is typically conducted for a five-day duration. The difference in the dissolved oxygen concentrations represents the oxygen consumption. The biological reactions may require in excess of thirty days to reach completion, in which case the five-day oxygen consumption (BOD_5) would represent a fraction of the total oxygen demand. The total oxygen demand, assuming complete conversion of organics to carbon dioxide, theoretically occurs at time equal to infinity. The total demand is also referred to as the ultimate BOD (BOD_u).

Example Problem 4.1

A 300 ml BOD bottle was filled with 5 ml of municipal wastewater and 295 ml of dilution water. The initial and final (after 5 days of reaction) dissolved oxygen concentrations were 9.02 mg/L and 4.13 mg/L, respectively. Determine the BOD_5.

Solution

The dilution factor is calculated as the ratio of the diluted sample volume (300 ml) to the environmental sample volume (5 ml). The BOD_5 is calculated as the oxygen consumption in the BOD test bottle multiplied by the dilution factor:

$$BOD_5 = (9.02 \text{ mg/L} - 4.13 \text{ mg/L}) (300 \text{ ml})/(5 \text{ ml}) = 293 \text{ mg/L}$$

Mathematical equations have been developed to describe the time series progression of substrate and dissolved oxygen during the batch BOD test. It is customarily assumed that the rate of reaction is proportional to the amount of substrate remaining at any time (first-order kinetics):

$$\frac{dL}{dt} = -k_d L \tag{4.3}$$

where:

L = substrate concentration measured as BOD_u (mg/L)

t = time (days)

k_d = deoxygenation rate constant (days^{-1})

The differential equation (4.3) can be solved by separation of variables and integration. Definite integrals are used, corresponding with an initial condition ($t = 0$ and $L = L_0$) and a general condition ($t = t$ and $L = L_t$):

$$\int_{L_0}^{L_t} \frac{dL}{L} = \int_0^t -k_d \, dt \tag{4.4}$$

$$\ln \frac{L_t}{L_0} = -k_d \, t \tag{4.5}$$

$$L_t = L_0 \exp(-k_d \, t) \tag{4.6}$$

The oxygen consumed at any time during the BOD test is equivalent to the amount of substrate which has reacted:

$$y_t = L_0 - L_t \tag{4.7}$$

where: y_t = BOD exerted at time t (mg/L)

Combining equations (4.6) and (4.7) yields the following general equation to define the BOD progression as a function of time:

$$y_t = L_0 \left[1 - \exp(-k_d \, t) \right] \tag{4.8}$$

From equation (4.8), it is observed that the oxygen consumed at time = 0 and time = infinity equals zero and L_0, respectively:

$$y_\infty = L_0 \tag{4.9}$$

Thus in the limit as time approaches infinity, the ultimate BOD is shown to equal the initial substrate concentration present in the environmental sample. The time series progression for substrate remaining (equation 4.6) and BOD (equation 4.8) are illustrated in Figure 4.1.

The numerical value for the deoxygenation rate constant is dependent on the ease of degradation of the organic materials that are present. For municipal wastewaters, a range of values from 0.10 to 0.23 days^{-1} has been reported (Metcalf and Eddy, 1979; Sawyer, McCarty, and Parkin, 1994). Values for industrial wastewaters would be expected to exhibit greater variability.

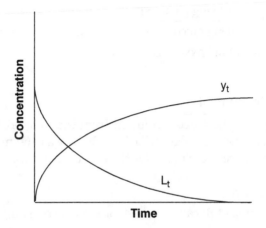

Figure 4.1
BOD progression.

Example Problem 4.2

Determine the ultimate BOD of a 200 mg/L solution of glucose. The chemical formula for glucose is $C_6H_{12}O_6$.

Solution

The ultimate BOD is defined as the oxygen consumed during complete oxidation of the organic compound. It is necessary to write a balanced equation, assuming glucose is oxidized completely to carbon dioxide. This particular reaction can be balanced by inspection by sequential determination of the molar quantities of carbon dioxide (to balance carbon), water (to balance hydrogen), and molecular oxygen (to balance oxygen):

$$C_6H_{12}O_6 \rightarrow 6\ CO_2$$

$$C_6H_{12}O_6 \rightarrow 6\ CO_2 + 6\ H_2O$$

$$C_6H_{12}O_6 + 6\ O_2 \rightarrow 6\ CO_2 + 6\ H_2O$$

A more general procedure may also be used to balance this reaction using half-reactions for the oxidation of glucose to carbon dioxide and water and the reduction of oxygen to water.

Half-reactions are written for the oxidation and reduction reactions:

$$C_6H_{12}O_6 + 6\ H_2O \rightarrow 6\ CO_2 + 24\ H^+ + 24\ e^-$$
$$+ 6 \times [\ O_2 + 4\ H^+ + 4\ e^- \rightarrow 2\ H_2O]$$

$$\overline{C_6H_{12}O_6 + 6\ O_2 \rightarrow 6\ CO_2 + 6\ H_2O}$$

A stoichiometry calculation is completed to determine the quantity of oxygen required to react with the referenced glucose concentration of 200 mg/L:

$$L_0 = \frac{200\ \text{mg glucose/L}}{180\ \text{mg glucose/mmole}} \times \frac{6\ \text{mmole oxygen}}{1\ \text{mmole glucose}} \times 32\ \text{mg oxygen/mmole}$$

$$= 213\ \text{mg oxygen/L}$$

Example Problem 4.3

Suppose the glucose solution from example problem 4.2 was analyzed to determine the 5-day BOD. The BOD_5 was 150 mg/L. Determine the value, with appropriate units, for the rate constant k_d.

Solution

Equation (4.8) can be solved for the rate constant:

$$k_d = \frac{-\ln\left(1 - \dfrac{y_t}{L_0}\right)}{t} = \frac{-\ln\left(1 - \dfrac{150\ \text{mg/L}}{213\ \text{mg/L}}\right)}{5\ \text{days}} = 0.244\ \text{day}^{-1}$$

Example Problem 4.4

Determine the fraction of the ultimate BOD which is exerted during a 5-day BOD test for a domestic wastewater with a rate constant = 0.23 day^{-1}. Repeat for a slowly degradable industrial wastewater with a rate constant = 0.05 day^{-1}.

Solution

Equation (4.8) is solved for the ratio of BOD_5 to BOD ultimate:

$$\frac{y_5}{L_0} = 1 - \exp(-5\,k_d)$$

For the domestic wastewater, this ratio is 0.68. For the less degradable industrial wastewater, the ratio is only 0.22.

The BOD test is widely used to characterize the organic content in wastewaters and to establish criteria for effluent discharge into the environment. Typical municipal wastewaters contain 200 to 300 mg/L of BOD_5 (Metcalf and Eddy, 1991). Typical effluent standards for a treated municipal wastewater effluent range from 5 to 30 mg/L of BOD_5.

The BOD test has proved to be quite useful as a measure of substrate in the design and operation of wastewater treatment facilities, especially for domestic and readily degradable industrial wastewaters (for example, food processing wastewaters). Certain limitations of the test have been documented, however, including:

1. Simultaneous oxidation of inorganic compounds, such as ammonia, also exerts an oxygen demand. Interpretation of BOD results is complicated by competing reactions with oxygen.

2. The oxidation does not go to completion within the normal five-day test period. Knowledge of the rate constant is therefore required to relate five-day and ultimate BOD values.

3. The test period is excessive (five days); consequently, the results have little value for process control.

4. The presence of toxic materials will interfere with biological activity during the test. If toxic effects are experienced, the measured BOD will underestimate the true organic content in the sample. Many industrial wastewaters contain compounds which may be inhibitory or toxic; therefore BOD data may not be reliable for these wastewaters.

5. Characteristics of the seed organisms may influence the rate of oxidation of organics. Because different seeds may produce different BOD results, reproducibility of the test may be difficult. Many industrial wastewaters contain organic compounds that are not readily degraded by seed organisms derived from a municipal wastewater. Interpretation of BOD results for such industrial wastewaters is difficult unless the seed used has been acclimated to the specific wastewater.

4.4   Chemical Oxygen Demand Test

The chemical oxygen demand test (COD) may be used as an alternative to the BOD test for quantification of substrate. There is one fundamental similarity between the two analytical tests—both tests measure the quantity of electron acceptor required to oxidize organic material present in an environmental sample. In the BOD test, the quantity of oxygen required during aerobic biological oxidation is measured. Under COD test conditions, potassium dichromate ($K_2Cr_2O_7$) serves as the oxidizing agent (electron acceptor). The quantity of organic material is expressed in terms of the equivalent amount of electron acceptor, assuming oxygen is the electron acceptor. Because very strong oxidizing conditions are provided during the COD test (elevated temperature, 70% sulfuric acid, silver catalyst), complete oxidation of most organic compounds to carbon dioxide is achieved within several hours.

The COD test offers some decided advantages over the BOD test for quantification of substrate, particularly for industrial wastewaters which may contain toxic compounds. The COD test also can be completed in a matter of hours, and a stoichiometric endpoint is reached so that knowledge of a rate constant is unnecessary. Unfortunately, the COD test is not able to distinguish between organic materials that are biodegradable and those compounds that are either toxic or recalcitrant (recalcitrant compounds are not toxic but are not biodegradable). In addition, some inorganic compounds may be oxidized during the COD test. This limitation of the COD test is very significant for description of biological processes, since a successful measurement of substrate must not be inflated by the presence of compounds such as toxic organics, recalcitrant organics, or inorganic species that are not suitable as an energy source for the organisms.

Example Problem 4.5

Determine the COD of a 200 mg/L solution of glucose ($C_6H_{12}O_6$).

Solution

The COD is defined as the oxygen consumed during complete oxidation of the organic compound. It is necessary to write a balanced equation, assuming glucose is oxidized completely to carbon dioxide. Half-reactions are written for the oxidation and reduction reactions:

$$C_6H_{12}O_6 + 6\ H_2O \rightarrow 6\ CO_2 + 24\ H^+ + 24\ e^-$$
$$\underline{+\ 6 \times [O_2 + 4\ H^+ + 4\ e^- \rightarrow 2\ H_2O]}$$
$$C_6H_{12}O_6 + 6\ O_2 \rightarrow 6\ CO_2 + 6\ H_2O$$

A stoichiometry calculation is completed to determine the quantity of oxygen required to react with the referenced glucose concentration of 200 mg/L:

$$COD = \frac{200 \text{ mg glucose/L}}{180 \text{ mg glucose/mmole}} \times \frac{6 \text{ mmole oxygen}}{1 \text{ mmole glucose}} \times 32 \text{ mg oxygen/mmole}$$
$$= 213 \text{ mg/L}$$

Comparison of example problems 4.2 and 4.5 indicates that the ultimate BOD and COD are equal for this glucose solution. This relation may be generalized, provided that the environmental sample does not contain significant quantities of organic materials that are not biodegradable or toxic materials that prevent biological activity. In most wastewater samples, the COD is somewhat greater than the ultimate BOD due to the presence of nonbiodegradable organics.

4.5   Nitrogenous Oxygen Demand

It was noted in the preceding discussion that certain bacteria are capable of oxidizing reduced nitrogen compounds, such as ammonia, to supply energy for cellular functions. The biological oxidation of ammonium ion to nitrate is referred to as **nitrification**.

$$NH_4^+ + 2\,O_2 \rightarrow 2\,H^+ + NO_3^- + H_2O \qquad (4.10)$$

This reaction, equation (4.10), is important in surface waters that receive nitrogen-containing wastewater discharges because the consumption of oxygen during the reaction, coupled with oxygen consumption due to carbonaceous BOD, may deplete oxygen levels and interfere with propagation of desirable sport fish species. The oxygen consumption associated with nitrification is defined as the **nitrogenous oxygen demand**. The oxygen consumption associated with oxidation of organic material to carbon dioxide is defined as the **carbonaceous oxygen demand**. Nitrification is often provided during wastewater treatment to minimize the depletion of dissolved oxygen in surface waters which receive treated effluents.

Nitrification may occur during the BOD test, especially if the duration of the BOD test exceeds five days. The oxygen consumption may exhibit two distinct phases, corresponding with exertion of the carbonaceous and nitrogenous oxygen demands, respectively. The effect of nitrification in the BOD test is illustrated in Figure 4.2. Reagents are available which can be added to the BOD dilution water to suppress nitrification (Standard Methods, 1992). Oxidation of ammonia does not occur during COD test conditions.

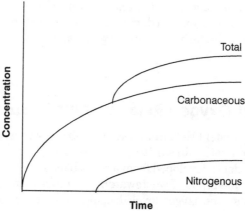

Figure 4.2

Carbonaceous and nitrogenous oxygen demand progression.

Example Problem 4.6

Bacterial cell mass is reported to have the following empirical formula: $C_5H_7O_2N$. Determine the carbonaceous, nitrogenous, and total oxygen demand of one pound of cell mass. Determine the COD of one pound of cell mass.

Solution

The carbonaceous oxygen demand is defined by the biochemical reaction for oxidation of the organic material with complete conversion to carbon dioxide and release of nitrogen as ammonia:

$$C_5H_7O_2N + 5\ O_2 \rightarrow 5\ CO_2 + NH_3 + 2\ H_2O$$

The carbonaceous oxygen demand is determined by stoichiometry:

$$O_2 = \frac{1 \text{ pound cells}}{113 \text{ lb cells/lb-mole}} \times \frac{5 \text{ lb-mole oxygen}}{1 \text{ lb-mole cells}} \times 32 \text{ lb oxygen/lb-mole}$$

$$= 1.42 \text{ lb carbonaceous oxygen demand per pound cell mass}$$

The total oxygen demand (carbonaceous plus nitrogenous) is defined by the biochemical reaction for oxidation of the cellular material with complete conversion to carbon dioxide and nitrate:

$$C_5H_7O_2N + 7\ O_2 \rightarrow 5\ CO_2 + H^+ + NO_3^- + 3\ H_2O$$

$$O_2 = \frac{1 \text{ pound cells}}{113 \text{ lb cells/lb-mole}} \times \frac{7 \text{ lb-mole oxygen}}{1 \text{ lb-mole cells}} \times 32 \text{ lb oxygen/lb-mole}$$

$$= 1.98 \text{ lb total oxygen demand per pound of cell mass}$$

The nitrogenous oxygen demand is determined as the difference between the total oxygen demand and the carbonaceous oxygen demand:

$$O_2 = 1.98 - 1.42 = 0.56 \text{ lb nitrogenous oxygen demand per pound of cell mass}$$

Under COD test conditions, the oxidation products are identical as reported for the carbonaceous oxygen demand: carbon dioxide and ammonia. Thus the COD is equal to the carbonaceous oxygen demand: 1.42 lb oxygen per lb of cell mass.

4.6   Dissolved Oxygen Relationships in Surface Waters

A simplified model (Metcalf and Eddy, 1991) for determining the dissolved oxygen concentration in surface waters is presented in the following discussion. The model is developed from a mass balance on a section of a river that receives a single input of wastewater. The model considers only two factors which influence the oxygen level: decomposition of organics and atmospheric reaeration. Other processes also are known to influence dissolved oxygen values, including photosynthesis, respiration, and nitrification.

The wastewater discharge is introduced into the river and complete mixing is assumed to occur within an abbreviated mixing zone. The concentration of ultimate BOD (S, mg/L), concentration of dissolved oxygen (DO, mg/L), temperature (T, °C), and flow rate (Q, cfs, or MGD) are determined by steady-state material or energy balances around the mixing zone. In the following balance equations, the subscripts R, W, and 0 refer to the river just upstream from the point of discharge, the wastewater discharge, and the river immediately downstream from the point of discharge, respectively. A schematic diagram of the system is presented in Figure 4.3.

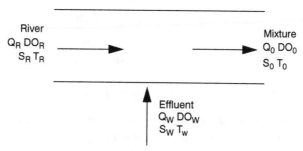

Figure 4.3
Discharge mixing zone.

$$S_0 = \frac{Q_R\,S_R + Q_W\,S_W}{Q_R + Q_W} \tag{4.11}$$

$$DO_0 = \frac{Q_R\,DO_R + Q_W\,DO_W}{Q_R + Q_W} \tag{4.12}$$

$$T_0 = \frac{Q_R\,T_R + Q_W\,T_W}{Q_R + Q_W} \tag{4.13}$$

$$Q_0 = Q_R + Q_W \tag{4.14}$$

The rate of oxidation of organics was defined previously in equation (4.3):

$$\frac{dS}{dt} = -k_d\,S \tag{4.3}$$

Because the concentration of organics is expressed as the ultimate BOD, the rates of oxygen depletion and oxidation of organic materials are equivalent:

$$\frac{d\,DO}{dt} = \frac{dS}{dt} = -k_d\,S \tag{4.15}$$

To facilitate solution of the subsequent differential equations, it is useful to introduce the concept of a **dissolved oxygen deficit**. The deficit is defined as the difference between the concentration of dissolved oxygen at saturation (that is, the concentration in water in equilibrium with the atmosphere) and the actual concentration. Saturation values are sensitive to temperature and salinity; values are provided in Table 4.1.

$$D = DO_{sat} - DO \tag{4.16}$$

where:

D = dissolved oxygen deficit (mg/L)

DO_{sat} = saturation dissolved oxygen concentration (mg/L)

Taking the derivative of equation (4.16), and recognizing that DO_{sat} is a constant:

$$\frac{dD}{dt} = -\frac{dDO}{dt} \tag{4.17}$$

Thus, oxygen depletion due to oxidation of organics is:

$$\frac{dD}{dt} = k_d S \tag{4.18}$$

The transfer of oxygen from the atmosphere into the surface water is referred to as **reaeration**. This rate of oxygen transfer is first-order in terms of the dissolved oxygen deficit, so deficit increase due to reaeration is:

$$\frac{dD}{dt} = -k_a D \tag{4.19}$$

where: k_a = reaeration rate constant (days^{-1})

The combined effects of decomposition of organic material and reaeration may be evaluated simultaneously by adding equations (4.18) and (4.19):

$$\frac{dD}{dt} = k_d S - k_a D \tag{4.20}$$

Table 4.1 Saturation values for dissolved oxygen (mg/L) in water (zero salinity) at 1 atmosphere pressure as a function of temperature (°C).

T	DO_{sat}	T	DO_{sat}	T	DO_{sat}	T	DO_{sat}
1	14.20	11	11.02	21	8.90	31	7.41
2	13.81	12	10.77	22	8.73	32	7.29
3	13.45	13	10.53	23	8.56	33	7.17
4	13.09	14	10.29	24	8.40	34	7.05
5	12.76	15	10.07	25	8.24	35	6.93
6	12.44	16	9.86	26	8.09	36	6.82
7	12.13	17	9.65	27	7.95	37	6.72
8	11.83	18	9.45	28	7.81	38	6.61
9	11.55	19	9.26	29	7.67	39	6.51
10	11.28	20	9.08	30	7.54	40	6.41

Source: Adapted from Metcalf and Eddy (1991).

Solving the differential equation in equation (4.20) requires specification of the initial condition for deficit (D_0) and ultimate BOD (S_0). In these equations, time = zero corresponds with the conditions leaving the mixing zone. Positive values of time correspond with the travel time downstream from the point of discharge. The initial deficit is defined using equations (4.12) and (4.16):

$$D_0 = DO_{sat} - DO_0 \qquad (4.21)$$

The initial condition for ultimate BOD is defined with equation (4.11):

$$S_0 = \frac{Q_R S_R + Q_W S_W}{Q_R + Q_W} \qquad (4.11)$$

The differential equation (4.20) may be solved to provide a general equation for predicting the deficit at downstream locations from the point of discharge. This equation was first reported in the early 1900s and is known as the Streeter-Phelps Equation:

$$D = \frac{k_d S_0}{k_a - k_d} \left[\exp\left(-k_d\, t\right) - \exp\left(-k_a\, t\right) \right] + D_0 \exp\left(-k_a\, t\right) \qquad (4.22)$$

For a river flowing at a constant velocity (u), travel time (t) and distance (x) are related as noted in equation (4.23):

$$x = u\, t \qquad (4.23)$$

The resulting dissolved oxygen profile downstream from the point of discharge is presented in Figure 4.4 (recall that $DO = DO_{sat} - D$). If the rate of oxygen consumption exceeds the rate of reaeration, the general response to introduction of organic material in the wastewater effluent is to decrease the dissolved oxygen in the region immediately downstream from the point of discharge. After decomposition of a portion of the organic material, the rates of deoxygenation and reaeration become equal and the dissolved oxygen reaches a minimum value. Downstream from the minimum dissolved oxygen concentration, the rate of reaeration exceeds the rate of deoxygenation and the oxygen levels recover, ultimately approaching saturation values.

If the minimum dissolved oxygen concentration is too small (commonly, a critical value of 5 mg/L is used), desired sport fish species may not survive. If the concentration of organic material is excessive, dissolved oxygen levels may approach zero, resulting in anaerobic (without oxygen) conditions and foul odors.

The time at which the minimum dissolved oxygen concentration occurs is determined by differentiating equation (4.22), setting the derivative equal to zero, and solving for time.

$$t_c = \frac{1}{(k_a - k_d)} \ln\left[\frac{k_a}{k_d} \left(1 - \frac{D_0\, (k_a - k_d)}{k_d\, S_0} \right) \right] \qquad (4.24)$$

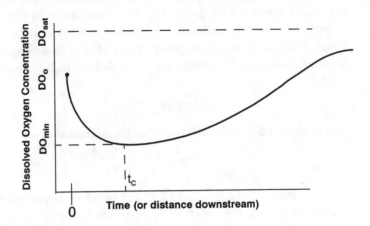

Figure 4.4
Dissolved oxygen sag curve.

where: t_c = time to achieve minimum dissolved oxygen concentration (days).

The rate constants for decomposition of organics and reaeration are sensitive to temperature (Metcalf and Eddy, 1991):

$$k_{d_T} = k_{d_{20}} (1.135)^{(T-20)} \text{ where } T \leq 20°C \tag{4.25}$$

$$k_{d_T} = k_{d_{20}} (1.056)^{(T-20)} \text{ where } 20°C < T \leq 30°C \tag{4.26}$$

$$k_{a_T} = k_{a_{20}} (1.024)^{(T-20)} \tag{4.27}$$

Equation (4.25) is valid for temperatures from 4 to 20 °C, and equation (4.26) is valid for temperatures from 20 to 30 °C. In these equations, a subscript of T indicates the rate constant is valid for a temperature of T °C, while a subscript of 20 indicates the rate constant is valid for a reference temperature of 20 °C.

Determining the dissolved oxygen in a river downstream from the point of wastewater discharge requires completion of the following calculations (Dietz, 1996):

1. Characterization of the conditions in a mixing zone at the point of discharge. A mass balance must be completed to determine the initial BOD ultimate and the initial dissolved oxygen (equations 4.11 and 4.12). A thermal balance must be completed to determine the initial temperature (equation 4.13). All BOD$_5$ values must be converted to BOD ultimate values using equation (4.8). Flow, temperature, dissolved oxygen, and ultimate BOD must be known (or assumed) for the river upstream from the point of discharge and for the effluent.

2. Based on the temperature in the river after discharge, values can be determined for the dissolved oxygen saturation concentration (Table 4.1), deoxygenation coefficient (equation 4.25 or 4.26), and the reaeration coefficient (equation 4.27). The initial deficit is then calculated (equation 4.21).

3. For a known value of time, the deficit is calculated (equation 4.22).

4. To determine the minimum dissolved oxygen concentration, the maximum deficit is determined (equation 4.24, followed by equation 4.22). The minimum dissolved oxygen is determined with equation (4.16).

The value for the deoxygenation coefficient may be specific to the wastewater due to differences in the inherent rate of degradation of the organic compounds present. The reaeration coefficient is highly dependent on local river conditions (depth and velocity), with a range of values reported from 0.10 per day for small ponds to 1.15 per day for swift streams (Metcalf and Eddy, 1979). Site-specific values for these coefficients should be obtained for accurate evaluation of receiving water dissolved oxygen.

In many practical situations, dissolved oxygen sag calculations are completed by state regulatory agencies to determine an allowable concentration of BOD_5 in an effluent discharge which will not degrade water quality criteria (typically 5 mg/L of dissolved oxygen). For this wasteload allocation exercise, the concentration of BOD_5 in the effluent is unknown. An iterative calculation procedure is necessary in which successive values of the effluent BOD_5 are assumed until the calculated minimum dissolved oxygen concentration is satisfactory. A discharge permit is then granted for operation of a wastewater treatment facility.

Example Problem 4.7*

For a wastewater with a flow rate of 4.00 MGD, determine the minimum dissolved oxygen concentration downstream from a treatment facility which achieves secondary treatment ($BOD_5 < 30$ mg/L). The characteristics of the receiving stream upstream from the point of discharge are noted below:

Minimum flow = 20.0 cfs   Temperature = 32.0 °C

Ultimate BOD = 3.0 mg/L   Dissolved oxygen = 7.0 mg/L

The wastewater temperature is 25.0 °C and the wastewater dissolved oxygen is 5.0 mg/L. The deoxygenation coefficient for the wastewater is 0.17 days^{-1} at 20 °C. The reaeration coefficient for the stream is 0.40 days^{-1} at 20 °C.

Solution

The initial conditions (ultimate BOD, DO, and temperature) in the stream at the point of discharge must be determined using mass or energy balances based on the defined inputs. The concentration and temperature of the mixture would be a flow–weighted average of the inputs. All BOD_5 values must be converted to ultimate BOD using equation (4.8).

The resulting inputs and output mixture are summarized below:

Parameter	Effluent	Stream	After Mixing
Flow, MGD	4.00	12.93	16.93
S, mg/L	52.4	3.0	14.7
DO, mg/L	5.0	7.0	6.5
T, °C	25.0	32.0	30.3

The rate coefficients must be corrected for temperature using equations (4.26) and (4.27):

$$k_d = (0.17 \text{ days}^{-1})(1.056)^{(30.3-20)} = 0.30 \text{ days}^{-1}$$

$$k_a = (0.40 \text{ days}^{-1})(1.024)^{(30.3-20)} = 0.51 \text{ days}^{-1}$$

The dissolved oxygen saturation concentration at 30.3 °C is 7.50 mg/L (see Table 4.1). The initial deficit is calculated with equation (4.21):

$$D_0 = 7.5 - 6.5 = 1.0 \text{ mg/L}$$

The time at which the maximum deficit occurs is calculated with equation (4.24):

$$t_c = \frac{1}{(0.51 - 0.30)} \ln\left[\frac{0.51}{0.30}\left(1 - \frac{1.0(0.51 - 0.30)}{(0.30)(14.7)}\right)\right] = 2.30 \text{ days}$$

The maximum deficit is calculated with equation (4.22):

$$D = \frac{(0.30)(14.7)}{0.51 - 0.30}\left[\exp\left(-(0.30)(2.30)\right) - \exp\left(-(0.51)(2.30)\right)\right]$$
$$+ 1.0 \exp\left(-(0.51)(2.30)\right) = 4.3 \text{ mg/L}$$

The minimum dissolved oxygen is calculated with equation (4.16):

$$DO = DO_{sat} - D = 7.5 - 4.3 = 3.2 \text{ mg/L}$$

*(From Dietz, 1996)

Example Problem 4.8[*]

For the situation described in example problem 4.7, determine the maximum allowable effluent BOD_5 concentration which can be discharged while maintaining compliance with a water quality standard of 5.0 mg/L of dissolved oxygen in the receiving stream.

Solution

A trial-and-error solution is required in which successive values of effluent BOD_5 are assumed and a corresponding minimum dissolved oxygen is determined. The methodology is identical to the procedure presented in the solution to example problem 4.7. Results are tabulated below. The assumed value for the third trial was determined by linear interpolation based on the results of the first two trials.

Trial	BOD_5 mg/L	BOD_{ult} mg/L	S_0 mg/L	t_c days	D mg/L	Minimum DO mg/L
1	30	52.4	14.7	2.30	4.31	3.19
2	0	0.0	2.3	0.84	1.04	6.46
3	13	22.7	7.7	2.08	2.40	5.10
4	14	24.5	8.1	2.11	2.51	4.99

The maximum allowable BOD_5 in the effluent is between 13 and 14 mg/L.
*(From Dietz, 1996)

4.7   Bacterial Growth

Bacterial metabolism may be described by the following generalized oxidation/reduction equation:

Energy Source (Electron Donor) + Carbon Source + Electron Acceptor

+ Nutrients → Bacterial Cell Mass + Metabolic End-Products   (4.28)

The energy source has been defined previously for this biological process as the substrate, typically measured as BOD_5 (or COD). For nitrification processes, the substrate was defined previously as ammonia. The source of carbon for cell synthesis can be either carbon dioxide or the same organic material which is the substrate, depending on the bacterial group. Some compound must serve as the electron acceptor to balance the electrons lost during oxidation of substrate. For aerobic processes, oxygen is the electron acceptor. For other applications, the electron acceptor may be nitrate, sulfate, carbon dioxide, or even an organic compound. The principal inorganic nutrients required for bacterial growth are nitrogen and phosphorus, although trace quantities of many elements are necessary for growth.

The products of the reaction include bacterial cell mass, commonly characterized by the empirical formula $C_5H_7O_2N$ (Sawyer, McCarty, and Parkin, 1994). Cell mass is measured analytically by a gravimetric procedure that includes filtration of a known sample volume, drying, and determination of the dry weight of the filter residue. Results are reported as suspended solids (SS) or volatile suspended solids (VSS), depending on the temperature employed for the drying process (Standard Methods, 1992). The metabolic end products indicated in equation (4.28) may include carbon dioxide, water, nitrate, nitrogen gas, methane gas, and various organic compounds.

Various biological processes are significant in wastewater treatment operations for removal of organics, ammonia, nitrogen, and/or phosphorus. Selected examples of these reactions are described below (unbalanced reactions). For purposes of this presentation, the organic material in domestic wastewater is assumed to have the following empirical formula: $C_{10}H_{19}O_3N$ (Sawyer, McCarty, and Parkin, 1994). Of course, many other organic compounds could also be represented as the substrate.

1. Secondary treatment for BOD removal

$$C_{10}H_{19}O_3N + O_2 \rightarrow \text{cell mass} + CO_2 + NH_3 + H_2O \qquad (4.29)$$

2. Nitrification (oxidation of ammonia to nitrate)

$$NH_3 + CO_2 + O_2 \rightarrow \text{cell mass} + HNO_3 + H_2O \qquad (4.30)$$

3. Denitrification (reduction of nitrate to nitrogen gas)

$$HNO_3 + C_{10}H_{19}O_3N \rightarrow \text{cell mass} + CO_2 + N_2 \qquad (4.31)$$

4. Anaerobic treatment (conversion of organic compounds to methane)

$$C_{10}H_{19}O_3N \rightarrow \text{cell mass} + CH_4 + CO_2 \qquad (4.32)$$

Environmental conditions (pH, temperature, dissolved oxygen concentration, salinity, and other factors) are known to influence the metabolism of specific bacterial species. Therefore, these conditions must be maintained at proper levels to promote the desired biological reactions in natural or engineered systems.

Specifying the kinetic relationships for bacterial growth requires description of the rate of substrate removal and the rate of cell synthesis. Knowledge of the substrate removal kinetics is necessary to predict the fate of organic compounds in an aquatic environment or to design facilities for removal of BOD. Since bacterial cell mass is produced as a result of biological reactions, excess cellular material (biomass) is produced which requires proper disposal. The quantity of this biomass, also referred to as residuals (or sludge), can be determined with knowledge of the cell synthesis kinetics.

The terminology and nomenclature used throughout this presentation are summarized in Table 4.2; definitions of selected parameters are provided in equations (4.33) to (4.36). These kinetic relationships must be determined by conduct of laboratory studies. Knowledge of the kinetics may be used during design of treatment facilities (chapter 6).

$$r_s = \frac{dS}{dt} \qquad (4.33)$$

Table 4.2 Bacterial kinetics terminology.

Parameter	Symbol	Analytical Test	Units
Substrate	S	BOD, COD, NH_3	mg/L
Biomass	X	VSS, SS	mg/L
Substrate removal rate	$-r_s$		mg S/L–day
Biomass production rate	r_x		mg X/L–day
Specific substrate removal rate	q		mg S/mg X–day
Specific growth rate	μ		days^{-1}

$$r_x = \frac{dX}{dt} \qquad (4.34)$$

$$q = \frac{-r_s}{X} \qquad (4.35)$$

$$\mu = \frac{r_x}{X} \qquad (4.36)$$

In order to specify kinetic relationships we must identify the reaction order and estimate rate constants. Experimental studies must be completed in order to define the reaction order and estimate parameters. The modeling process is largely an empirical exercise, with a goal to identify a model that accurately describes the experimental data.

Many kinetic models for specific substrate utilization rate have been used with success. Since the process is empirical in nature, there is no expectation that any single reaction order would always provide the best description of a particular system. Several of the most common kinetic models are presented in Table 4.3. Of these models, the variable-order model, also commonly referred to as Monod kinetics, exhibits flexibility and can accurately describe many biological processes. It is also noted that the inhibitory kinetic expression collapses to yield the variable-order expression for large values of K_I.

The specific growth rate is normally described with the relationship in equation (4.37). It is noted that the growth rate and substrate removal rate are coupled by the yield parameter. The endogenous energy requirements include all cell maintenance functions which do not contribute to synthesis of new cellular material.

$$\mu = Y_{max}\, q - k_e \qquad (4.37)$$

where:

Y_{max} = maximum yield (mg X/mg S)
k_e = endogenous decay rate (days^{-1})

Table 4.3 Kinetic expressions for specific substrate utilization rate.

Order	Rate Expression	
Zero	$q = k$	
First	$q = kS$	
Variable	$q = \dfrac{kS}{K_S + S}$	K_S = saturation constant
		K_I = inhibition constant
Inhibitory	$q = \dfrac{kS}{K_S + S + \dfrac{S^2}{K_I}}$	

Example Problem 4.9

The following steady-state data were obtained by operation of completely mixed, no-recycle reactors to support identification of the kinetics for substrate removal and cell production. The influent substrate concentration for all reactors was 600 mg COD/L. The influent biomass concentration for all reactors was negligible. The volume of each reactor was 20.0 Liters. Determine kinetic expressions for q and μ. Provide numerical estimates for all parameters in the models, including units.

Reactor	Effluent S mg COD/L	Effluent X mg VSS/L	Flow Rate L/day
1	160	206	15.0
2	110	223	10.0
3	60	225	5.0
4	35	202	2.5
5	20	145	1.0

Solution

A steady-state material balance for substrate can be written to determine q.

$$0 = Q\,S_{influent} - Q\,S_{effluent} + Vr_s$$

$$q = \frac{-r_s}{X} = \frac{Q\,(S_{influent} - S_{effluent})}{V\,X_{effluent}}$$

A similar material balance for biomass can be developed to determine μ:

$$0 = Q\,X_{influent} - Q\,X_{effluent} + Vr_X$$

Since the influent biomass concentration is negligible:

$$\mu = \frac{r_X}{X} = \frac{Q\,X_{effluent}}{V\,X_{effluent}} = \frac{Q}{V}$$

Calculated values for q and μ are summarized for each reactor below:

Reactor	q mg S/mg X–day	μ days^{-1}	S mg COD/L
1	1.602	0.750	160
2	1.099	0.500	110
3	0.600	0.250	60
4	0.350	0.125	35
5	0.200	0.050	20

A plot of q as a function of S is useful to suggest the reaction order. In this case, the data plot is a straight line which passes through the origin. The data are consistent with first-order kinetics. The rate constant, k, is defined as the slope of the line on this plot of q versus S. The rate constant (k) is estimated to equal 0.010 L/mg VSS–day.

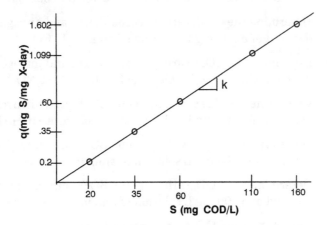

A plot of μ as a function of q may be used to estimate Y_{max} and k_e. Based on examination of equation (4.37), this plot would be expected to have a slope of Y_{max} and an intercept of $-k_e$. Analysis of the data (linear least squares parameter estimation) provides estimates for Y_{max} and k_e of 0.50 mg VSS/mg COD and 0.050 days^{-1}, respectively.

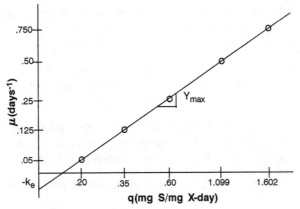

End-of-Chapter Problems

4.1   An industrial wastewater was analyzed to determine the BOD_5. Four different dilutions were completed, with initial and final dissolved oxygen concentrations as noted below. Determine the BOD_5 concentration.

Dilution Factor	Initial DO (mg/L)	Final DO (mg/L)
1.0	9.20	0.00
10.0	9.20	0.00
100.0	9.20	2.05
1000.0	9.20	8.47

4.2   In order for BOD results to be valid, it is desirable to maintain a minimum dissolved oxygen concentration at the end of the incubation period of at least 1.0 mg/L. If an industrial wastewater is known to have a COD of 1000 mg/L, recommend a dilution factor for conducting a BOD_5 test.

4.3   Determine the COD, carbonaceous BOD_u, and nitrogenous oxygen demand of a 100 mg/L solution of urea (N_2H_4CO).

4.4   Determine the COD, carbonaceous BOD_u, and nitrogenous oxygen demand of a 100 mg/L solution of alanine ($C_3H_7NO_2$).

4.5   Determine the COD, carbonaceous BOD_u, and nitrogenous oxygen demand of a 100 mg/L solution of carbohydrate (CH_2O).

4.6   Determine the COD, carbonaceous BOD_u, and nitrogenous oxygen demand of a 100 mg/L solution of ammonia (NH_3).

4.7   Determine the COD, carbonaceous BOD_u, and nitrogenous oxygen demand of a 100 mg/L solution of ammonium nitrate (NH_4NO_3).

4.8   Determine the COD, carbonaceous BOD_u, and nitrogenous oxygen demand of a 100 mg/L solution of sodium nitrate ($NaNO_3$).

4.9   A biodegradable industrial wastewater (paper production) has a COD of 600 mg/L. If the BOD progression follows first-order kinetics with a rate constant = 0.10 per day, determine the BOD_5.

4.10 A biodegradable industrial wastewater (petrochemical) has a COD of 600 mg/L. If the BOD progression follows first-order kinetics with a rate constant = 0.20 per day, determine the BOD_5.

4.11 A biodegradable industrial wastewater (pharmaceutical) has a COD of 600 mg/L. If the BOD progression follows first-order kinetics with a rate constant = 0.30 per day, determine the BOD_5.

4.12 A biodegradable industrial wastewater (paper production) has a BOD_5 of 600 mg/L. If the BOD progression follows first-order kinetics with a rate constant = 0.10 per day, determine the BOD_u.

4.13 A biodegradable industrial wastewater (petrochemical) has a BOD_5 of 600 mg/L. If the BOD progression follows first-order kinetics with a rate constant = 0.20 per day, determine the BOD_u.

4.14 A biodegradable industrial wastewater (pharmaceutical) has a BOD_5 of 600 mg/L. If the BOD progression follows first-order kinetics with a rate constant = 0.30 per day, determine the BOD_u.

4.15 Based on the answers to questions 4.12, 4.13, and 4.14, do you believe that knowledge of the BOD at a single point on the BOD progression curve is sufficient to define the impact of organics discharge to a surface water? If not, what additional information is needed?

4.16 An industrial wastewater has a 5-day BOD of 370 mg/L and a 10-day BOD of 500 mg/L. If the BOD progression follows first-order kinetics, determine the ultimate BOD.

4.17 Discuss adverse environmental impacts associated with discharge of nitrate into surface waters.

4.18 Discuss adverse environmental impacts associated with discharge of ammonia into surface waters.

4.19 Discuss adverse environmental impacts associated with discharge of phosphate into surface waters.

4.20 A municipal wastewater treatment facility currently achieves secondary treatment. The existing facility discharge is characterized as follows:

BOD_5 = 20 mg/L NH_3–N = 20 mg/L
SS = 20 mg/L NO_3–N = 0 mg/L

An upgrade of the facility to add biological nitrification is proposed. Describe any expected benefits to the water quality in the receiving stream after the facility modification.

4.21 A municipal wastewater treatment facility currently achieves secondary treatment. The existing facility discharge is characterized as follows:

BOD_5 = 20 mg/L NH_3–N = 20 mg/L
SS = 20 mg/L NO_3–N = 0 mg/L

An upgrade of the facility to add biological nitrification and denitrification is proposed. Describe any expected benefits to the water quality in the receiving stream after the facility modification.

4.22 Repeat example problem 4.8 for winter conditions in Florida. Assume all conditions are the same as originally stated in example problem 4.7 except as noted below:

Receiving stream temperature = 15 °C

Wastewater temperature = 20 °C

Discuss the seasonal effects (temperature) on the assimilative capacity of surface waters.

4.23 Repeat example problem 4.8 for drought conditions. Assume all conditions are the same as originally stated in example problem 4.7 except as noted below:

Receiving stream flow = 2.0 cfs

Discuss the effect of reduced flow conditions on the assimilative capacity of surface waters.

4.24 Repeat example problem 4.8 for flood conditions. Assume all conditions are the same as originally stated in example problem 4.7 except as noted below:

Receiving stream flow = 2000 cfs

Discuss the effect of elevated flow conditions on the assimilative capacity of surface waters.

4.25 Repeat example problem 4.7 for flood conditions. Assume all conditions are the same as originally stated in example problem 4.7 except as noted below:

Receiving stream flow = 2000 cfs

Receiving stream ultimate BOD = 0.1 mg/L

Discuss the effect of elevated flow conditions on the assimilative capacity of surface waters.

4.26 The following steady-state data were obtained by operation of completely mixed, no-recycle reactors to support identification of the kinetics for substrate removal and cell production. The influent substrate concentration for all reactors was 500 mg/L of BOD_5. The influent biomass concentration for all reactors was negligible. The volume of each reactor was 10.0 Liters. Calculate q and μ for each reactor. Plot q versus S. Determine kinetic expressions for q and μ. Provide numerical estimates for all parameters in the models, including units.

Reactor	Effluent S mg BOD_5/L	Effluent X mg VSS/L	Flow Rate L/day
1	500	0	15.0
2	424	36	10.0
3	224	123	5.0
4	124	152	2.5
5	64	136	1.0

4.27 The following steady-state data were obtained by operation of completely mixed, no-recycle reactors to support identification of the kinetics for substrate removal and cell production. The influent substrate concentration for all reactors was 500 mg/L of BOD_5. The influent biomass concentration for all reactors was negligible. The volume of each reactor was 10.0 Liters. Calculate q and μ for each reactor. Plot q versus S. Determine kinetic expressions for q and μ. Provide numerical estimates for all parameters in the models, including units.

Reactor	Effluent S mg BOD₅/L	Effluent X mg VSS/L	Flow Rate L/day
1	500	0	15.0
2	65	251	10.0
3	23	272	8.0
4	8	273	5.0
5	3	257	2.5
6	1	214	1.0

4.28 The following steady-state data were obtained by operation of completely mixed, no-recycle reactors to support identification of the kinetics for substrate removal and cell production. The influent substrate concentration for all reactors was 500 mg/L of BOD_5. The influent biomass concentration for all reactors was negligible. The volume of each reactor was 10.0 Liters. Calculate q for each reactor. Plot q versus S. Determine a kinetic expression for q. Provide numerical estimates for all parameters in the model, including units.

Reactor	Effluent S mg BOD₅/L	Effluent X mg VSS/L	Flow Rate L/day
1	500	0	15.0
2	104	230	10.0
3	38	241	9.0
4	13	228	8.0
5	4	210	7.0

4.29 Verify equation (4.24) by taking the derivative of equation (4.22), setting the derivative equal to zero, and solving for time.

4.30 Verify equation (4.22) as the solution to the differential equation (4.20) with initial conditions of $S = S_0$ and $D = D_0$ using integrating factor solution techniques.

4.31. Verify equation (4.22) as the solution to the differential equation (4.20) with initial conditions of $S = S_0$ and $D = D_0$ using Laplace domain solution techniques.

4.32 Repeat example problem 4.7 using a numerical solution method. Use a time increment of one hour for the simulation. Plot the oxygen sag curve for the river downstream from the discharge for the problem conditions obtained with the numerical solution. Compare this plot with a plot of the analytical solution obtained with equations (4.16) and (4.22).

4.33 For the critical discharge condition reported in example problem 4.8 (effluent $BOD_5 = 14$ mg/L), determine the dissolved oxygen concentration in the river downstream from the discharge using a numerical simulation method. Use a time increment of one hour for the simulation. Plot the oxygen sag curve for the river downstream from the discharge for the problem conditions. Compare this plot with a plot of the analytical solution obtained with equations (4.16) and (4.22).

4.34 For the critical discharge condition determined in end-of-chapter problem 4.22, determine the dissolved oxygen concentration in the river downstream from the discharge using a numerical simulation method. Use a time increment of one hour for the simulation. Plot the oxygen sag curve for the river downstream from the discharge for the problem conditions. Compare this plot with a plot of the analytical solution obtained with equations (4.16) and (4.22).

4.35 For the critical discharge condition determined in end-of-chapter problem 4.23, determine the dissolved oxygen concentration in the river downstream from the discharge using a numerical simulation method. Use a time increment of one hour for the simulation. Plot the oxygen sag curve for the river downstream from the discharge for the problem conditions. Compare this plot with a plot of the analytical solution obtained with equations (4.16) and (4.22).

4.36 For the critical discharge condition determined in end-of-chapter problem 4.24, determine the dissolved oxygen concentration in the river downstream from the discharge using a numerical simulation method. Use a time increment of one hour for the simulation. Plot the oxygen sag curve for the river downstream from the discharge for the problem conditions. Compare this plot with a plot of the analytical solution obtained with equations (4.16) and (4.22).

4.37 For the discharge condition described in end-of-chapter problem 4.25, determine the dissolved oxygen concentration in the river downstream from the discharge using a numerical simulation method. Use a time increment of one hour for the simulation. Plot the oxygen sag curve for the river downstream from the discharge for the problem conditions. Compare this plot with a plot of the analytical solution obtained with equations (4.16) and (4.22).

References

Dietz, John D. 1996. "Wastewater Treatment." In *Environmental Engineering P.E. Examination Guide & Handbook*, edited by W. Christopher King. Annapolis, MD: American Academy of Environmental Engineers.

Metcalf and Eddy, Inc. 1979. *Wastewater Engineering*. 2nd ed. New York: McGraw-Hill.

Metcalf and Eddy, Inc. 1991. *Wastewater Engineering*. 3rd ed. New York: McGraw-Hill.

Sawyer, C. N., P. L. McCarty, and G. F. Parkin. 1994. *Chemistry for Environmental Engineering*. 4th ed. New York: McGraw-Hill.

Standard Methods. 1992. *Standard Methods for the Examination of Water and Wastewater*. 18th ed. Washington, DC: American Public Health Association.

Chapter 5

Air Resources

5.1 Management of Air Resources

When our early ancestors first discovered fire they also first experienced air pollution, especially if that first fire was built inside a cave without good ventilation! Air pollution is tied directly to the burning of fuels, and can be pervasive in large, densely populated urban areas. Air pollution has caused local problems for many centuries (the use of coal in London was banned for several years starting in 1307 by King Edward I), but it did not become a global problem until the advent of the industrial revolution. Now, air pollution is a serious problem in many countries, with the highest concentrations being in the world's megacities, including Jakarta (Indonesia), São Paulo (Brazil), Bangkok (Thailand), Cairo (Egypt), Santiago (Chile), Los Angeles (United States), Mexico City (Mexico), and others.

Since the early 1800s, human population has increased from about 500 million to over 5 billion, or about one order of magnitude. World energy consumption has increased perhaps by two or three orders of magnitude, with much greater than average increases occurring in the industrialized centers. With such great increases in energy consumption, air pollution emissions have increased enormously since the 1800s, not withstanding the efforts of the last twenty-five years to control pollution. Based on the principles of material balance, with increased **emissions,** and with no change in the natural **rates of removal** from the atmosphere, it should not be surprising that **accumulations** of pollutants in the air have occurred frequently, resulting in harmful levels of air pollution. In this chapter we briefly address some major air pollution problems facing the world and review some modern approaches to the management and control of these problems.

Air resource management (ARM) involves several steps, including identifying the problem pollutants, determining appropriate air quality standards, identifying and controlling major sources of emissions, conducting air quality modeling, monitoring actual air quality, and assessing program effectiveness. The purpose of ARM is to provide good air quality to all citizens while balancing other social, cultural, political, and economic needs. Because the atmosphere knows no boundaries, pollutants emitted in one location can easily be transported to another. This same transport dilutes pollution and makes it difficult to identify polluters, so in the past, many industries simply discharged their wastes into the air, and counted on the atmosphere to carry them away. However, in very large urban areas, there is no "away," so it becomes much more important to reduce emissions and prevent pollution.

There are several well-known episodes where very high air pollution concentrations in cities caused thousands to get sick and resulted in many deaths (Donora, Pennsylvania in 1948 and London, England in 1952, for example), but air resource management is more than just emergency action planning. Modern ARM involves making plans and taking actions to reduce everyday concentrations and to provide healthy air on a long-term basis.

An important prerequisite to ARM is to know what the problem is. That is, an **emissions inventory** is needed to identify problem pollutants and polluters. An emissions inventory (EI) is a comprehensive and quantitative listing or estimation of major sources of pollution within a geographic area. An EI is very helpful in deciding where to focus the efforts of a city or state in controlling air pollution. The EI is developed from measured or estimated emissions data about all major pollutants and for all sources, including industry, mobile (or transportation) sources, area sources, and natural sources.

The agencies in charge of air quality must organize and present their data effectively to encourage political support and public participation. They must analyze the emissions data in conjunction with the ambient air quality data to propose the best strategies for reducing certain pollutants. These strategies might include emissions limits on certain industries, on-site inspections and/or fines, vehicle inspections, and traffic reduction ordinances and/or programs. Such agencies must then follow up and measure or estimate the results of their efforts.

In this chapter, we will learn about the major components of air resource management. This knowledge will enable us to identify and quantify the major pollutants (and their causes, sources, and effects), to be familiar with the legislative and regulatory standards that have been developed for effective management of local and national air quality, to distinguish among engineered equipment for air pollution control (both from stationary sources and from mobile sources), and to conduct dispersion modeling of air pollution from proposed or existing sources to predict pollutant concentrations in ambient air at locations downwind of those sources.

5.2   The Major Air Pollutants

Air pollution can be defined as foreign matter contained in the air in high enough concentrations to cause harm to people, plants, animals, or things.

Primary air pollutants (those emitted directly to the air) and **secondary** pollutants (those formed by reactions in the atmosphere) are both serious problems. **Major** pollutants are defined in this text to mean those emitted in very large quantities (millions of metric tons per year in the United States) or those of national concern because of their health effects. The six major primary pollutants are particulate matter (PM), sulfur oxides (SO_x), nitrogen oxides (NO_x), volatile organic compounds (VOCs), hazardous air pollutants (HAPs), and carbon monoxide (CO). The one major secondary pollutant is ground-level ozone (O_3). All of the above except VOCs and HAPs are also called **criteria** pollutants, because the U.S. EPA in the 1970s established ambient standards based on measurable health effects (the criteria for the standards).

Particulate matter (PM) is the term used to describe very small diameter solids or liquids which remain suspended in the atmosphere. PM-10 and PM-2.5 refer to particulate matter less than 10 and 2.5 μm in diameter, respectively. Particles are emitted from a variety of sources, including fossil-fuel combustion, metals and mineral processing, fugitive dusts from roads, agricultural fields, and many others (U.S. EPA, 1995). In the 1970s, the EPA established health-based air quality standards for total PM, but changed the standards to PM-10 in the late 1980s, and changed them again in 1997 to PM-2.5 in recognition of the more serious health effects of smaller particles. The effects of PM-2.5 include an adverse impact on human health (mainly related to the respiratory system), reduction in visibility due to haze (small particles scatter light and make things appear hazy), and soiling of buildings and other materials.

Nitrogen oxides are formed whenever any fuel is burned in air at a high enough temperature. In high-temperature combustion gases, the nitrogen and oxygen in the air combine to form NO and NO_2. Also, nitrogen atoms present in some fuels can contribute to emissions of nitrogen oxides (NO_x). Total U.S. emissions of NO_x are about equally distributed between stationary sources and mobile sources (vehicles). Nitrogen oxides can contribute to a reduction in visibility (NO_2 absorbs light), can be injurious to plants and animals, and can have adverse effects on human health. Nitrogen oxides also contribute to acid deposition. Perhaps most importantly for urban areas, NO_x reacts with VOCs in the presence of sunlight to form ground-level ozone, a major secondary pollutant.

Sulfur oxides are produced whenever a fuel that contains sulfur is burned. The main source is fossil-fuel combustion, although nonferrous metal smelting is also a major source. The main sulfur oxide is SO_2; however, some SO_3 is formed during combustion, and some SO_3 is formed by oxidization of SO_2 in the atmosphere. SO_2 and SO_3 form acids when they combine with water, and these acids can have detrimental effects on aquatic and terrestrial ecosystems, and on statues and buildings. In addition, gaseous SO_2 has been associated directly with human health problems and can cause damage to plants and animals.

Volatile organic compounds (VOCs) include any organics with an appreciable vapor pressure such that they vaporize when exposed to air. A major source of VOCs are automobiles and other mobile sources, from which small amounts of unburned fuel are exhausted to the air. Petrochemicals production, petroleum refining, transport, storage, and the pumping of gas into consumers' cars account for substantial VOC emissions, as does evaporation of solvents (such as those in oil-based paints, or printing inks). Some VOCs are carcinogenic, but the major problem with VOCs is that they react photochemically with NO_x in the atmosphere to form ozone.

Hazardous air pollutants (HAPs) are certain compounds (such as benzene, formaldehyde, vinyl chloride, lead, mercury, and many others) that were specifically identified by the U.S. EPA in the 1990 Clean Air Amendments. HAPs are emitted from a wide variety of sources, both combustion and noncombustion. Most HAPs are not emitted in very large quantities, but are considered serious because of their potential to severely damage human health.

The category of HAPS has only recently come into the spotlight, so the emission rates by source categories are not known with the same degree of certainty as with the criteria pollutants. The uncertainty is compounded by the fact that many HAPs are created as unwanted by-products (for example, during the combustion of various fuels). The emission rates of a few HAPS have been estimated (U.S. EPA, 1995). For example, benzene (a known carcinogen) is a component of gasoline and is emitted from mobile sources at the rate of about 300,000 tons per year. But it is also created during forest fires, prescribed burning, residential wood combustion, and by many different industrial furnaces and waste incinerators. The total benzene emissions from these combustion sources is in the range of another 100,000 tons per year.

Carbon monoxide (CO) is a colorless, odorless, tasteless gas that results from the incomplete combustion of any carbonaceous fuel. Usually, power plants and other large furnaces do not emit much CO because they are designed and operated carefully enough to ensure fairly complete combustion. Thus, the major sources of CO are on-road and off-road mobile sources. Automobiles, trucks, buses, airplanes, trains, boats, construction equipment, lawn and garden equipment, and farm vehicles exhausted over 76 million tons of CO in 1994 (U.S. EPA, 1995). CO reacts with the hemoglobin in blood to block oxygen transfer. Depending on the concentration of CO and length of exposure, effects of polluted air on humans may range from slight impairment of some psychomotor functions to dizziness and nausea.

Ozone and other oxidants are not emitted from a source per se, but are formed by complex photochemical reactions in the atmosphere involving mainly VOCs, NO_x, and sunlight. The classical term "smog" is defined as a mixture of smoke and fog, and stems from the early part of the twentieth century when the major problems were PM and SO_x. The term "smog" as used today describes the complex mix of air pollutants found in many cities, including high levels of ozone, but also PM, NO_x, VOCs, SO_x, and many other compounds; it often appears from a distance as a visible layer of material hanging over the city. Thus, urban smog is caused both by direct emissions of pollutants and by photochemical reactions in the air, and is a problem during the warm months of the year. The photochemical reactions also form small particles that contribute to haze.

The ozone and other oxidants attack plants and materials, and cause serious health effects including eye, nose, and throat irritation, and premature "aging" of the lungs. Reduction of ozone problems focuses on controlling emissions of VOCs or NO_x, depending on which compound is present in the air in the "critical" concentration. Detailed emission inventories and modeling are required to determine the best strategy for each urban area.

The preceding brief discussion of the causes, sources, and effects of air pollution is summarized in Table 5.1. Recent estimates of U.S. annual emissions of air pollutants are presented in Table 5.2.

5.3   Global Issues

There are three global air pollution issues discussed in this section: acid deposition, ozone layer depletion, and global climate change. The reader must recognize in advance that the following discussions are brief, and that there have been volumes written about each of these issues. Although we cannot devote much space to these topics in a general text such as ours, this in no way diminishes their level of importance to the world community.

5.3.1 Acid Deposition

Acid deposition refers to the fallout of acid rain, acid snow (or any other material that is acidic) from the atmosphere onto the earth. The acids are usually sulfuric or nitric acids that are formed in the atmosphere as a result of SO_x and NO_x emissions. Acid deposition is a serious issue and has been studied and debated for more than three decades due to two important characteristics: (1) acid deposition has seriously damaged lake and forest ecosystems, and (2) the acids tend to be transported long distances before they are deposited. For example, acids originating in the upper midwestern region of the United States have acidified lakes in Canada as well as damaged forests from New England to Virginia; acids from many countries in western Europe have impacted lakes in Scandinavia, as well as destroyed forests in Germany. Because of this characteristic, there have been acrimonious debates between countries about how to control acid deposition and who should do what to mitigate its effects. It is clear that acid deposition can be reduced only if acid gas emissions are reduced. The United States and other developed countries have made a major commitment to this goal. However, emissions of sulfur oxides and nitrogen oxides continue to increase in many countries in the world.

Table 5.1 Summary of major air pollutants—causes, sources, and effects.

Pollutant	Cause	Source	Effects On:			
			Human Health	Plants, Animals	Materials	Other
Particulate Matter PM-2.5	Burning coal, crushing, grinding	Power plant boilers, construction, industrial processes	Bronchitis, emphysema, cancer, etc.	Damage to leaf structure, toxic effects	Soiling, corrosion	Haze, smog
NO_x	Reaction of N_2 with O_2 at high temperature	Boilers, furnaces, vehicles	Respiratory irritant	Small	Corrosion	Reacts with VOCs to form photochemical smog and forms HNO_3 in atmosphere
SO_x	Burning any fuel with sulfur; processes using liquid SO_2 or H_2SO_4	Boilers and furnaces, industrial processes, smelting	Acts with particulates (synergism)	Necrosis, chlorosis	Corrosion	Forms H_2SO_4 in atmosphere; adds to haze and smog
VOCs	Incomplete burning of fuels; evaporation of solvents	Motor vehicles, industrial processes	Certain VOCs are toxic or carcinogenic	Small	None	Reacts with NO_x to form photochemical smog
HAPs	Similar to causes for VOC emissions, plus others	HAPs are emitted from various industrial and commercial processes and furnaces	HAPs are suspected or known carcinogens or have other toxic health effects	Certain HAPs have similar toxic effects on animals	Some HAPs are acids	Some HAPs bio-accumulate in the environment
CO	Incomplete combustion of carbon fuels	Vehicles and industrial processes	Poisonous; reacts with blood hemoglobin	None to plants	None	Eventually oxidizes to CO_2. Contributes to global warming
O_3	Chemical reaction of VOCs, NO_x, and sunlight	N.A.	Irritating and damaging to lungs, eyes, nose and throat	Severe damage to plants	Corrosion, oxidation, bleaching	Photochemical smog formation

Table 5.2 National U.S. emissions estimates of criteria pollutants, 1994 (millions of tons/yr).

Source Category	CO	SO$_X$	NO$_X$	VOC	PM-10
Stationary Sources					
Fuel Combustion					
Electric Utilities	0.32	14.87	7.80	0.04	0.27
Industrial Furnaces	0.67	3.03	3.21	0.14	0.24
Other combustion	3.89	0.60	0.73	0.71	0.53
Subtotal: fuel combustion	4.88	18.50	11.74	0.89	1.04
Chemical & Allied Manuf.	2.05	0.46	0.29	1.58	0.06
Metals Processing	2.17	0.69	0.08	0.08	0.14
Petroleum & Related Ind.	0.39	0.41	0.09	0.63	0.03
Other Industrial Processes	0.75	0.43	0.33	0.41	0.39
Solvent Utilization	0.00	0.00	0.00	6.31	0.00
Storage & Transport	0.06	0.00	0.00	1.77	0.06
Waste Disposal/Recycling	1.75	0.04	0.08	2.27	0.25
Mobile Sources					
On-Road Vehicles					
Light Duty Vehicles	54.44	-	5.18	5.58	-
Heavy Duty Vehicles	6.63	-	2.34	0.71	-
Subtotal (on-road)	61.07	0.30	7.52	6.29	0.31
Non-Road Vehicles					
Constr., lawn, farm, recr. boats, etc.	14.46	0.20	2.00	-	-
Aircraft	1.06	0.01	0.15	-	-
Railroads	0.12	0.07	0.95	-	-
Subtotal (non-road)	15.64	0.28	3.10	2.26	0.41
Subtotal Mobile Sources	76.71	0.58	10.62	8.55	0.72
Miscellaneous	9.25	0.01	0.37	0.68	42.74*
TOTAL ALL SOURCES	98.01	21.12	23.60	23.17	45.43

* Note: Miscellaneous PM-10 distributed approximately as follows:

natural sources—wind erosion	2.59
agriculture and forestry	7.12
wildfires/managed burns	1.02
fugitive dust	
unpaved roads	12.88
paved roads	6.36
other	12.77

Source: U.S. EPA, 1995.

5.3.2 Stratospheric Ozone Depletion

The ozone layer, located about 15 to 30 km above the earth's surface, is a portion of the atmosphere containing relatively high concentrations of ozone. This stratospheric ozone is formed naturally by photo-dissociation of oxygen atoms, and is chemically identical to the ground-level ozone that pollutes many of our cities. However, ozone in the stratosphere is extremely important to life on earth because ozone absorbs ultraviolet rays, preventing this harmful radiation from reaching the surface. The small fraction of the solar ultraviolet radiation that makes it to ground level is responsible for sunburns, skin cancer, and cataracts.

In the mid-1970s, scientists discovered that certain chlorine-containing compounds that had been emitted into the atmosphere were slowly working their way up through the lower layers of air into the ozone layer. These compounds are collectively known as **chlorofluorocarbons** (CFCs). Once in the stratosphere, with its higher levels of ultraviolet radiation, the CFCs break down—releasing free chlorine atoms that decompose ozone. The process is catalytic, meaning that one free chlorine atom can break down thousands of ozone molecules before eventually becoming inactivated.

For years, CFCs were used in aerosol spray cans, in automobile air conditioners, as foaming agents in making plastics, as fire suppressants, as degreasing compounds for certain metal parts before electroplating, and in many other uses. The high usage level resulted in emissions of CFCs to the atmosphere with significant long-term damage to the earth's protective ozone layer. In the late 1970s scientific evidence was gathered to prove the thinning of the ozone layer, especially in the Antarctic region. In 1987, representatives from thirty-six countries met in Montreal and signed an international accord aimed at limiting the production and use of these compounds. This accord, known as the Montreal Protocol, has resulted in significant decreases during the past dozen years in the use and manufacture of these and other ozone-depleting chemicals.

5.3.3 Global Climate Change

Global climate change (GCC) is perhaps the greatest environmental threat yet faced by humankind. GCC refers to the change in climate caused by emissions of various compounds into the atmosphere and the resulting change in the earth's energy balance. It is well known that the sun provides energy to the earth that warms the surface during the day, fuels the winds and hydrologic cycle, and is the basis for plant photosynthesis. It is also known that the earth cools off during the night by radiating infrared energy (heat) out to space. Based on a simple energy balance, we know that over any substantial period of time, the amount of energy given off during the nights must exactly equal the amount of energy absorbed during the days or else the heat content of the earth will change.

During the past two hundred years, human activity has disrupted this energy balance. Emissions of fine particles and sulfuric acid mists have contributed to more reflection of sunlight, preventing some energy from reaching the earth's surface. On the other hand, gases in the atmosphere (mainly carbon dioxide, methane, CFCs, and nitrous oxide) absorb infrared radiation, thus retaining heat and preventing the earth from cooling as much as it otherwise would. This latter effect seems to be outweighing the former, and GCC is popularly called the **greenhouse effect**.

Carbon dioxide (CO_2) is responsible for the majority of the greenhouse effect, but the other gases—methane, CFCs, and nitrous oxide—all contribute substantially. The latter three are each more "powerful" than CO_2 in that they absorb infrared radiation more effectively. However, the production rate of CO_2 is much greater, and it continues to be emitted in great quantity by fuel combustion. Because CO_2 has no short-term toxic or irritating effects, and because it is abundant in the atmosphere and is necessary to plant life, it is normally not considered a pollutant. However, from material balance considerations, we know that excess emissions of CO_2 into the atmosphere will result in an increase in its overall concentration. Scientists and engineers for years have had proof of the steady increases in the atmospheric concentration of CO_2 (see Figure 5.1).

Example Problem 5.1

In chapter 1, some data were given as to global energy usage. Assume that fossil fuel burning accounts for an energy use of 300 quadrillion kJ/year. Also, assume that all fossil fuels can be represented by the formula C_3H_5 with energy content of 40,000 kJ/kg. Finally, assume that air is 79% N_2 and 21% O_2, and has a "molecular weight" of 29 kg/kmole.

(a) Estimate the annual release of CO_2 to the atmosphere by burning fossil fuels, in kg/yr.

(b) If all that CO_2 was excess (that is, it entered the atmosphere and none was removed), estimate the increase in atmospheric concentration, in ppm by volume. Assume the atmosphere contains $5 (10)^{18}$ kg of air.

Solution

(a) The mass of fossil fuels combusted per year is:

$$\frac{300 (10)^{15} \text{ kJ/yr}}{40,000 \text{ kJ/kg}} = 7.5 (10)^{12} \text{ kg/yr}$$

Since the carbon mass fraction of the fuel is 36/41 = 0.878, then the carbon combusted is

$$= 0.878 \times 7.5 (10)^{12} = 6.58 (10)^{12} \text{ kg/yr}$$

A balanced reaction for the combustion of carbon is

$$C + O_2 \rightarrow CO_2$$

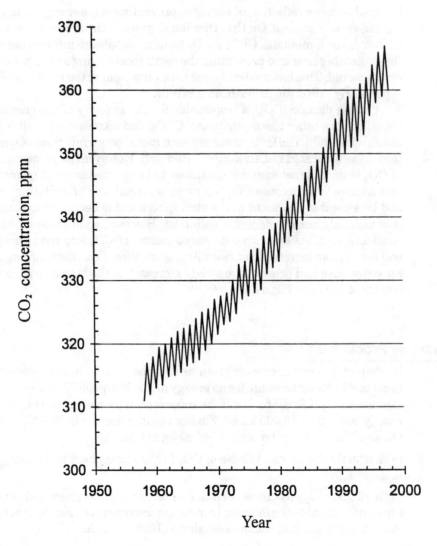

Figure 5.1
Monthly average concentrations of atmospheric CO_2 at Mauna Loa, Hawaii. (*Keeling, Whorf, Wahlen, and Van der Plicht, 1995. Updated figure based on personal communication with C.D. Keeling and T. P. Whorf at Scripps Institute of Oceanography, La Jolla, California.*)

which shows that CO_2 is produced mole for mole. Therefore, the moles of CO_2 produced are:

$$6.58 \ (10)^{12} \ \text{kg/yr} \ / \ 12 = 5.49 \ (10)^{11} \ \text{kmole/yr},$$

which translates into:

$$44 \times 5.49 \ (10)^{11} = 2.41 \ (10)^{13} \ \text{kg of } CO_2 \ \text{per year}$$

(b) The moles of air in the atmosphere are:

$$5 \, (10)^{18} / 29 = 1.72 \, (10)^{17} \text{ kmoles}$$

Therefore the increase in concentration is:

$$\frac{5.49 \, (10)^{11} \text{ kmole}}{1.72 \, (10)^{17} \text{ kmole}} \times 10^6 = 3.2 \text{ ppm}$$

Of course, the carbon balance for the earth is much more complicated than implied by example problem 5.1. There are many sources of CO_2 (including human and animal respiration, burning of trees and brush, dissolution of CO_2 from ocean water, volcanic eruptions, and biodegradation of organic matter), and many sinks for CO_2 (such as photosynthesis, dissolution of CO_2 into ocean water, mineralization into carbonates, and storage in living or dead organic matter). All the sources and sinks are interconnected, many in nonlinear ways. Modeling the system is difficult and uncertain, but the net result of this biogeochemical cycle is the steady increase of atmospheric CO_2 as shown in Figure 5.1. In fact, evidence from analyzing gases trapped in old ice cores indicates that the carbon dioxide content of the atmosphere has varied between about 190 and 290 ppm for the last 160,000 years, and has remained relatively stable at 260–290 ppm for the last 10,000 years. Through analysis of the ice core gases and carbon isotope ratios, a strong correlation has been demonstrated between low CO_2 levels and the ice ages, and high CO_2 levels and warmer interglacial periods. CO_2 was about 280 ppm as recently as 1750, had increased to 315 ppm in 1958, and is now above 365 ppm. The change from 1958 to the present is an amazing 16% in just 40 years, an enormous change over a tiny interval of time.

The greenhouse effect means much more than just a gentle warming of the earth as the name might imply. As the atmospheric heat content increases, it seems inevitable that weather and climate will change, perhaps significantly. Rainfall patterns will change, hurricanes may become more frequent and more severe, crop-growing regions may shift northward, plant and animal (especially insect) habitats may be affected, and sea level will increase. Predictions of an average warming of 1 to 2 °C within 50 years are not uncommon. While this might not sound alarming, remember that during the great ice ages, the average temperature was only about 7 to 9 °C less than it is today.

It has been argued for years that such increases in CO_2 could increase global temperature and change the climate, both regionally and globally. In fact, such an effect was first predicted in a technical publication by the famous Swedish chemist Svante Arrhenius one hundred years ago. However, the earth is a massive system; changes are slow and are not well understood. The effects of change in one parameter are masked and counteracted by hundreds of other factors. One thing is clear—increased heat retention should result in warmer temperatures. One common measure is the average global temperature (AGT). Unfortunately, this is not the most sensitive measure and the record is very "noisy" (erratic), making conclusive trends more difficult to spot (see Figure 5.2). Thus, a number of people have remained unconvinced that global climate change is a real concern. Hence, arguments and debates over the validity of GCC have continued for the last hundred years. However, in the mid-1990s most technical experts have come to agree that global climate change is definitely occurring (Hileman, 1995) and may cause dramatic and unpleasant changes in the environments of many regions of the world. It has been projected that an increase in the mean atmospheric CO_2 concentration from its present value of about 365 ppm to about 600 ppm will occur by the early part of the next century. The potential consequences of the global warming that may result from such an increase include dramatic effects on the earth's climate and weather conditions, such as altered rainfall patterns and crop-growing regions, more severe tropical storms, increased insect and bacterial infestations, and a rise in sea level and subsequent flooding in low-lying coastal areas. For more information, students can visit the web sites www.climate.org and www.ncdc.noaa.gov.

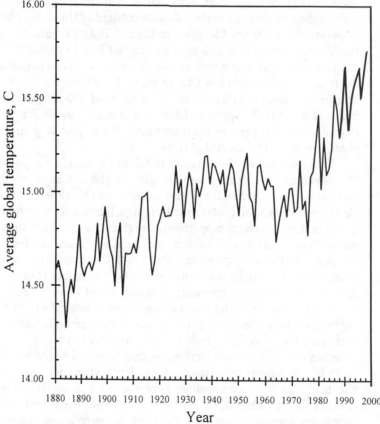

Figure 5.2
Average global temperature for the last century. (*Adapted from NASA Goddard Institute for Space Studies and National Climatic Data Center.*)

5.4   Legislative and Regulatory Standards

It has been said that something that is everybody's responsibility often ends up being nobody's responsibility. So it was with our air resources up until the early 1950s, when the federal government began to recognize its role and to assume its responsibility. Because air knows no political boundaries, air quality management appropriately begins at the national level. But, in addition to national management, *state and local efforts are essential* to the attainment and maintenance of good air quality.

Air-pollution standards were mandated by Congress and established by the U.S. EPA to protect the health and promote the well-being of individuals and communities. These standards were set by government with input from professional organizations as a result of increased awareness of pollutants and their effects upon living organisms, especially people. Federal legislation and regulations have been developed over a period of more than three decades with input from many interested groups. Some of these laws were the Air Pollution Control Act of 1955, the Motor Vehicle Air Pollution Control Act of 1965, the far-reaching Clean Air Act Amendments (CAAA) of 1970, the CAAA of 1977, and the comprehensive CAAA of 1990. Compliance with these laws requires not only proper environmental engineering design and operation of pollution-abatement equipment, but careful analysis and accurate measurements of specified pollutants and environmental quality parameters.

There are two types of standards: **ambient air quality standards** (AAQSs) deal with concentrations of pollutants in the outdoor atmosphere, while **source performance standards** (SPSs) apply to emissions of pollutants from specific sources. Ambient air quality standards are always written in terms of concentration ($\mu g/m^3$ or ppm), while SPSs are written in terms of mass emissions per unit of time or unit of production (g/min or kg of pollutant per ton of product produced).

National ambient air quality standards (NAAQS) were set by the EPA for several of the major pollutants at levels to protect the public health. The current standards are presented in Table 5.3. It should be noted that some states have set their own standards, which are stricter than those listed. Note also that some pollutants have more than one standard (depending on the averaging time, or time of exposure). Source performance standards (or emissions standards) are numerous because of the variety of sources. Some examples are given in Table 5.4. The new source performance standards listed in Table 5.4 were derived either from materials balance considerations or from actual field tests at a number of industrial plants. One use of these standards is as emissions factors to estimate emissions from plants yet to be constructed. Another use is to estimate emissions from sources that are difficult to measure in the field. The next few example problems illustrate some calculations with these standards.

Table 5.3 National ambient air quality standards.

Pollutant	Averaging Time	Primary Standard
PM-10	Annual arithmetic mean	50 $\mu g/m^3$
	24-hour average	150 $\mu g/m^3$
PM-2.5	Annual arithmetic mean	15 $\mu g/m^3$
	24-hour average	65 $\mu g/m^3$
CO	1-hour average	35 ppm
	8-hour average	9 ppm
SO$_2$	Annual arithmetic mean	80 $\mu g/m^3$
	24-hour average	365 $\mu g/m^3$
NO$_2$	Annual arithmetic mean	0.053 ppm
O$_3$	3-year average of annual 4th highest daily 8-hour maximum	0.08 ppm

Example Problem 5.2

The 1-hour National Ambient Air Quality Standard for CO is 35 ppm. Calculate the corresponding concentration in mg/m³, at 25° C and atmospheric pressure.

Solution

For gases in air, ppm refers to volume fraction. So 35 ppm is equivalent to 35 ml of CO per million ml of polluted air, or 35 ml of CO per m³ of air. The mass density of pure CO in mg/ml can be derived from the ideal gas law:

$$\frac{n}{V} = \frac{P}{RT} \quad \text{or} \quad \frac{m}{V} = \frac{P(M.W.)}{RT}$$

$$\frac{m}{V} = \frac{1.0 \text{ atm} (28 \text{ g/mol})}{.08209 \dfrac{L\text{-atm}}{mol\text{-}K} 298K} \quad \text{or} \quad \frac{m}{V} = 1.145 \text{ g/L or } 1.145 \text{ mg/ml}$$

Thus, the *concentration* of CO in air (in mg/m³) that equates to 35 ppm is:

$$CO = \frac{35 \text{ ml CO}}{m^3 air} \times \frac{1.145 \text{ mg CO}}{ml CO} = 40 \text{ mg/m}^3$$

Table 5.4 Selected examples of new source performance standards (NSPS).

Steam electric power plants
   a. Particulates: 0.03 lb/million Btu of heat input (13 g/million kJ)
   b. NO_x: 0.20 lb/million Btu (86 g/million kJ) for gaseous fuel
      0.30 lb/million Btu (130 g/million kJ) for liquid fuel
      0.60 lb/million Btu (260 g/million kJ) for anthracite or bituminous coal
   c. SO_2: 0.20 lb/million Btu (86 g/million kJ) for gas or liquid fuel. For coal-fired
      plants, the SO_2 standard requires a scrubber that is at least 70% efficient and may
      be more than 90% efficient depending on the percent sulfur in the coal. The max-
      imum permissible emissions rate is 1.2 lb SO_2 per million Btu of heat input; the
      permissible emissions rate may be less depending on the coal sulfur content and
      the scrubber efficiency required.
Solid waste incinerators
   The particulate emission standard is a maximum 3-hr. average concentration of 0.18
   g/dscm* corrected to 12 percent carbon dioxide.
Nitric acid plants
   Standard is a maximum 3-hr. average NO_x emission of 1.5 kg NO_x (expressed as
   nitrogen dioxide)/metric ton of acid produced.
Sulfuric acid plants
   Applies to plants employing the contact process. Standard is a maximum 3-hr. aver-
   age SO_2 emission of 2 kg/metric ton of acid produced. An acid mist standard is a
   maximum 3-hr. emission of 0.075 kg/metric ton of acid produced.
Wet-process phosphoric acid plants
   The standard for total fluorides emissions is 10.0 g/metric ton of P_2O_5 feed.
Iron and steel plants
   Particulate discharges may not exceed 50 mg/dscm, and the opacity must be 10%
   or less except for 2 minutes in any hour.
1995 passenger cars
   a. CO: 3.4 g/mile
   b NO_x: 0.4 g/mile
   c. VOC: 0.41 g/mile
   d. PM-10: 0.08 g/mile

Note: dscm = dry standard cubic meter.

Example Problem 5.3

Calculate the daily SO_2 emissions that a 200-ton-per-day sulfuric acid plant
would emit if emitting at the maximum allowable rate.

Solution

The standard is a maximum of 2 kg SO_2 per metric ton of sulfuric acid.

$$\frac{2 \text{ kg } SO_2}{\text{MT acid}} \times \frac{1 \text{ MT}}{1.102 \text{ T}} \times 200 \text{ T/day} = 363 \text{ kg } SO_2/\text{day}$$

Example Problem 5.4

Calculate the daily emissions of particulates and SO_2 from a 1000-MW coal-fired power plant which meets the performance standards listed in Table 5.4, including an SO_2 standard of 1.2 lb/million Btu heat input. Assume the coal has a heating value of 11,000 Btu/lb and that the plant has an overall efficiency of 39%.

Solution

First calculate the heat input rate for a 39% efficient plant:

$$E_{in} = \frac{1000 \text{ MW}}{0.39} \times \frac{1000 \text{ kw}}{\text{MW}} \times \frac{24 \text{ hr}}{\text{day}} \times \frac{3412 \text{ Btu}}{\text{kwh}} = 2.10 \times 10^{11} \text{ Btu/day}$$

$$\text{Particulates emitted} = \frac{0.03 \text{ lb}}{10^6 \text{ Btu}} \times \frac{2.10 \times 10^{11} \text{ Btu}}{\text{day}} \times \frac{1 \text{ ton}}{2000 \text{ lb}} = 3.2 \text{ tons/day}$$

$$SO_2 \text{ emitted} = \frac{1.2 \text{ lb } SO_2}{10^6 \text{ Btu}} \times \frac{2.10 \times 10^{11} \text{ Btu}}{\text{day}} \times \frac{1 \text{ ton}}{2000 \text{ lb}} = 126 \text{ tons/day}$$

5.5   Air Pollution Control: Stationary Sources

The most effective control often is simply a step or steps to *prevent* pollution from being formed. In recent years, industries have increasingly taken such steps. As discussed previously, since the mid-1980s industries have worked hard to find replacement cleaning techniques and have significantly reduced their use of (and emissions of) CFCs. Oil refineries have continually increased their standards of maintenance over the last twenty years and have significantly reduced leaks of petroleum products (this not only prevents air pollution but results in the recovery of more products that can be sold). Nevertheless, no process can be made 100% efficient. Therefore, there will always be some air pollution emissions that must be controlled. Engineers have developed several large, interesting, and important pollution control devices for industrial sources. These devices fall into two broad categories—those that remove particles and those that remove gases.

5.5.1 Particulate Matter Control Devices

There are several control devices for removing particulate matter from exhaust gases before they are emitted into the atmosphere. These include cyclones, electrostatic precipitators, baghouses, and scrubbers. In the following few pages we give a brief description of each device, then compare their advantages and disadvantages for removing particulate matter. In comparing particulate control devices we should consider collection efficiency and cost (both capital and operating) simultaneously. Collection efficiency is defined as:

$$\eta = \frac{\text{Mass rate of particles collected}}{\text{Mass input rate of particles}} \times 100\% \qquad (5.1)$$

A **cyclone** is designed to remove particles by causing the entire gas stream to spin in a vortex. The centrifugal force acting on the larger particles flings them toward the wall of the cyclone where they impinge and then fall to the bottom of the cyclone. The gas flows out through the top of the cyclone and the particles can be removed from the bottom. Advantages of cyclones are that they are simple, rugged, and inexpensive. Also, they collect the PM in a dry form so that the solids can be re-used. The major disadvantage is that the efficiency is usually not high enough to be able to use the cyclone as a final control device. Also, moving the gas through a cyclone at high enough velocities to collect a reasonable fraction of the PM creates a substantial pressure drop (which means an increase in operating costs). Figure 5.3 presents a schematic diagram of a cyclone.

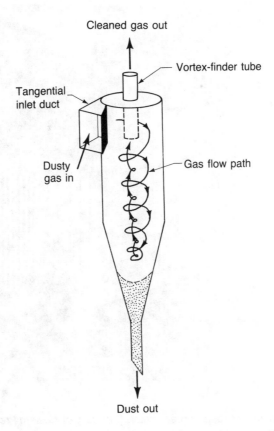

Figure 5.3
Schematic flow
diagram of a cyclone.

An **electrostatic precipitator** (ESP) removes particulate matter from a gas stream by creating a high voltage drop between electrodes. A cutaway view of an electrostatic precipitator is presented in Figure 5.4. A gas stream carrying particles flows into the ESP and between sets of large plate electrodes; gas molecules are ionized, the resulting ions stick to the particles, and the particles acquire a charge. The charged particles are attracted to and collected on the oppositely charged plates while the cleaned gas flows through the device. While the gas flows between the plates at velocities in the range of meters per second, the particles move towards the plates at a velocity (called the drift velocity) which is in the range of a meter per minute. During the operation of the device, the plates are rapped periodically to shake off the layer of dust that builds up. The dust is collected and disposed of or recycled.

ESPs are large and expensive, but collect particles with very high efficiencies. A major advantage is that they present very little resistance to gas flow and therefore cause only a slight pressure drop even when operating on flows as large as a million cubic feet per minute. Because of this advantage,

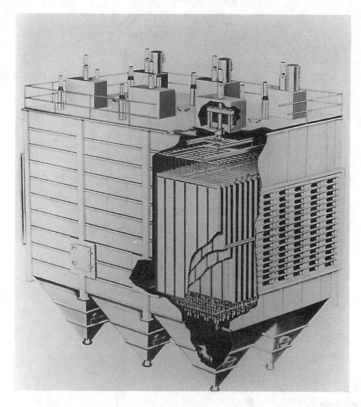

Figure 5.4
Cutaway view of an electrostatic precipitator. (*Courtesy of Western Precipitation Division, Joy Manufacturing Co., Los Angeles, CA.*)

their operating costs are not as great as one might expect. Many coal-fired power plants use ESPs. The design of ESPs is complicated, but the collection efficiency is reasonably well-modeled by the Deutsch equation:

$$\eta = 1 - \exp\left(-Aw/Q\right) \qquad (5.2)$$

where:

A = area of the collection plates, ft^2

w = effective drift velocity of particles towards the plates, ft/min

Q = gas volumetric flow through the ESP, ft^3/min

A **baghouse** can be thought of as a giant multiple-bag vacuum cleaner. Gas containing the particulates flows through cloth filter bags. The dust is filtered from the gas stream, while the gas passes through the cloth and is exhausted to the atmosphere. The bags are periodically cleaned (usually by shaking the bags or by blowing clean air backwards through them) to knock the dust down to the bottom hoppers where it can be removed to be either recycled or disposed. A cutaway view of a baghouse is presented in Figure 5.5.

The capital costs of baghouses are high, but their efficiencies are so high that they have become very popular as final control devices. Many power plants, and a variety of dry-process industries, use baghouses. Recently, baghouses have been used at hospitals to control incinerator emissions. When powdered lime and activated carbon are injected into the gases before flowing into the baghouse, the system will control not only particles, but also HCl gases and mercury fumes. The biggest operating cost comes from forcing large volumetric flows of air or combustion gases through the bags, which creates a substantial pressure drop. An equation relating pressure drop to energy consumption is:

$$W = Q \, \Delta P / \, \eta \qquad (5.3)$$

where:

W   = the energy consumed by a fan, kw

Q   = gas flow rate, m^3/s

ΔP = pressure drop, kPa

η = fan and motor efficiency (not particle collection efficiency)

Wet scrubbers operate on the principle of collision between particles and water droplets, collecting the particles in the larger, heavier water drops. The water falls through the upward-flowing gases, collides with and removes particles, and accumulates in the bottom of the scrubber. Later, the "dirty" water is treated to remove the solids.

Advantages of wet scrubbers include being able to handle flammable or explosive dusts, providing cooling of the gases, and neutralizing acid mists and vapors. Disadvantages include a high potential for corrosion, a high use of water, and a liquid or sludge effluent that must be treated and/or disposed.

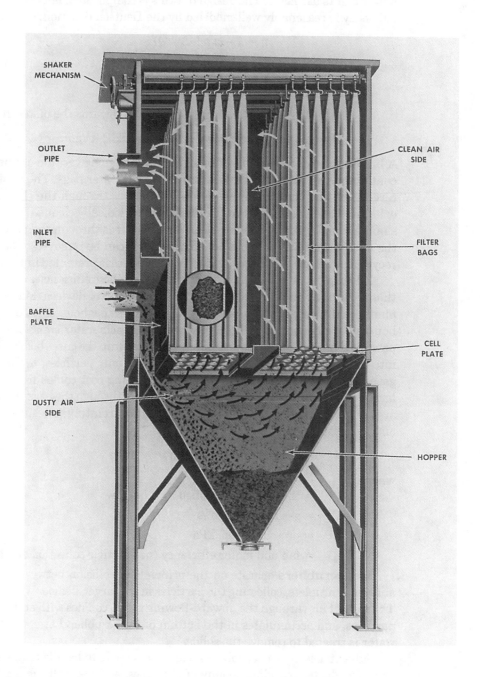

Figure 5.5
Cutaway view of a shaker-type baghouse.(*Courtesy of Wheelabrator Frye, Inc., Mishawaka, IN.*)

The capital and operating costs of wet scrubbers vary considerably with type of scrubber, efficiency desired, and local availability of water and local wastewater discharge standards. Figure 5.6 presents a cutaway view of one of the many types of spray tower scrubbers.

In summary, there are several different types of particulate matter control devices, with varying efficiencies and costs. Each has its own advantages and disadvantages; site-specific engineering is needed to make the best choice. Figure 5.7 compares the collection efficiencies of these devices over a range of particle sizes.

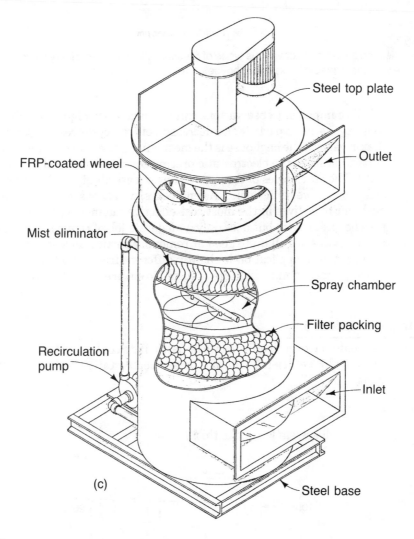

Figure 5.6
Cutaway view of a spray tower scrubber. (*Courtesy of Midwest Air Products Co., Owosso, MI.*)

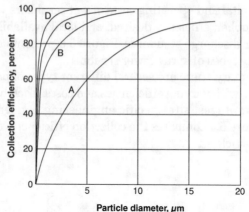

Figure 5.7
Typical collection efficiencies for various types of particle collectors: A–ordinary cyclone; B–spray tower; C–electrostatic precipitator; D–baghouse.

Two important observations can be made from Figure 5.7. First, notice that in general, as particle size decreases, efficiency decreases. Second, notice that in general the baghouse is the most efficient device over the widest range of particle sizes; the electrostatic precipitator is second, and the scrubber is third. The efficiencies of these three devices are all high, while cyclones typically have lower efficiencies. As engineers, we always want to try to accomplish our objectives in the most cost-effective manner. The cyclone is by far the cheapest device but does only a moderate job of removing particles. The other devices are all quite good at removing particulates but are expensive. However, each application is slightly different from all others. Proper engineering analysis and design are required to ensure the right choice for the job.

Example Problem 5.5

Calculate the overall efficiency of a particulate control system composed of a cyclone (75% efficient) followed by an electrostatic precipitator (95% efficient).

Solution

The overall system looks like this:

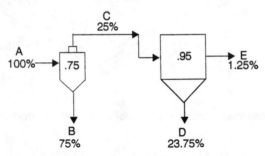

where streams A, B, C, D, and E represent particulate flow rates into and out of the equipment. The ESP collects 95% of the particles that pass through the cyclone. The overall collection efficiency is the sum of B and D, or 98.75%.

In example problem 5.5, we analyzed each piece of equipment, then added streams B and D to get the total collection efficiency. Let us define fractional penetration as one minus the fractional efficiency:

$$Pt = 1 - \eta/100 \tag{5.4}$$

Then, it should be obvious that penetration is the fraction of particle pass-through and that overall penetration for two devices in series is

$$Pt_{overall} = Pt_1 \times Pt_2 \tag{5.5}$$

Thus, the overall efficiency of collection for two devices in series is

$$\eta_{overall} = (1 - Pt_{overall}) \times 100\% \tag{5.6}$$

Equations (5.5) and (5.6) allow us to solve example problem 5.5 directly.

5.5.2 Gaseous Emissions Control Devices

As with particulate matter, there are several control devices and processes for removing or destroying gaseous pollutants before they are emitted into the atmosphere. These include carbon adsorbers, vapor incinerators, and scrubbers. In the next few pages we give a brief description of each device.

Volatile organic compounds usually are controlled by one of two methods: carbon adsorption or vapor incineration. Both are effective, and while carbon adsorption is a bit more complicated, it usually allows for recovery and reuse of the organic compounds. Incineration may be preferred for highly toxic or very odorous compounds.

In the **carbon adsorption** process, the contaminated air stream is passed through a bed of granular activated charcoal. The organic molecules are adsorbed onto the surface of the highly porous carbon pellets, while the cleaned air flows through. When all the surfaces of the carbon have been covered, the air stream is switched to an identical bed. While the air is being treated on the new bed, the old bed is regenerated by passing steam through it. The hot steam desorbs the organics and carries them out of the bed, thus renewing the carbon for another cycle of adsorption. The steam and organic vapors are separated by condensing the vapors, and then decanting the two immiscible liquids. The costs include the capital cost of the system, the fan power to move air through the bed, the amount of steam used for regeneration, and the costs for disposal/replacement of carbon. If enough VOC is recovered and if it has a high value, this type of system can actually generate revenue. Figure 5.8 presents a schematic diagram of a carbon adsorption system.

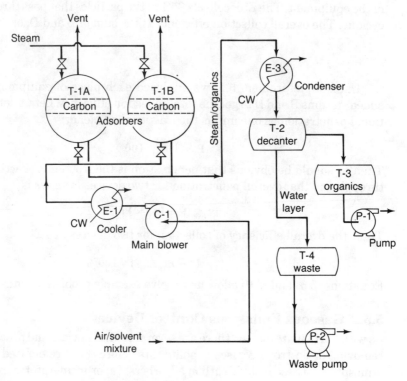

Figure 5.8
Process flow schematic of carbon adsorption system.

Vapor incineration (sometimes called thermal oxidation) is a simple process. The contaminated air stream is heated by burning a fuel gas in air. In the hot air stream, the VOCs are oxidized by reaction with the oxygen already present in the air stream. Figure 5.9 presents a cutaway view of a small unit. The desired end products of the reactions are CO_2 and water vapor; however, other products are possible. For example, if the organic waste is a chlorinated compound, HCl will be formed as a product of combustion. It is only necessary to provide enough temperature, time, and turbulence (the "three Ts" of incinerator design) to ensure good performance of the thermal oxidizer. Because the operating temperatures are 650 to 1100 °C, good designs almost always provide for heat recovery. Catalytic oxidizers also are used on relatively clean gas streams. The catalyst allows the reactions to occur at a lower temperature, saving fuel.

Another major use of **wet scrubbers** (besides removing particles) is to absorb a pollutant gas from a mixture of gases. The rate and extent of absorption are commonly assisted by a chemical reaction in the absorbing medium. A widespread example is the scrubbing of SO_2 from power-plant combustion

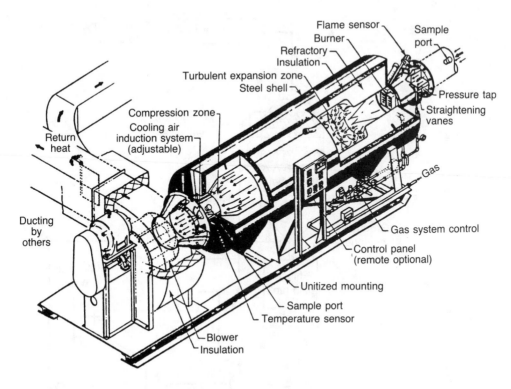

Figure 5.9
A cutaway view of a small thermal oxidizer. (*Courtesy of KTI Gas Processors, Inc., Santa Ana, CA.*)

gases by an alkaline solution. The most widely used method in this country uses a lime or limestone slurry to scrub the SO_2, producing a $CaSO_3$ plus $CaSO_4$ sludge that is usually discarded.

The chemistry of limestone scrubbing is complex and only a brief synopsis is presented here. The absorption of SO_2 gas is a two-step process: first, the SO_2 gas is absorbed into the liquid, forming bisulfite ions (HSO_3^-), and second, the bisulfite ions react with calcium ions to precipitate solid calcium sulfite (Henzel et al., 1981). The net reaction in the presence of limestone is

$$CaCO_3(s) + SO_2 + H_2O \rightarrow CaSO_3(s) + CO_2 + H_2O \qquad (5.7)$$

The pH in the scrubber must be kept fairly low (less than about 6) to prevent precipitation in the scrubber itself. The possible plugging of the scrubber is the reason that a separate vessel, the effluent hold tank (shown in Figure 5.10), is provided. More limestone is added here, and the precipitation reactions occur in this tank. With excess oxygen in the scrubber there is some oxidation of sulfite, so some of the sludge produced is $CaSO_4$. In fact, because $CaSO_4$ can be dewatered much more easily than $CaSO_3$, and because it is a

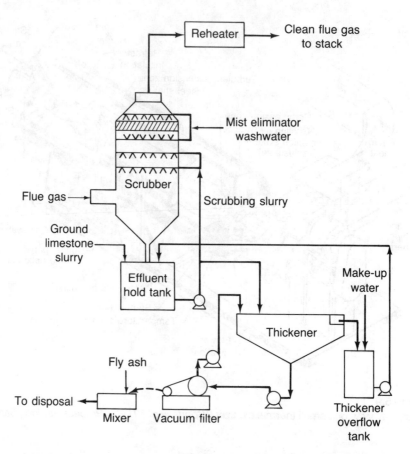

Figure 5.10
Schematic process flow diagram for limestone-based SO_2 scrubbing system. (*Cooper and Alley, 1994.*)

more stable sludge in a landfill, a popular process modification, called forced oxidation, has been developed to purposely oxidize the sludge before disposal. To summarize, the sulfur enters as a gas and leaves as a solid.

Lime-limestone scrubbing is a "throwaway" process because of the discarded sludge. This process accounts for more than 90% of all scrubbing systems in the United States because it is the cheapest and easiest to use. However, there are many other processes, some of which recover the SO_2 as either H_2SO_4 or elemental sulfur, both of which are valuable products.

Other countries use recovery-type processes. In Japan, for example, SO_2 scrubbing with recovery is mandated by the government and results in significant volumes of sulfuric acid, which is marketed within that country as well as exported. As sulfur supplies decline and become more expensive, these regenerative processes should become more widely used in the future. No matter what type of process is used, SO_2 scrubbing is a complex, large-scale, expensive undertaking. This type of chemical processing is unfamiliar to most power-plant operators, and most utilities have been slow to accept SO_2 scrubbing. However, federal law requires it on all new coal-fired power plants, and it certainly does reduce acid gas discharges to the atmosphere. The Clean Air Act Amendments of 1990 mandated significant reductions in SO_2 from many existing power plants. It is highly likely that many more scrubbers will be built in the near future.

Gases such as SO_2 can be scrubbed out of exhaust gases with alkaline aqueous solutions, and this is currently being done at a number of power plants throughout the world. But a gas like NO_2 or NO, which is relatively insoluble, is very difficult to remove once it has been formed. Therefore, the principal control mechanism for NO_x in the United States has been to minimize its formation through proper design and operation of burners and furnaces. However, catalytic reduction is also used, and is popular in Japan.

5.6   Air Pollution Control: Mobile Sources

5.6.1 Mobile Sources—A Global Problem

As stated earlier, mobile sources emit all the criteria pollutants, but especially large amounts of CO. In addition, mobile sources are major producers of VOCs and NO_x, which participate in the formation of ground-level (tropospheric) ozone, a significant problem in many of the world's largest urban areas. Mobile sources include cars, trucks, buses, airplanes, boats, locomotives, construction equipment, lawn and garden equipment, and other items. The only criteria for inclusion as a mobile source is the presence of a fossil-fuel fired engine and the ability to move around. Their mobility and their individually small sizes make air pollution control from this source category difficult. However, the extremely large and growing numbers of mobile sources makes such control imperative.

Although mobile-source pollution is a major problem for the United States and other industrialized countries, it is also a serious problem in many less developed countries. One problem facing these developing countries is that as their economies grow, their people tend to want more cars—clearly, an automobile is a status symbol around the world! Moreover, in developing countries, the affordable cars are usually older and not as well equipped for pollution control. Table 5.5 displays the numbers of road vehicles in various countries and regions of the world, and Table 5.6 displays vehicle emissions of the major air pollutants for several countries, as well as a world total.

5.6.2 Characteristics of Engines and Fuels

The internal combustion engine operating on gasoline or diesel is the prime mover of most mobile sources. A well-operated gasoline engine brings in precisely controlled amounts of fuel and air into the cylinder, ignites the mixture

Table 5.5 On-road vehicles around the world, 1988.

Country or Region	Millions of vehicles			
	Cars	Trucks & Buses	2- & 3-Wheelers	Total
U.S.A.	141.3	43.1	7.1	191.5
Canada	11.9	4.0	0.4	16.3
Japan	30.5	21.7	18.2	70.7
Europe	138.5	18.0	22.2	178.5
Latin America	27.3	7.9	4.3	45.6
Africa	8.0	4.4	1.2	13.6
Asia	10.2	10.0	33.3	53.5
World total	413.4	126.8	91.4	631.6

Table 5.6 Air pollution from vehicles for selected countries, late 1980s.

Country	Millions of metric tons/year					Thou. MT/yr
	CO	VOC	NO_x	PM	SO_x	Pb
U.S.A.	40.1	6.4	7.7	1.5	1.0	3.0
Canada	6.6	0.8	1.1	0.2	0.1	0.2
France	4.3	0.8	1.0	0.08	0.08	8.0
United Kingdom	4.3	0.6	1.2	0.2	0.05	3.1
Poland	0.4	0.1	0.6	n.a.	0.1	1.0
Russia	28.8	5.6	1.8	4.2	n.a.	n.a.
South Korea	0.8	0.1	0.6	0.04	0.04	n.a.
Thailand	0.4	0.02	0.05	0.01	0.04	1.5
World total	160	21	30	8	5	30

with a spark from the spark plug, and combusts the gasoline almost completely. An idealized equation showing the stoichiometric combustion of octane in air is shown in equation (5.8).

$$C_8H_{16} + 12\,O_2 + 45.1\,N_2 \rightarrow 8\,CO_2 + 8\,H_2O + 45.1\,N_2 \qquad (5.8)$$

However, even in an extremely well-designed car, the combustion reactions are not perfect, and small amounts of CO and NO_x are formed and are emitted from the cylinders along with small amounts of unburned gasoline fragments (VOCs). The actual air-to-fuel ratio, the mixing, and other factors greatly influence the relative amounts of these pollutants that escape from the engine. Furthermore, adjustments to reduce one pollutant often result in increasing another. For example, tuning an engine to run "richer" (more fuel and less air) may reduce NO_x but will increase CO and VOCs (see Figure 5.11).

Modern cars have a variety of pollution controls: computer control of the engine operation (air-to-fuel ratio, spark timing, etc.) to minimize the forma-

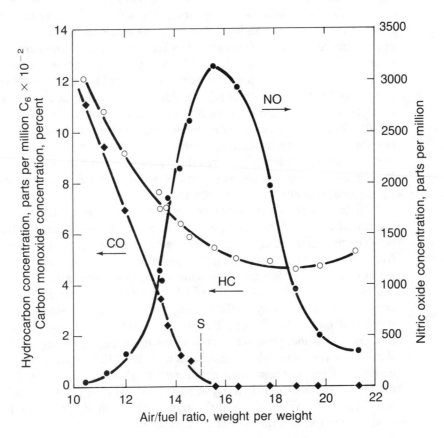

Figure 5.11
Effects of air/fuel ratio on air pollution from a gasoline engine. (*Adapted from Agnew, 1968.*)

tion of pollutants, catalytic converters (a small catalytic oxidizer) to further
react pollutants while they are still in the vehicle exhaust system, carbon can-
isters to capture evaporative emissions of VOCs, and others. However, to reap
these benefits the equipment must not be tampered with and must be properly maintained. In addition to equipment wearing out, there are many older
cars without these pollution-control devices still being driven—everyone has
seen "old clunkers" going down the street emitting visible smoke (as well as
invisible CO and NO). In many U.S. urban areas, it has been found that more
than half the vehicle pollution is caused by less than 15% of the vehicles. Of
course, in many countries pollution-control equipment either was never
installed, or was broken and then removed.

Fuel quality is very important. Based on mass balance, we know that
what goes in must come out. So any sulfur and/or chlorine contained in the
fuel comes out as acid gases, and any metal impurities are emitted as metal

vapors or metal salts. Diesel fuel is lower quality fuel than gasoline and has more sulfur. Also, because of the nature of the fuel and the design of diesel engines, a lot more smoke is produced during acceleration. In the less developed countries, drivers often can only afford the cheaper, lower quality fuels (more sulfur and other impurities). Moreover, many countries still use leaded gasoline, which is a cheap way to gain octane. Lead is a toxic metal that has been linked with reduction of IQ in children. Also, lead quickly deactivates a catalytic converter and renders it useless in controlling other pollutants. Finally, there is a high usage of motor scooters and small three-wheeled vehicles that use two-stroke engines (in which the oil and gasoline are pre-mixed and burned together), which results in more pollution per vehicle mile. Higher densities of smaller vehicles and more crowded cities also result in higher local concentrations of pollutants.

Environmental conditions (such as temperature and altitude) and driving conditions (such as average speed and smoothness of traffic flow) can affect engine operation and thus the rates of emissions. (In general, emissions from vehicles are measured in grams per mile driven; these emissions go up rapidly as the average speed goes down, and as there are more stops and starts.) The U.S. EPA has developed a large computer program to predict the emissions from roadway traffic for any set of conditions—speeds, temperatures, and many other factors. The program, MOBILE5a, requires user input for speed, altitude, year, temperature, vehicle type, and many other conditions; it contains much built-in data as well. MOBILE5a predicts an average emission factor for these input conditions, in g/veh-mile, that can be used to estimate total emissions in a city or region, or just along one particular roadway. This model is widely used throughout the United States and has been modified for use in a number of other countries.

5.6.3 Strategies for Pollution Prevention/Control from Mobile Sources

The vehicle manufacturer has the responsibility to produce vehicles that meet all applicable federal and state laws with regard to air pollution emissions. As soon as vehicles are sold, however, it becomes the owners' responsibility to operate them properly and to maintain the vehicles such that they will continue to drive cleanly. Oftentimes this goal is not achieved. Over the past two decades, it has become recognized that organized efforts by state and local governments are necessary to ensure that we achieve clean air in our urban areas. Several strategies have evolved for controlling or preventing pollution from vehicles, and are discussed in the next few paragraphs.

Inspection and maintenance programs have been implemented in metropolitan areas where emissions levels violate the national ambient air quality standards. Such programs involve annual or biannual vehicle inspections in which a computer-linked probe is inserted into the exhaust pipe to measure emissions. Vehicles that pass get a driving permit (sticker). Vehicles that do not pass are required to be repaired and then must get reinspected.

Transportation control measures (TCMs) and **traffic management techniques** (TMTs) can be effective. TCMs include providing mass transit options (more buses, exclusive bus or car-pool lanes, park and ride lots), economic incentives and penalties (subsidies for van pools, higher priced parking in downtown areas), and regulatory steps (parking bans, restricted driving zones, or even restricted driving days). Traffic management techniques refer to the control of traffic flow on the streets and the management of travel demand. A very successful TMT is the proper timing and sequencing of traffic signals on the main thoroughfares of a downtown area. Other techniques to reduce congestion include creating one-way street pairs, adding protected left turn lanes, planning truck routes, and others. Travel demand can be altered by staggered work hours, working four-day work weeks, or allowing for employee work-at-home programs (electronic commuting).

Changes in motor vehicle fuels can have a significant effect on air pollution. One of the great success stories in this regard is the EPA's mandating the phase-out of tetraethyl lead from gasoline in the mid-1970s. In 1970, lead emissions from motor vehicles were 200,000 metric tons per year; in 1990 they were only 2,000—a 99% decrease! Other recent fuel changes that have reduced air pollution are (1) reducing the vapor pressure of gasoline (cuts down on evaporative and engine emissions of VOCs) and (2) using some oxygenated fuels, like ethanol or ether (reduces CO emissions). In the near term, we might look for electric cars, and compressed natural gas and hydrogen fueled vehicles. Over the long term, solar-powered vehicles might provide a substantial answer to this vexing problem of mobile source emissions.

5.7 Meteorology and Atmospheric Dispersion of Pollutants

The release of pollutants into the atmosphere is a time-honored technique for "disposing" of them. One of the reasons that the air did not become completely unusable long ago is the atmosphere's ability to quickly disperse high concentrations of pollutants to lower, relatively harmless concentrations. Of course, the atmosphere's ability to disperse pollutants is not infinite and varies from quite good to quite poor, depending on meteorological and geographical conditions. Thus, with the onset of the industrial revolution and the subsequent exponential growth of human population and energy consumption, we have had to begin installing pollution-control equipment to prevent indiscriminate abuse of the atmosphere. However, we still rely heavily on atmospheric dispersion for final disposal of many pollutants.

5.7.1 Meteorology and Atmospheric Stability

Harmful effects occur when pollutants build up to high concentrations in a local area. The accumulation of pollutants in any localized region is a function of emission (input) rates, transport and dispersion (output) rates, and generation or destruction rates (by chemical reaction). The dispersion of pollutants is almost entirely due to natural conditions like geography and local meteorological conditions such as wind, rainfall, and atmospheric stability. Therefore, meteorology is extremely important. Meteorology includes both horizontal and vertical movements of the atmosphere, on a local, regional, and global scale.

Atmospheric stability is a term used to describe the mixing tendencies of the air. Air is termed **unstable** when there is good vertical mixing. This occurs when there is strong insolation and consequent heating of the layers of air near the ground. The warm air rises and is replaced by cooler air, which in turn is heated and rises. This process is good for diluting pollutants and carrying them away from the ground.

Stable air results when the surface of the earth is cooler than the air above it (such as on a clear, cool night). The layers of air next to the earth are thus cooled and no vertical mixing can occur. In stable air, pollutants tend to stay near the ground.

If there is a strong wind or good vertical mixing in the atmosphere, pollutants will be dispersed quickly into a large volume of air, resulting in low concentrations. If the wind is weak and there is an **inversion layer** (defined as a layer of warm air above a layer of cooler air, which prevents the vertical mixing of air in the region), then pollutant concentrations can increase. The modeling of air pollution dispersion is an important part of managing our air resources.

For convenience, atmospheric stability has been broken into six classes, arbitrarily labeled A through F, with A being the most unstable. Stability classes are correlated to the solar insolation and to the wind speed near the ground; Table 5.7 depicts these relationships. Researchers have also correlated values of the dispersion coefficients (see discussion in section 5.7.3) with atmospheric stability and distance away from the source. These correlations are presented in Figures 5.12 and 5.13. Once the stability classification is determined, this can be combined with other information to predict the concentration anywhere downwind from the source (as shown in section 5.7.3).

5.7.2 The Box Model

The simplest model that can be developed to predict pollutant concentrations under an inversion is called a box model. In this model we picture an area being covered by a rectangular box (see Figure 5.14). The height of the box is determined by the inversion layer and the box is aligned with the wind direction. Pollutants can be carried into the box from upwind or be emitted from ground-based sources. Pollutants are carried out of the box by the wind. If the air in the box is assumed to be well mixed (like a CSTR), then the analysis is straightforward. One input of pollutant is the material carried into the box with the incoming wind. The volumetric flow into the box is visualized as

Table 5.7 Atmospheric stability classifications.

Surface Wind Speed[a] m/s	Day Incoming Solar Radiation			Night Cloudiness[e]	
	Strong[b]	Moderate[c]	Slight[d]	Cloudy (≥4/8)	Clear (≤3/8)
<2	A	A–B[f]	B	E	F
2–3	A–B	B	C	E	F
3–5	B	B–C	C	D	E
5–6	C	C–D	D	D	D
>6	C	D	D	D	D

Notes:
a. Surface wind speed is measured at 10 m above the ground.
b. Corresponds to clear summer day with sun higher than 60° above the horizon.
c. Corresponds to a summer day with a few broken clouds, or a clear day with sun 35–60° above the horizon.
d. Corresponds to a fall afternoon, or a cloudy summer day, or clear summer day with the sun 15–35°.
e. Cloudiness is defined as the fraction of sky covered by clouds.
f. For A–B, B–C, or C–D conditions, average the values obtained for each.
A = Very unstable   D = Neutral
B = Moderately unstable   E = Slightly stable
C = Slightly unstable   F = Stable
Regardless of wind speed, Class D should be assumed for overcast conditions, day or night.

Source: Adapted from Turner, 1970.

the wind velocity times the front face area of the box, so the mass input of pollution is $uWHC_0$. Another input comes from source emissions within the box, which can either be identifiable point sources or pictured as a flux of pollutants coming from the "floor" of the box. Reaction is certainly possible, but first consider the steady-state material balance on a nonreactive pollutant:

$$0 = uWHC_0 + qLW - uWHC_e \qquad (5.9)$$

where terms are as depicted in Figure 5.14. Equation (5.9) can be solved for the outlet concentration as follows:

$$C_e = C_0 + \frac{qL}{uH} \qquad (5.10)$$

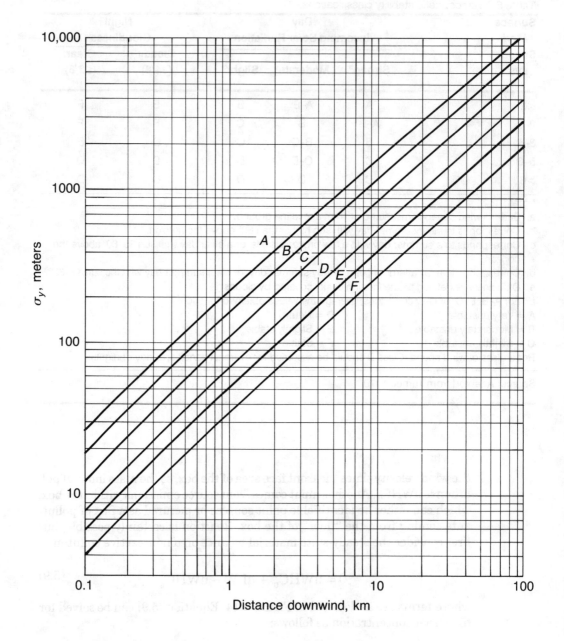

Figure 5.12
Horizontal dispersion coefficient as a function of downwind distance from the source and
atmospheric stability class. (*Adapted from Turner, 1970.*)

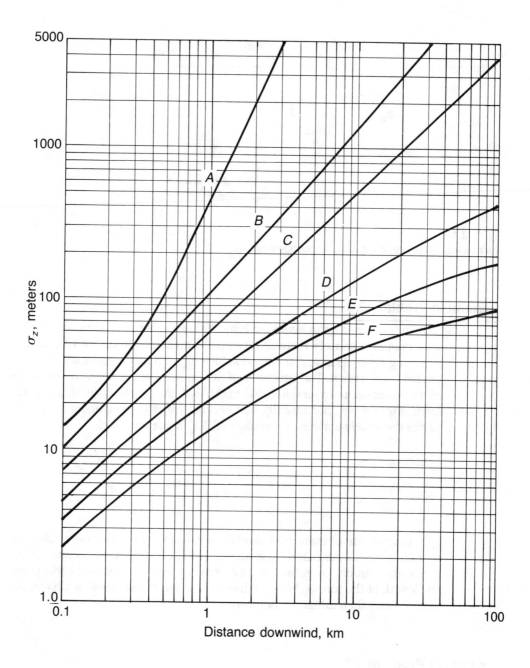

Figure 5.13
Vertical dispersion coefficient as a function of downwind distance from the source and atmospheric stability class. (*Adapted from Turner, 1970.*)

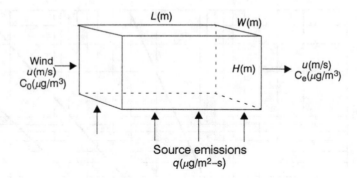

Figure 5.14
The box model for predicting pollutant concentrations.

Example Problem 5.6

A town is covered by an inversion layer 500 m above the ground. The town's area is approximately rectangular with dimensions 9 km by 12 km. The incoming air is clean and the wind is blowing at 3 m/s parallel with the 12-km side of the town. The emissions of SO_2 in this town come from a number of small sources and average 0.0008 mg/m²-s. Assuming that the air above the town is well-mixed horizontally, and vertically up to the inversion layer, calculate the steady-state concentration of SO_2 in the town's air.

Solution

$$C_e = 0 + \frac{0.8 \times 12,000}{3 \times 500} = 6.4 \, \mu g/m^3$$

Another use of material balances to calculate air pollutant concentrations is given in the following example. Note, however, that because of the physical situation we cannot assume that the air is well mixed throughout the length of the tunnel. We shall have to make our balance on a differential element and then integrate as shown.

Example Problem 5.7

A 2000-foot, one-way, two-lane rectangular tunnel filled with cars is blocked by an accident at the exit. The longitudinal ventilation system involves large exhaust fans blowing ambient air out of the exit end of the tunnel. Assume that the tunnel fills with cars, all with engines idling. One car occupies 20 linear feet of space in one lane in the tunnel and emits 2.8 grams of CO per minute with the engine idling. The tunnel is 30 feet wide and 12 feet tall. The fans blow 1,000,000 ft³/min of fresh air (which contains 2.0 mg/m³ of CO) into the tunnel entrance. Derive an equation to predict the steady-state concentration of CO as a function of position in the tunnel. Calculate the distance where the CO concentration surpasses 40 mg/m³.

Solution

The problem is diagrammed schematically below.

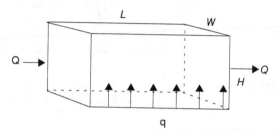

Assume that emissions q can be represented as a continuous line source along the length of the tunnel.

$$q = \frac{1\ car}{lane\text{-}20\ ft} \times 2\ lanes \times \frac{2.8\ g\ CO}{minute\text{-}car} = \frac{0.28\ g\ CO}{ft\text{-}minute}$$

Choose an arbitrary volume element in the tunnel located x feet from the entrance.

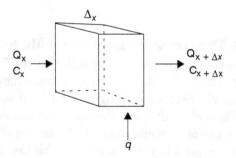

A steady-state material balance for CO on this volume element (assuming it is well-mixed vertically and horizontally) is

$$0 = Q_x C_x + q\Delta x - Q_{x+\Delta x} C_{x+\Delta x}$$

Since Q is essentially constant, we get

$$-Q(C_x - C_{x+\Delta x}) = q\Delta x$$

Dividing by Δx, we get

$$Q\left(\frac{C_{x+\Delta x} - C_x}{\Delta x}\right) = q$$

Taking the limit as $\Delta x \rightarrow 0$ yields the definition of a derivative:

$$Q\left(\frac{dC}{dx}\right) = q$$

This equation is easily integrated as follows:

$$Q\int_{C_0}^{C_L} dC = q\int_0^L dx$$

or

$$Q(C_L - C_0) = qL \text{ or } C_L = C_0 + \frac{q}{Q}L$$

Solving for L,

$$L = \frac{Q(C_L - C_0)}{q}$$

and substituting, we get:

$$L = \frac{1,000,000\,\dfrac{ft^3}{min}(40-2)mg\,/\,m^3 \times \dfrac{1\,m^3}{35.3\,ft^3}}{\dfrac{0.28\,g}{ft-min} \times \dfrac{1000\,mg}{g}}$$

$$L = 384 \text{ ft}$$

5.7.3 The Gaussian Dispersion Model

Releases of large amounts of pollution (as from a large power plant) are directed through tall stacks to allow some time and space for the pollutants to disperse before reaching ground level where we live (and where the NAAQS apply). Although the equations to describe the dispersion of pollutants in the open atmosphere can be developed from materials-balance considerations, it is beyond the scope of this text to do so. Rather, we will qualitatively describe this process and then present the equation that is widely used to calculate concentrations at points downwind from the source.

Pollutants released into the atmosphere travel with the mean wind. They also spread out in both the horizontal and vertical directions from the center line of the plume. The rate of spread in each direction is a complex function of micrometeorological conditions, the characteristics of the pollutants, and local geographical features. Figure 5.15 schematically portrays the time-averaged behavior of a plume being emitted into the air and bent over by a steady wind.

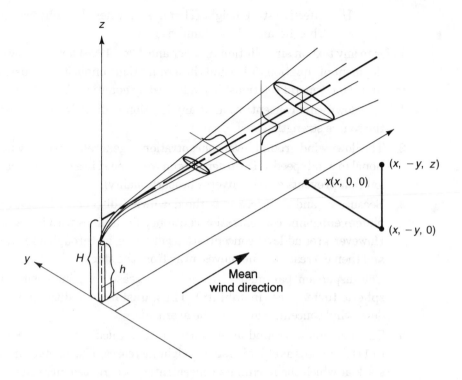

Figure 5.15
The spreading of a bent-over plume, with x, y, z axes. (*Adapted from Turner, 1970.*)

It has been shown that the spread of pollutants can be approximated by a Gaussian or normal distribution (Turner, 1970). The equation that models the normal dispersion of a gaseous pollutant from an elevated source is given below in a form that predicts the steady-state concentration at a point (x, y, z):

$$C = \frac{Q}{2\pi\, u\sigma_y\sigma_x} \exp\left(-\frac{1}{2}\frac{y^2}{\sigma_y^{\,2}}\right)\left\{\exp\left(-\frac{1}{2}\frac{(z-H)^2}{\sigma_z^2}\right) + \exp\left(-\frac{1}{2}\frac{(z+H)^2}{\sigma_z^2}\right)\right\}$$

(5.11)

where:

C = steady-state concentration at a point (x, y, z), $\mu g/m^3$

Q = emissions rate, $\mu g/s$

σ_y, σ_z = horizontal and vertical dispersion parameters, m

u = average wind speed at stack height, m/s

y = horizontal distance from plume center line, m

z = vertical distance from ground level, m

H = effective stack height ($H = h + \Delta h$, where h = physical stack height, and Δh = plume rise), m

Refer to any text on air pollution (Cooper and Alley, 1994) for a review of the development of equation (5.11) and the assumptions implicit in its use. Keep in mind some general relationships indicated by equation (5.11):

1. The downwind concentration at any location is directly proportional to the source strength, Q.

2. The downwind ground-level concentration is generally inversely proportional to wind speed. (H also depends on wind speed in a complicated fashion that prevents a strict inverse proportionality).

3. Because σ_y and σ_z increase as the downwind distance x increases, the plume center-line concentration continuously declines with increasing x. However, ground-level concentrations increase, go through a maximum, and then decrease as one moves away from the stack.

4. The dispersion parameters σ_y and σ_z increase with increasing atmospheric turbulence (instability). Thus, unstable conditions decrease downwind concentrations (on the average).

5. The maximum ground-level concentration calculated from equation (5.11) decreases as effective stack height increases. The distance from the stack at which the maximum concentration occurs also increases.

The Gaussian dispersion equation is extremely important in air pollution modeling work, and is used in almost all of the computer programs developed by the U.S. EPA for atmospheric dispersion modeling. The relationships in Figures 5.12 and 5.13 have been curve-fit for use in computer programs (Martin, 1976); the curve fit equations and constants are presented in Table 5.8.

Table 5.8 Curve fit equations and constants for dispersion parameters.

$\sigma_y = a\, x^b$; $\sigma_z = c\, x^d + f$			(x must be in units of km in both equations)					
			x<1 km			x>1 km		
Stability	a	b	c	d	f	c	d	f
A	213		440.8	1.941	9.27	459.7	2.094	−9.6
B	156		106.6	1.149	3.3	108.2	1.098	2.0
C	104		61.0	0.911	0	61.0	0.911	0
D	68		33.2	0.725	−1.7	44.5	0.516	−13.0
E	50.5		22.8	0.678	−1.3	55.4	0.305	−34.0
F	34		14.35	0.740	−0.35	62.6	0.180	−48.6

Note: b = 0.894 for all stability classes and values of x.

Source: Adapted from Martin, 1976.

Example Problem 5.8

(a) Using a hand calculator, calculate the ground-level downwind center-line concentration of SO_2 10 km from the source. The effective stack height is 100 m, the wind speed is 6 m/s, and the emissions rate is 0.25 kg/s. The atmosphere has a neutral stability.

(b) Using a spreadsheet, calculate the concentration at enough downwind distances to be able to draw a curve depicting how the ground-level concentration changes with distance from the source. Find the maximum ground-level concentration, and where it occurs.

Solution

(a) For D stability at x = 10 km, the values of σ_x and σ_z (as read from Figures 5.12 and 5.13) are 500 m and 130 m, respectively. In air modeling parlance, "ground-level" means z = 0 and "center-line" means y = 0. Substituting into equation (5.11) we get:

$$C = \frac{0.25(10)^9\ \mu g/s}{2\pi\ 6\dfrac{m}{s}\ 500m\ 130m}\ (1) \left\{ \exp\left(-\frac{1}{2}\frac{100^2}{130^2} \right) + \exp\left(-\frac{1}{2}\frac{100^2}{130^2} \right) \right\} = 102\ (1.49)$$

$$= 152\ \mu g/m^3$$

(b) A portion of the spreadsheet is shown in Figure 5.16. Increments of 100 m were chosen over the range of 100 m to 20 km. This choice gives 200 points with which to plot the curve. If, after developing the spreadsheet, the resulting curve is deemed not acceptable, then either smaller increments or a wider range can be implemented easily. From Figure 5.16, note that the calculation formula for the sigma values switches at x = 1 km. Also, note that at 10 km, the spreadsheet answer is 141 $\mu g/m^3$ as compared with the 152 $\mu g/m^3$ obtained in part (a) above. The difference is due to errors incurred in visually reading Figures 5.12 and 5.13.

Figure 5.17 displays the plot that was obtained directly from the spreadsheet. The plot clearly shows how, for an elevated point source, the downwind ground-level concentrations start low, increase through a maximum, and then decline with distance. From the curve, the maximum is estimated as 345 $\mu g/m^3$ and occurs 2.7 km downwind.

5.8 Indoor Air Quality

Up to this point, we have focused on ambient (outdoor) air quality. However, in modern society, many people spend 90% or more of their time inside: in their homes, in their workplaces, in stores, or inside vehicles going to or from those places. Indoor air quality (IAQ) thus is very important to people's health. Indeed, there have been a number of instances of "sick building" syndrome in the news in recent years, resulting in a variety of lawsuits against building owners and operators.

Define input values		H =	100 m			
		u =	6 m/s			
		Q =	0.25 kg/s			
		Class =	D			
		xstart =	100 m			
		x inc =	100 m			
		x end =	20 km			

Stability Class	For x < 1 km			For x > 1 km		
	a	c	d	f	c	d	f
A	213	440.8	1.941	9.27	459.7	2.094	-9.6
B	156	106.6	1.149	3.3	108.2	1.098	2
C	104	61	0.911	0	61	0.911	0
D	68	33.2	0.725	-1.7	44.5	0.516	-13
E	50.5	22.8	0.678	-1.3	55.4	0.305	-34
F	34	14.35	0.74	-0.35	62.6	0.18	-48.6

Note: b = 0.894 for all classes and all distances

Calculations and results

For a more general solution, pick coefficients for each class and distance using IF statements. For this example problem, coeffs were hand-selected and copied into the right cells

Calc No.	x dist, km	c	d	f	sigma y	sigma z	Conc
1	0.1000	33.2	0.725	-1.7	9	5	0.0
2	0.2000	33.2	0.725	-1.7	16	9	0.0
3	0.3000	33.2	0.725	-1.7	23	12	0.0
4	0.4000	33.2	0.725	-1.7	30	15	0.0
5	0.5000	33.2	0.725	-1.7	37	18	0.0
6	0.6000	33.2	0.725	-1.7	43	21	0.2
7	0.7000	33.2	0.725	-1.7	49	24	1.8
8	0.8000	33.2	0.725	-1.7	56	27	7.4
9	0.9000	33.2	0.725	-1.7	62	29	19.8
10	1.0000	44.5	0.5160	-13	68	32	40.1
11	1.1000	44.5	0.5160	-13	74	34	65.7
12	1.2000	44.5	0.5160	-13	80	36	95.2
13	1.3000	44.5	0.5160	-13	86	38	126.3
14	1.4000	44.5	0.5160	-13	92	40	157.3

Figure 5.16
A portion from the spreadsheet solution for example problem 5.8.

Indoor air pollutants include tobacco smoke, radon, formaldehyde, molds and mildews, and others. The concentrations of pollutants to which people are exposed can vary dramatically based on location, age of the building (but newer isn't always better), design and construction of the building, methods of ventilation, and whether control techniques are being used. This section presents information on the major pollutants of concern—where they come from, how they can accumulate in indoor air, and how we can predict and control IAQ.

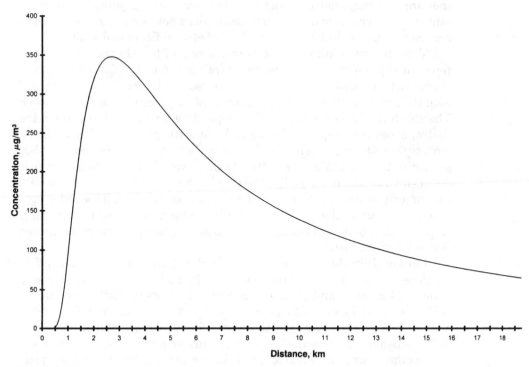

Figure 5.17
The spreadsheet plot of concentration versus distance for example problem 5.8.

5.8.1 Some Pollutants of Concern

Tobacco Smoke. Smoke from cigarettes, cigars, and pipes contains a number of pollutants ranging from fine particulate matter and CO, to formaldehyde and nicotine, to a variety of known or suspected carcinogens such as benzene and benzo-a-pyrene. Smoking has been blamed for more than 10 million deaths in the last thirty years. While smokers voluntarily choose to pollute their lungs, nonsmokers who share the same indoor space with smokers have no choice. Smoke, in addition to circulating throughout the air within the home, restaurant, office, or shopping area, can deposit on surfaces (such as curtains) and the smell can linger for a long time. Fortunately, the public relations campaign against smoking appears to be successful, and the percentage of smokers in the population is declining.

Radon. Radon (Rn) gas is a radioactive decay product of radium (which is itself a decay product of uranium), and is found in rocks and soils in many parts of the country. Although radon is not chemically active, its decay products (polonium, lead, and bismuth) are, and can lodge in the lungs where they emit alpha particles. Radon gas migrates upward through porous soils, and relatively large amounts of the gas can enter homes and other buildings if they are in its migration pathway. The gas enters through paths of least resistance (such as cracks in the concrete slab under a house, through semi-porous basement walls or flooring, or loose joints between floors and walls).

The concentration of radon is measured with a Geiger counter and reported as pico-Curies per cubic meter (pCi/m^3). A Curie (Ci) is a large unit of radioactivity, and a pCi is 10^{-12} Ci. The radioactivity of radon, combined with the chemical effects and radioactivity of its progeny, cause lung cancer. The soil flux of radon varies widely depending on location in the United States; it has been reported between 0.1 and 100 pCi/m^2-s, with 1.0 being most typical. Many homes have indoor air concentrations in the range of 1500 pCi/m^3 or 1.5 pCi/L (Masters, 1991). At this level, the risk of lung cancer is about equal to that obtained by receiving 75 chest x-rays per year. (The EPA recommends remedial action to reduce radon levels in homes if the radon concentration reaches about 4 to 8 pCi/L.) It has been estimated that radon is responsible for between 5000 and 20,000 cancer deaths per year in the United States (Masters, 1991).

Formaldehyde. Formaldehyde (HCHO) frequently is found inside buildings; it can come from several sources. Formaldehyde is emitted from a variety of laminates and glues, is found in some foam insulation, and in the "off-gases" from new carpets, plywood, and particle board. Therefore, it can be a problem in brand new office buildings. The principal human health effect of formaldehyde is eye irritation, which occurs at very low concentrations.

Another source of formaldehyde is indoor combustion (HCHO is a relatively stable product of incomplete combustion of any carbonaceous fuel). Sources include wood stoves and fireplaces, natural gas stoves and water heaters, or oil or coal furnaces with slight leaks. Also, cigarettes are known to emit formaldehyde. Formaldehyde concentrations inside homes and offices range from 0.02 to 0.3 ppm, although in some new mobile homes HCHO has been reported as high as 1 to 2 ppm (Godish, 1985). Furthermore, formaldehyde concentrations have been shown to increase substantially with increasing indoor air temperature, and/or increasing inside air humidity.

Two other pollutants that come from combustion sources are CO and NO_x. Both are criteria pollutants and have been discussed previously. However, high concentrations can be reached indoors if there is little or no ventilation for the combustion products. It should be noted that people die each year from accidental CO poisoning when CO accumulates indoors owing to the operation of a kerosene space heater, faulty furnace, or even a charcoal grill inside a poorly ventilated home.

Molds, Mildews, and Allergens. In buildings with central air conditioning, the inside walls of the ductwork can become a breeding ground for mold, mildew, and bacteria. This problem can be especially bad in warm, humid climates, and has been exacerbated by energy-saving trends in the last decade. These energy-saving steps are (1) to use a higher percentage of recirculated air and a lower percentage of fresh air and (2) to keep the air moister when using air conditioning. Inspections have found millions of spores per square inch on the inside walls of ducts leading from the air conditioners in homes and commercial buildings. People may be allergic to mold and mildew spores and/or their gaseous metabolic products.

5.8.2 Ventilation and Infiltration

Ventilation is defined as fresh outside air coming in to replace inside air that is being exhausted to the outside. This process is called simply **air exchange**. Air exchange occurs in buildings in three ways: forced ventilation, natural ventilation, and infiltration. Forced ventilation uses fans or blowers to forcibly exchange the air, while natural ventilation permits natural air exchange through open windows or doors. Infiltration refers to the natural air exchange that occurs even when all windows and doors are closed. Air can leak into buildings through numerous gaps and openings in the building envelope such as the cracks around doors and windows, the gaps around pipes and electrical conduits, kitchen and bathroom vent pipes, floor-wall joints, mortar joints, and others. Ventilation and infiltration are depicted schematically in Figure 5.18.

Excess outside air that enters a building must be heated or cooled to keep the inside conditions comfortable, and infiltration accounts for most of this "excess" energy use. Infiltration occurs on a much larger scale than one might think—by some estimates, it accounts for up to 10% of total U.S. energy use, or about $10 billion per year (Masters, 1991).

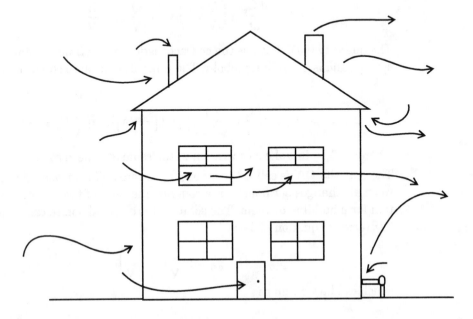

Figure 5.18
Ventilation and infiltration.

5.8.3 Material Balance Models for Indoor Air Quality

The indoor concentrations of air contaminants can be predicted by simple mathematical models; the key variables are emission rates and ventilation rates. In some buildings, the air flow is very simple, and we can assume the whole building acts like a single CSTR. For other situations, we might have to model the building as many stirred tanks connected in series and parallel. For the simple model of one well-mixed room, the material balance equation is similar to that for the box model (equation 5.9):

$$\frac{V d C_e}{dt} = Q C_i + E - Q C_e - k C_e V \qquad (5.12)$$

where:

V = volume of the room, m^3

C_e = concentration indoors, $\mu g/m^3$

C_i = concentration in the fresh air coming into the room, $\mu g/m^3$

Q = ventilation rate, m^3/hr

E = emission rate inside the room, $\mu g/hr$

k = reaction rate constant (here assumed to be first-order destruction), hr^{-1}

Equation (5.12) can be rewritten as

$$\frac{d C_e}{dt} + \left(\frac{Q}{V} + k\right) C_e = \frac{Q}{V} C_i + \frac{E}{V} \qquad (5.13)$$

The quantity $(Q/V + k)^{-1}$ is the characteristic time or time constant for this system, and is given the symbol τ. The general solution to equation (5.13) is:

$$C_e = \left\{ C_0 - \tau\left(\frac{Q}{V} C_i + \frac{E}{V}\right) \right\} \left\{ \exp[-t / \tau] \right\} + \tau\left(\frac{Q}{V} C_i + \frac{E}{V}\right) \qquad (5.14)$$

where: C_0 = initial concentration in the room at time zero.

The quantity Q/V is called the air exchange rate and is measured in units of room air changes per hour. It is a common measure of the degree of ventilation for a building or room. The solution to the steady-state case is just the last term of equation (5.14):

$$C_{ess} = \tau\left(\frac{Q}{V} C_i + \frac{E}{V}\right)$$

which can be written as

$$C_{ess} = \left(\frac{Q}{V} C_i + \frac{E}{V}\right) / \left(\frac{Q}{V} + k\right)$$

or

$$C_{e\ ss} = (AC_i + E/V) / (A + k) \tag{5.15}$$

where: $A = Q/V$, air changes per hour.
Obviously, as the air exchange rate increases, the indoor concentration approaches the outdoor concentration level.

Example Problem 5.9

Two friends go to a nightclub one evening to listen to a popular band. As the room begins to fill up, more and more smokers light up. The room fills quickly, and soon there are 600 people there, one-third of whom are smoking at any given moment. Each cigarette emits 1.4 mg of formaldehyde, and each smoker averages three cigarettes per hour. Formaldehyde decays to carbon dioxide with a first-order rate constant of 0.40 per hour. The room measures 20 m by 30 m by 5 m, the ventilation rate is 2.5 air changes per hour for this event, and the outdoor concentration is zero. (a) Assuming that the friends stay there until the steady-state concentration is reached, what is the maximum concentration of formaldehyde to which they are exposed? (b) Assuming the initial HCHO concentration is zero, how long does it take to reach 95% of the steady-state concentration?

Solution

(a) At steady state, equation (5.15) applies, namely:

$$C_{e\ ss} = (AC_i + E/V) / (A + k)$$

The emissions rate is:

$$E = 0.333 \times 600 \text{ people} \times 1.4 \text{ mg/cig} \times 3 \text{ cig/hr}$$

$$= 839 \text{ mg/hr of formaldehyde emissions}$$

The room volume is 3000 m^3, and since C_i is zero, then

$$C_{e\ ss} = (E/V) / (A + k)$$

or

$$= (839/3000) / (2.5 + 0.4) = 0.0965 \text{ mg/m}^3$$

This concentration converts to 0.079 ppm, more than enough to cause serious eye irritation.

(b) To estimate the time to reach 95% of the steady-state concentration, we must know how the emissions rate function behaves. A simple approach, since the room fills quickly, is to assume that the emissions jump up immediately to their final level (a so-called step increase). For $C_0 = C_i = $ zero, equation (5.14) reduces to:

$$C_e = \frac{E/V}{(A+k)}\left(1 - \exp\left[-t/\tau\right]\right)$$

$$= C_{e\,ss}\left(1 - \exp\left[-t/\tau\right]\right) \quad \text{(recall } \tau = 1/(2.5+0.4) = 0.345\,\text{hr)}$$

Setting $C_e/C_{e\,ss} = 0.95$, and solving for t, we find that t equals 1.03 hours.

The response of a first-order system (such as described in example problem 5.9) to a step-increase forcing function is always an exponential response curve (see Figure 5.19). For such response curves, 63% of the final response is reached in a time numerically equal to one time constant, 86% in two time constants, 95% in three time constants, and so forth. Steady state is effectively achieved after four time constants. In example problem 5.9, the time constant is 0.345 hours or about 20 minutes, so the time to reach 95% of the steady-state concentration is three time constants or about one hour, which confirms the answer to part (b).

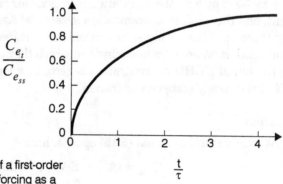

Figure 5.19
Dimensionless response of a first-order system to a step increase forcing as a function of dimensionless time.

The unsteady-state problem of example problem 5.9 can be solved easily using numerical methods as illustrated in example problem 5.10.

Example Problem 5.10

Solve example problem 5.9 using a spreadsheet.

Solution

We start by rewriting the mass balance (equation 5.12) as a finite difference equation:

$$V\,\Delta C_e/\Delta t = QC_i + E - Q\,C_e - k\,C_e\,V$$

Defining Δt as 1 minute, and setting C_e equal to C_0 at time zero, we set up a spreadsheet (see Figure 5.20). Then simply copy down any number of rows and watch the results until the calculated C_e approaches a constant value.

Rows		Columns				
	B	C	D	E	F	G

3

4 Example Problem 5.10   Spreadsheet solution for smoky nightclub

5

6 Define system parameters   K =   0.4 1/hr

7 Define initial conditions   Czero =   0 mg/L   Also, Cin=0

8   t zero =   0 minutes

9   Q =   7500 cu m/hour

10   E =   840 mg/hour

11   V =   3000 cubic meters

12 Define simulation params   deltime=   1 minute

13

14 Equation to be solved   VdC/d t = QCin + E - QCout - kCoutV

15   or   dC/dt = (Q/V)Cin + E/V - (Q/V)Cout - kCout

16 Also, new estimate for Cout: Cnew = Cold + dC/dt * del t

METHOD   Start in row 23 with Cold = Czero; calc dC/dt from above;

18   multiply by del t & add result of Czero to get Cnew

19   copy Cnew into next row cell for Cold and repeat

20

21 **Calculations and results**   With spreadsheets, a formula with the cell

Row	t, min	C old	dCdt	C new	
22	t, min	C old	dCdt	C new	address as, for example, B13, when copied
23	0	0.0000	0.2800	0.0047	will change to follow the cell position. But
24	1	0.0047	0.2665	0.0091	with the cell address given as $B$13, the cell
25	2	0.0091	0.2536	0.0133	address will not be indexed.
26	3	0.0133	0.2413	0.0174	
27	4	0.0174	0.2297	0.0212	
28	5	0.0212	0.2186	0.0248	
29	6	0.0248	0.2080	0.0283	
30	7	0.0283	0.1979	0.0316	
31	8	0.0316	0.1884	0.0347	
32	9	0.0347	0.1793	0.0377	Note break here; rows 33–80 are hidden
81	58	0.0911	0.0158	0.0914	
82	59	0.0914	0.0151	0.0916	
83	60	0.0916	0.0143	0.0918	**One hour**
84	61	0.0918	0.0136	0.0921	
85	62	0.0921	0.0130	0.0923	Note: rows 86–140 are hidden
141	118	0.0963	0.0008	0.0963	
142	119	0.0963	0.0008	0.0963	
143	120	0.0963	0.0007	0.0963	**Two hours**
144	121	0.0963	0.0007	0.0963	
145	122	0.0963	0.0007	0.0963	Note: rows 146–200 are hidden
201	178	0.0965	0.0000	0.0965	
202	179	0.0965	0.0000	0.0965	
203	180	0.0965	0.0000	0.0965	**Three hours**

Figure 5.20

Spreadsheet solution to example problem 5.10.

5.8.4 Solutions to IAQ Problems

Review of equation (5.15) shows that the only viable approaches to solving IAQ problems are to increase Q or reduce E. Increasing Q causes increased energy costs on a continuing basis, and has been discouraged as a solution (after all, it is only a "dilution" method). However, if we know exactly where the emissions are originating, and if we are smart about how and where to ventilate, this approach can be very effective. For example, to control cooking emissions, a local vent fan is installed over the stove. It exhausts only a small volume of air compared with the whole house, but that small volume may contain 90% of the emissions of certain pollutants. In another example, if the problem is radon emissions entering through the basement of a house, the basement could be isolated and ventilated continuously as shown in Figure 5.21.

For other problems, such as molds and mildews in the ductwork, a regular schedule of cleaning and disinfection may be required. Recently, smoking-related problems have been eased by bans on the smoking of tobacco within the confines of certain buildings. Each IAQ problem is unique and requires careful investigation, analysis, design, and implementation to solve the problem in a cost effective manner.

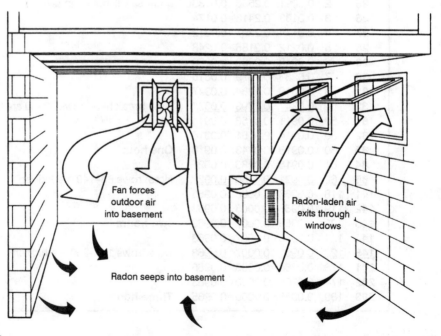

Fan forces outdoor air into basement

Radon-laden air exits through windows

Radon seeps into basement

Figure 5.21
One method of controlling radon entry into the living space of a house.

5.9   Noise Pollution*

Noise can be defined as unwanted sound. Accordingly, to investigate noise, some fundamental concepts of sound must first be understood. The perception of sound by an individual, whether it is from a tuning fork producing a pure tone or the complicated spectra from community noise, is truly an amazing process. An individual evaluates sound by at least four distinct criteria. These are loudness, frequency, duration, and subjectivity.

5.9.1 Loudness

Sound is propagated through the air as a cyclic wave of pressure fluctuations. The loudness or intensity of the sound is directly related to the amplitude of those pressure fluctuations. The pressure fluctuations cause the ear drum to be flexed, stimulating the auditory nerves and creating the sensation of sound. The human ear can sense pressure fluctuations as low as 2×10^{-5} Pa (the threshold of hearing) up to about 63 Pa (the threshold of pain). This range represents a pressure change by a factor of more than three million! Figure 5.22 shows typical sound pressure levels for various urban and rural noises.

The large range of pressure fluctuations that can be perceived by the ear makes for a clumsy scale. Additionally, as a protective mechanism the auditory response is not linearly related to pressure fluctuations. To overcome these two problems the sound pressure level (SPL), measured in decibels (dB), is used to describe loudness. The SPL(dB) is defined by:

$$ \text{SPL (dB)} = 10 \log_{10} (P^2/P_o^2) \tag{5.16} $$

where:

P_o = the reference pressure (2×10^{-5} Pa)

P = the sound pressure of concern (Pa)

The use of dB indicates that loudness is being reported as a sound pressure level, which is different from sound pressure. While sound pressures can be directly added, equation (5.16) indicates that decibels do not add in a linear fashion, but logarithmically. This means that if the sound pressure is doubled, the increase in the sound pressure *level* is only 3 dB. Computing the sum of several SPLs in dB may be accomplished using the equation:

$$ \text{SPL}_{total} = 10 \log_{10} \Sigma\ 10^{\ \text{SPL}/10} \tag{5.17} $$

Section 5.9 on noise pollution was contributed by Roger L. Wayson, Ph.D., P.E., Associate Professor, University of Central Florida.

Example Problem 5.11

A sound source is measured to be a constant 70 dB at a receiver location. If two more identical sources are making identical sounds, what is the resulting overall sound pressure level?

Solution

Since we are adding sources, we can use equation (5.17):

$$SPL_{total} = 10 \log_{10} \Sigma\, 10^{\,SPL/10}$$

$$SPL_{total} = 10 \log_{10} (10^{70/10} + 10^{70/10} + 10^{70/10})$$

$$= 10 \log_{10} (30{,}000{,}000)$$

$$= 74.8 \text{ dB}$$

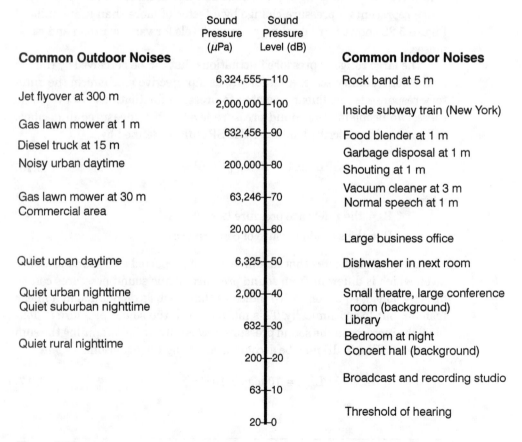

	Sound Pressure (μPa)	Sound Pressure Level (dB)	
Common Outdoor Noises			**Common Indoor Noises**
	6,324,555 — 110		Rock band at 5 m
Jet flyover at 300 m	2,000,000 — 100		Inside subway train (New York)
Gas lawn mower at 1 m	632,456 — 90		Food blender at 1 m
Diesel truck at 15 m			Garbage disposal at 1 m
Noisy urban daytime	200,000 — 80		Shouting at 1 m
Gas lawn mower at 30 m	63,246 — 70		Vacuum cleaner at 3 m
Commercial area			Normal speech at 1 m
	20,000 — 60		Large business office
Quiet urban daytime	6,325 — 50		Dishwasher in next room
Quiet urban nighttime	2,000 — 40		Small theatre, large conference room (background)
Quiet suburban nighttime			Library
	632 — 30		Bedroom at night
Quiet rural nighttime			Concert hall (background)
	200 — 20		
			Broadcast and recording studio
	63 — 10		
			Threshold of hearing
	20 — 0		

Figure 5.22
Typical community noises and their sound pressure levels.(*Federal Highway Administration.*)

A change of 10 dB is generally perceived to be a doubling or halving of the sound. In outdoor situations, it takes a change in SPL of greater than 3 dB to be noticeable, but changes in other components, such as the tone, can be perceived with no change in SPL.

5.9.2 Frequency

The human ear can distinguish among a large range of frequencies. The frequency is the number of pressure changes per second or oscillations per second, and has the unit of Hertz (Hz). The ear can detect frequencies extending from about 20 Hz to 20,000 Hz. These different rates of the pressure fluctuations provide the tonal quality of the sound. For instance, a flute has a much higher frequency than a bass guitar, and their tones are discerned as being quite different. The frequency and the wavelength of the sound wave are inversely related:

$$f = c/\lambda \qquad (5.18)$$

where:

f = frequency (Hz)

c = speed of sound (m/s)

λ = wavelength (m)

Example Problem 5.12

What is the wavelength of a 550 Hz tone?

Solution

Although the speed of sound changes slightly with environmental changes, a good approximation for the speed of sound in air is 343 m/s or 1100 ft/s. Rearranging equation (5.18):

$\lambda = c/f$

λ = 1100 ft/sec / 550 Hz

λ = 2 feet

The human ear does not detect all frequencies equally well. Low frequencies (less than 500 Hz) are not heard very well; neither are high frequencies (greater than 10,000 Hz). To incorporate the perceived loudness by the human ear, the frequency spectra must be described. The amplitude of each frequency could be reported and evaluated but this is not practical. Groups of frequencies called **octave bands** are used to describe sounds and provide a detailed description of the relative amplitudes of the frequency components as shown in Table 5.9. Octave bands are based on the way we perceive sound. Each octave band is described by a center frequency, which is the geometric mean. Each successive octave band center frequency is twice the last. Octave band center frequencies can be described by:

$$\frac{f_o}{f_1} = \frac{f_2}{f_o} = \left(\frac{f_2}{f_1}\right)^{1/2} \tag{5.19}$$

where:

f_2, f_1 = upper and lower frequency limit of range, respectively

f_o = center frequency

Sometimes a more detailed breakdown is required. In this case, one-third octave bands are used.

When evaluating community noise, a broad approach is used. In this approach, all frequency band contributions are first adjusted to approximate the way the ear hears each range, then the contributions are summed to a single number. Three common scales have been used. Figure 5.23 shows the A, B, and C weighting scales. The A scale is the way our ears respond to moderate sounds, the B scale is the response curve for more intense sound, and the C scale is the way our ears would respond to very loud sounds. The non-linear response of the ear at low and high frequencies is quite apparent from this graph.

Most regulations and evaluations applicable to community noise use the A scale and are reported as dB(A), decibels that have been A-weighted. It should be noted that dB are the units and are separated from the weighting scheme through the use of parentheses.

Table 5.9 Octave band ranges and their center frequencies.

Octave Frequency Range (Hz)	Geometric Mean (or Center) Frequency of Band (Hz)
22–44	31.5
44–88	63
88–177	125
177–355	250
355–710	500
710–1420	1000
1420–2840	2000
2840–5680	4000
5680–11360	8000
11360–22720	16000

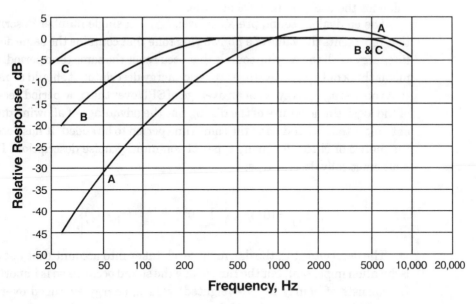

Figure 5.23
The decibel weighting scales.

5.9.3 Duration

A firecracker may be loud, but the sound lasts only a fraction of a second. Traffic noise may not be as intense, but it is continuous. Effective descriptors of how sound varies with time complete the description of noise, and several have been developed. Some of the more commonly used community noise descriptors are: maximum sound level, $[L_{max}(t)]$, statistical sound level $[L_{xx}(t)]$, equivalent sound level $[L_{eq}(t)]$, and day/night level $[L_{dn}]$. In each of these descriptors the capital L indicates that each is a sound pressure level and of course has the units of dB. The (t) indicates each is given for a specific period of time. By definition, L_{dn} is a 24-hour descriptor.

L_{max} represents the maximum noise level that occurs during a defined time period. This allows for a more complete description of the noise when combined with loudness and frequency description. For example, 60 dB(A); L_{max} (1 hr) means that the maximum sound pressure level that occurs during a one-hour period is 60 dB weighted by the A scale.

More description is possible with statistical descriptors (L_{xx}). The subscript xx indicates the percent of time that the listed level is exceeded. For instance, a reported sound level of 60 dB(A); L_{10} (1 hr), means that a sound pressure level of 60 dB on an A-weighted scale was exceeded 10 percent of the time in a one-hour time period. The numeric value may be any fraction of the time, but L_{10}, L_{50}, and L_{90} are most commonly used. L_{90} is the sound pressure level exceeded 90 percent of the time and is commonly used as the value for the background level of noise.

The equivalent sound pressure level, L_{eq}, is a single number descriptor that represents the value of a nonvarying tone that contains the same acoustic energy as all the varying sounds that occur over the same time period. One might think of L_{eq} as an average acoustic energy descriptor. It should be noted that the average energy is not an average of SPL over the time period because of the logarithmic nature of the dB. L_{eq} has the advantage of allowing different noises that occur during the same time period to be added. It has become the metric of choice for many types of community noise description. L_{eq} is mathematically described as:

$$L_{eq} = 10 \log_{10} \left\{ \left(\int_{t_1}^{t_2} 10^{SPL(t)/10} \right) / (t_2 - t_1) \right\}$$ (5.20)

The descriptor L_{dn}, the day/night level, takes into account that not only is duration important, but the time of day the sound occurs also is important. L_{dn} consists of hourly L_{eq} (A-weighted) values, energy-averaged over the entire 24-hour period, with a 10 dB (A-weighted) penalty added to the sound level during the time period from 10 P.M. until 7 A.M. Due to the logarithmic nature of dB, the 10 dB penalty in effect requires the sound pressure during nighttime hours to be one-tenth that of the daytime hours. Mathematically:

$$L_{dn} = 10 \log_{10} \left\{ 1/24 \left[15 \times 10^{(L_d/10)} + 9 \times 10^{(L_n+10/10)} \right] \right\}$$ (5.21)

where: L_d, L_n = SPL (L_{eq}) day and night, respectively.

Example Problem 5.13

You have been supplied ten sound level measurements, equally spaced in time. Approximate the L_{eq}.

Time	Sound Pressure Level (dB)
1	67.8
2	55.9
3	63.7
4	71.1
5	67.2
6	68.3
7	60.0
8	69.1
9	52.8
10	59.0

Solution

Since the measurements are evenly spaced in time, we can use a variation of equation (5.20). In this case, we will replace the integration with a simpler summation.

$$L_{eq} = 10 \log_{10} \left[1/n \ \Sigma \ 10^{(SPLi/10}} \right]$$

$$= 10 \log_{10} [1/10 \ (10^{6.78} + 10^{5.59} + 10^{6.37} + 10^{7.11} + 10^{6.72} + 10^{6.83}$$
$$+ 10^6 + 10^{6.91} + 10^{5.28} + 10^{5.9})]$$

$$= 66.4 \ dB$$

5.9.4 Effects and Subjectivity

The three components of sound discussed previously (loudness, frequency, duration) can be objectively described. However, individuals have different responses to various sounds, and the degree of pleasure or irritation felt when one hears a certain sound is quite subjective. Rock music may be a pleasing sound to one listener but only noise to another listener. Transportation noise is a common problem in urban areas, but the degree of annoyance caused by this noise is highly subjective. For this reason, evaluation criteria are usually based on attitudinal surveys.

Although a single very loud noise can result in an acute hearing loss, a single loud noise is not the typical problem with community noise. In our mechanized societies, noise tends to be more of a chronic problem, resulting in reduced hearing ability after long-term exposure. In the short-term, annoyance or irritation is of more importance. Studies have shown that excessive noise prevents deep sleep cycles needed for complete refreshment, causes increased tension, inhibits communication ability, and reduces the learning abilities of students.

Noise levels decrease with increased distance from a source; this is called **geometric spreading**. The attenuation due to geometric spreading may be characterized by the type of source. If noise is emitted from a single location, the source is referred to as a **point source**. An air conditioning unit, blower, or stationary vehicle could be identified as a point source. If a point source is extruded in space, a line is formed and the source is referred to as a **line source**. Highway traffic may be modeled as a series of moving point sources, or (for high-traffic highways) a line source.

For a point source the sound energy spreads out over the surface of a sphere (proportional to r^2). The intensity and the root-mean-square pressure decrease proportionally to the inverse of the square of the distance from the source (inverse-square law), as shown below:

$$\Delta SPL \ (dB) = 10 \log_{10} (r_1/r_2)^2 \qquad (5.22)$$

where:

Δ SPL (dB) = difference in SPL

r_1 = distance at point 1

r_2 = distance at point 2

For a point source, every time we double the distance we reduce the noise level by 6 dB as shown in example problem 5.14

Example Problem 5.14

What is the decrease in dB when the distance (point source to receiver) is doubled?

Solution

$$\Delta SPL \ (dB) = 10 \log_{10} (r_1/r_2)^2$$
$$\Delta SPL \ (dB) = 10 \log_{10} (1/2)^2$$
$$= 10 \log_{10} (1/4)$$
$$= \ ^-6 \ dB$$

Geometric spreading attenuation for a line source can be derived in a similar fashion to the point source. In this case, the energy is spread over the surface area of a cylinder, which is proportional to r. Using the same mathematical procedure as for a point source, a line source sound decreases with distance as:

$$\Delta SPL \ (dB) = 10 \log_{10} (r_1/r_2) \tag{5.23}$$

For line sources, the sound level decrease is proportional to the distance from the source, not the square of the distance. Solving as before for $r_2 = 2r_1$, a decrease of 3 dB occurs with each doubling of distance. Accordingly, it could be expected that the sound level would decrease by 3 dB for each doubling of distance from a highway. However, in reality the highway is not in free space but is close to the ground. As a result, the interaction of the sound wave with the ground surface causes excess attenuation above what would be expected from just geometric spreading. The excess attenuation effects are related to such factors as the type of soil, ground cover, and surface topography. Ground effects are difficult to predict. However, it has been determined that a value of 3 dB per doubling of distance is more typical of hard surfaces (like parking lots), and more attenuation occurs for soft surfaces (like loose soil with vegetative coverings).

Example Problem 5.15

If you measure a noise to be 72 dB(A) at 20 meters, what will be the resulting level at 100 meters if (a) the source is a point source, and (b) the source is a line source?

Solution

(a) For a point source, the inverse square law applies.

$$\Delta SPL \text{ (dB)} = 10 \log_{10} (r_1/r_2)^2$$
$$\Delta SPL \text{ (dB)} = 10 \log_{10} (20 \text{ m}/100 \text{ m})^2$$
$$\Delta SPL \text{ (dB)} = -14.0 \text{ dB}$$
$$SPL \text{ (dB) at 100 m} = 72 - 14 = 58 \text{ dB(A)}$$

(b) For a line source, the square does not apply

$$\Delta SPL \text{ (dB)} = 10 \log_{10} (r_1/r_2)$$
$$\Delta SPL \text{ (dB)} = 10 \log_{10} (20 \text{ m}/100 \text{ m})$$
$$\Delta SPL \text{ (dB)} = -7.0 \text{ dB}$$
$$SPL \text{ (dB) at 100 m} = 72 - 7 = 65 \text{ dB(A)}$$

The spatial relationship of the source location to the receiver not only determines the attenuation due to geometric spreading, but also determines any obstructions to the path of the noise. Obstructions in the noise path may cause diffraction or reflection of the sound. Diffraction, or the blocking of the sound, causes noise levels to be reduced. This region of decreased sound level is called the shadow zone (see Figure 5.24). Sound is attenuated the most just behind the object and the attenuation decreases further behind the object as the wave reforms. Diffraction is the reason properly designed highway noise barriers are effective. Obstructions may also reflect sound, causing a redirection of the sound energy. The angle of incidence equals the angle of reflection (see Figure 5.24).

5.9.5 Legislation and Regulations

Federal legislation for noise pollution was passed in the 1960s and 1970s and is still in effect. The Housing and Urban Development Act of 1965, reinforced with the Noise Control Act of 1972, mandated the control of urban noise impacts. The Control and Abatement of Aircraft Noise and Sonic Boom Act of 1968 led to noise standards being determined for aircraft. The Quiet Communities Act of 1978 better defined and added to the requirements of the Noise Control Act. This environmental legislation required noise pollution to be addressed at the community level. Analysis methodologies and regulations

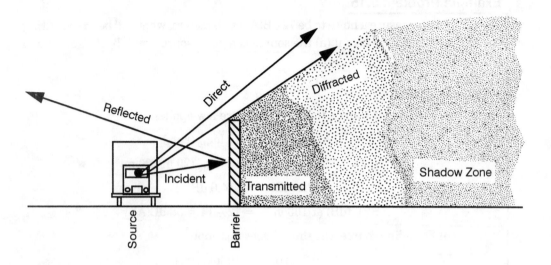

Figure 5.24
Noise attenuation behind a barrier. (*Federal Highway Administration, 1980.*)

to document and mitigate noise impacts resulted. To help ensure enforcement, the EPA created the short-lived Office of Noise Abatement and Control, which contributed significantly to determining noise sources and the necessary regulations. This office established the goal of 55 dB (A-weight); L_{dn} for neighborhoods.

Many discussions have considered the appropriate noise level and descriptor most applicable to various forms of transportation and land use. The U.S. Federal Highway Administration (FHWA) has defined procedures (23CFR772) that must be followed to predict the worst hourly noise levels where human activity normally occurs. Noise abatement criteria for various land uses are shown in Table 5.10. The legislation states that when the noise abatement criteria are approached or exceeded, noise mitigation must be considered. If abatement is considered feasible (possible) and reasonable (cost effective), then abatement measures must be implemented. Abatement may not occur if it is infeasible or unreasonable even though the criteria are exceeded. This leads to the requirement that each project be documented and considered individually. In addition to the noise abatement criteria, substantial increases also trigger abatement analysis for projects on new alignment or drastic changes to existing highways, even though the noise abatement criteria are not exceeded.

Aircraft noise also is controlled by federal legislation. The Control and Abatement of Aircraft Noise and Sonic Boom Act of 1968 mandated noise emission limits on aircraft beginning in 1970. The standards for new aircraft created classifications of aircraft based on noise emissions called Stage I, II,

Table 5.10 Noise abatement criteria.

Description of Activity	One-hour sound levels	
	L_{eq}	L_{10}
Lands on which serenity and quiet are of extraordinary significance and serve an important public need, and where the preservation of quiet is required for the area to continue to serve its intended purpose.	57 (ext)	60 (ext)
Picnic areas, recreation areas, playgrounds, active sports areas, residences, motels, hotels, schools, churches, libraries, and hospitals.	67 (ext)	70 (ext)
Other developed lands, properties, or activities.	72 (ext)	75 (ext)
Undeveloped lands.	–	–
Residences, hotels, motels, public meeting rooms, schools, churches, libraries, hospitals, and auditoriums.	52 (int)	55 (int)

Note: Exterior and interior sound levels are indicated by (ext) and (int), respectively.

Source: Title 23 United States Code, Chapter 772.

or III. The Stage I, noisier aircraft have all but been phased out in the United States. New regulations such as 14CFR91 (transition to an all stage-III fleet operating in the 48 contiguous states and District of Columbia) and 14CFR161 (notice and approval of airport noise and access restrictions) call for the fast phase-in of the quieter Stage-III aircraft.

In 1979, the Aviation Safety and Noise Abatement Act placed more responsibility on local and regional airport authorities. The Airport Noise Control and Land Use Compatibility (ANCLUC) planning process included in Part 150 of the Federal Aviation Regulations (FARs) allows federal funds to be allocated for noise abatement purposes. This process is often referred to as a "Part 150 study." FAA has also implemented a program that requires computer modeling for environmental analysis and documentation. Impacts are defined to occur if the L_{dn} is predicted to be above 65 dB (A-weighted).

In response to a lawsuit by the Association of American Railroads, the Federal Railroad Administration (FRA) has released standards as 40CFR Part 201. The lawsuit was necessary to circumvent hindrances to interstate commerce caused by inconsistent local ordinances.

In addition to U.S. Department of Transportation regulations, other criteria or regulations may be applicable such as the guidelines established by the Department of Housing and Urban Development (HUD) to protect housing areas. The HUD Site Acceptability Standards use L_{dn} (A-weighted) and are "acceptable" if less than 65 dB; "normally unacceptable" from 66 to 75 dB; and "unacceptable" if above 75 dB. In addition, state and/or local governments have also issued guidelines. The analyst should carefully review all applicable requirements before beginning any study.

End-of-Chapter Problems

5.1   Assume that gasoline is combusted with 99% efficiency in a car engine with 1% remaining in the exhaust gases as VOC. If the engine exhausts 16 kg of gases (M.W. = 30) for each kilogram of gasoline (M.W. = 100), calculate the fraction VOC in the exhaust. Give your answer in ppm.

5.2   What fraction of 1994 U.S. NO_x emissions were contributed by highway transportation? By electric utilities?

5.3   Wind blows down a trapezoidal valley at 8 m/s. The valley depth is 800 m, the floor width is 1000 m, and the width at the top is 2000 m. A smelter emits SO_2 at a steady rate of 10,000 kg/day. The valley is capped by an inversion. Calculate the steady-state concentration a long way downwind from the smelter, where the pollutant is uniformly spread across the width and height of the valley.

5.4   In problem 5.3 above, consider the same smelter to be on level, open ground. The stack is 200 m tall and the plume rise is 100 m. The wind is 4 m/s and the stability class is C. Estimate the ground-level SO_2 concentration 6000 m directly downwind. What is the concentration 300 m off the centerline at this same x value?

5.5   The secondary AAQS for SO_2 is 1300 $\mu g/m^3$ (for a 3-hour averaging time). Calculate the equivalent concentration in ppm.

5.6   A power plant uses 10,000 kg/hr of 3.00% sulfur coal. Approximately 90% of the SO_2 formed must be removed prior to release of the stack gases. A limestone system is to be used to remove the SO_2. Estimate the minimum limestone requirements for this plant (kg/hr).

5.7   A particulate removal system consists of a cyclone followed by an electrostatic precipitator. The cyclone is 65% efficient and the ESP is 95% efficient. Calculate the overall efficiency of the system.

5.8   A particulate removal system must achieve 99.4% overall efficiency. Calculate the required efficiency of a baghouse if it is preceded by an 80% efficient cyclone.

5.9   Assuming compliance with federal NSPS, predict the rate of emissions of NO_x and particulates from a coal-fired power plant producing 800 MW of electrical power at an overall thermal efficiency of 40%.

5.10 Which emits more SO_x per unit amount of electricity produced, a coal-fired plant with a 90% efficient SO_2 scrubber or an oil-fired plant with no scrubber? Assume the coal is Illinois bituminous with 3.5% sulfur and 11,500 Btu per pound heating value. Assume the oil has a heat content similar to that of a crude oil and contains 0.9% sulfur, and has a specific gravity of 0.92.

5.11 Calculate the annual emissions of CO, NO_x, and VOCs from an older car which travels 10,000 miles in a year. Assume it emits at 10 times the federal standards for 1995 cars.

5.12 Estimate the daily emissions of particulates from a solid waste incinerator emitting at 0.15 g/standard m^3. The incinerator burns 50 tons/day and exhausts gases in a ratio of 20 kg gases per kilogram of feed. The gases exit at 200 °C and 1 atm. Assume that the gases have an average molecular weight of 30.

5.13 Estimate the daily emissions of SO_2 (in kg/day) from a sulfuric acid plant emitting at 80% of the maximum standards. The production rate is 300 metric tons/day.

5.14 The diagram given below illustrates a coal-fired power plant with an SO_2 scrubber in which an overall SO_2 removal efficiency of 85% is required. Rather than treat the entire stream to 85% removal, the company proposes to treat part of the flue gas to 95% removal, and to bypass the remainder. The reblended stream must still satisfy the overall removal requirement. Calculate the fraction of the flue gas stream that can be bypassed around the scrubber.

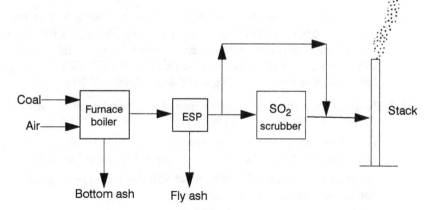

5.15 Referring to the diagram in problem 5.14, the gas temperature required in the stack for good buoyancy is at least 80 °C. The gases bypassing the scrubber are at 200 °C and the gases exiting the scrubber are at 50 °C. For a bypass percentage of 10%, calculate the temperature of the mixed stack gases, ignoring the water which evaporates in the scrubber. Will the blended stack gases have good buoyancy?

5.16 A carbon adsorption system has been proposed to treat a stream of contaminated air flowing at 17,090 ft^3/min (T = 104 °F, P = 1 atm) that contains 1000 ppm hexane vapor. Assume that we can recover 95% of the inlet hexane as liquid hexane, which can be reused. (Some hexane vapor is lost into the exhaust air when we switch beds, and some hexane liquid is lost into the wastewater stream). If hexane costs $1.50 per liquid gallon (which has a density of 5.50 lb/gal), calculate how much money we can potentially save by this operation. Give your answer in $/year assuming that the plant runs 24 hours per day, 240 days per year.

5.17 An older home was built in an area that has a high radon flux (10 pCi/m²-s). The concrete slab did not have a plastic vapor barrier installed, so assume that 85% of the Ra flux passes through the slab and into the house. The house has a slab area of 2400 ft² and an enclosed volume of 24,000 ft³ and averages 1.2 ach (air changes per hour). The half-life of radon is 3.82 days. Assuming the ambient concentration is negligible, estimate the steady-state indoor concentration of radon.

5.18 A person with no knowledge of the potential for carbon monoxide poisoning (and no common sense) brings his charcoal grill into his apartment when a sudden rain storm begins. The apartment has a volume of 7,000 ft³ and a ventilation rate of 0.5 ach. The ambient CO concentration and the initial indoor concentration of CO are both zero. After one hour, the CO concentration is 80 mg/m³. Calculate the emission rate of CO from the charcoal grill in mg/minute. What would the concentration be after two hours?

5.19 Solve problem 5.18 numerically, using spreadsheets, but assume that the emission rate of CO is 3.0 grams per minute. Produce a figure to show how the CO concentration in the apartment varies with time. How long will it take before the CO concentration reaches the NAAQS of 40 mg/m³? How long before it reaches the risk of fatality level of 800 mg/m³?

5.20 Noise from a roadway is measured at 100 feet from the centerline and found to be 72 dB(A); L_{eq}(1 hr).

(a) If the traffic volume increases by a factor of 1.5 and all else remains constant, what will the sound level be at the same location?

(b) With the increased traffic, what will the sound level be at 225 feet from the centerline of the roadway?

(c) At what distance will the FHWA noise abatement criteria for residences be exceeded with the increased traffic?

5.21 Approximate the L_{eq} value for the following sound measurements. Assume that all measurements are evenly spaced in time.

Measurement	SPL, dB(A)
1	67.3
2	71.6
3	81.0
4	71.9
5	65.4
6	62.1

5.22 Explain each piece of the notation: 65 dB(A); L_{dn} .

5.23 A large air compressor has been temporarily located 175 feet away from a home which is also 100 feet away from a highway. The compressor noise is rated at 61.2 dB(A);L_{eq} (at 50 feet), while the roadway noise (at 50 feet) was measured at 66.1 dB(A);L_{eq}. Calculate the combined noise level at the home.

5.24 Sulfur dioxide is emitted at 0.10 kg/s from a stack with a physical height of 75 meters and a plume rise of 25 meters. The wind speed at 100 m above the ground is 6 m/s on an overcast day. Calculate the ground level center-line concentrations at the following distances (all directly downwind from the stack): 2 km, 5 km, and 10 km.

5.25 Rework problem 5.24 with the following changes. Plume rise is 75 meters, and it is mid-afternoon on a warm sunny summer day (Class B stability). Wind speed at 150 m is 6 m/s. Calculate answers only for the 2 km and 5 km downwind distances.

5.26 Assume that 140,000 vehicles per day travel on an interstate highway that passes through an urban area. The length of the section of highway inside the urban area boundaries is 25 miles and at the speeds traveled the annual average CO emission factor is 30 grams per mile for the average vehicle on the road. Calculate the annual emissions of carbon monoxide emitted from vehicles on this interstate. Give your answer in metric tons.

References

Agnew, W. G. 1968. *Research Publication* GMR-743. Warren, MI: General Motors Corporation.

Code of Federal Regulations. 1993. Title 40, Part 50: National Primary and Secondary Ambient Air Quality Standards. Washington, DC: Office of Federal Register.

___. Title 23, Part 772: Noise Abatement Criteria. Washington, DC: Office of Federal Register.

Cooper, C. David, and F. C. Alley. 1994. *Air Pollution Control: A Design Approach.* 2nd ed. Prospect Heights, IL: Waveland Press.

Federal Highway Administration (FHWA). 1980. "Fundamentals and Abatement of Highway Traffic Noise." Federal Highway Administration (September).

Godish, Thad. 1985. *Air Quality.* Chelsea, MI: Lewis Publishers.

Hansen, J., D. Johnson, A. Lacis, S. Lebedeff, P. Lee, D. Rind, and G. Russell. 1981. "Climate Impact of Increasing Atmospheric Carbon Dioxide." *Science* 213:4511.

Henzel, D. S., B. A. Laseke, E. O. Smith, and D. O. Swenson. 1981. *Limestone FGD Scrubbers: User's Handbook.* EPA-600/8-81-017 (August).

Hileman, B. 1995. "Climate Observations Substantiate Global Warming Models." *C&E News* 18 (November 27).

Keeling, C. D., T. P. Whorf, M. Wahlen, and J. van der Plicht. 1995, updated May 1998. "Interannual Extremes in the Rate of Rise of Atmospheric Carbon Dioxide since 1980." *Nature* 375: 666–670.

Martin, D. O. 1976. "The Change of Concentration Standard Deviation with Distance." *Journal of the Air Pollution Control Association* 26 (2).

Masters, Gilbert M. 1991. *Introduction to Environmental Engineering and Science*. Englewood Cliffs, NJ: Prentice Hall.

Turner, D. B. 1970. *Workbook of Atmospheric Dispersion Estimates*. Washington, DC: Environmental Protection Agency Publication AP-26.

U.S. Environmental Protection Agency. 1995. *National Air Pollutant Emission Trends, 1900–1994*. EPA-454/R-95-011 (October). Research Triangle Park, NC.

Chapter 6

Water Resources

6.1   Hydrology

6.1.1 Water Quantity and Quality Issues

As we learned from an early age, the hydrologic cycle is the movement of water through the biosphere. Water is evaporated (and purified) by the sun, transported by the wind, falls to earth as rain (and snow, etc.), and travels "downhill" on top of and underneath the ground. This movement is highly nonuniform, both in time and place, and can lead, in the extremes, to droughts and floods. These nonuniformities are what makes the management of water resources so important and worthy of engineering study.

Distribution of water on planet earth is dominated by the oceans, which comprise 97% of the available water. Much of the remaining water is incorporated into ice caps and glaciers, and is therefore not available for societal development. The water resources that can be developed economically for domestic, industrial, and agricultural use include surface waters (lakes, rivers, and impoundments) and shallow groundwaters (subject to pumping costs). These potential water sources represent less than 1% of the total global quantity; therefore, proper management of this resource is critical to assure that people have access to reliable supplies of adequate quantity and quality water for all their needs.

Issues related to *excessive* water quantity must also be addressed. The occurrence of extreme precipitation events may result in flooding episodes with risk of loss of life and property damage. Engineering studies are routinely completed to assess the risks of flooding by identifying floodplains. This information is used to regulate development in flood-prone locations. In all cases where land development occurs, there is a potential that conversion from one land use (for example, agricultural) to a developed land use (for example, a shopping center) will result in a greater wet weather discharge downstream from the developed site than would have occurred prior to development. Mitigation facilities (e.g., stormwater retention basins) to address this increase in flooding potential must be included in the site development.

In recent years, the surface water pollution potential associated with stormwater runoff has been acknowledged and targeted by federal legislation (Clean Water Act Amendments). Urban runoff has been characterized to include toxic metals from highway runoff, organic materials that collect in storm sewers, and nutrients associated with fertilizer application. Agricultural runoff may increase nutrient, sediment, and pesticide loadings on surface water resources. Proper stormwater management must also consider measures for water quality improvement as well as flood mitigation.

6.1.2 Precipitation Quantities

Complete characterization of precipitation events must include specification of intensity (typically expressed as inches per hour), duration (recording period), and recurrence interval or frequency. The duration of the storm event is an important determinant of the resulting peak discharge because surface runoff rates do not attain maximum values instantaneously in response to precipitation. The frequency of an event can be related to the associated probability of occurrence and accepted risk of failure of stormwater management systems. Maximum recorded rainfall intensities have been compiled by Dominguez (1996) as summarized in Table 6.1. These "world record" rainfall amounts would not be representative of normal expectations for these locations, and are not practical for use in designing facilities.

Based on mass balance principles, precipitation on a specific area of land results in water that can accumulate on, infiltrate into, evaporate from, or flow off of (as runoff) that piece of land. Ignoring accumulation and evaporation, the quantity of runoff is simply:

$$R = P-I \qquad (6.1)$$

where:

R = runoff amount, inches

P = precipitation amount, inches

I = infiltration amount, inches

The amounts of each variable in equation (6.1) can be converted to volume units simply by multiplying by the area of land involved.

The recurrence interval defines the probability of the occurrence of an event in a one-year time frame. For example, a ten-year event (an event that recurs once every ten years) would have an annual probability of occurrence of one-tenth (0.1). The corresponding annual probabilities of a 25-year and 100-year event are 0.04 and 0.01, respectively. Selection of a design storm for engineering projects will reflect the acceptable risk of system failure (flooding). For residential development, it is common to require that all structures be located outside of the 100-year floodplain. This criterion is established

Table 6.1 Maximum recorded precipitation events.

Duration	Intensity (in/hour)	Amount (in.)	Location
8 min	30	4	Fussen, Bavaria
15 min	28	7	Plumb Point, Jamaica
20 min	24	8	Curtea de Arges, Romania
45 min	13	9.8	Holt, MO
2 hour	8	16	Rockport, WV
3 hour	7	21	D'Hanis, TX
5 hour	6	30	Smithport, PA
18 hour	2	36	Thrall, TX
1 day	1.7	40.8	Baguio, Philippines
5 day	1.3	156	Silver Hill Plantation, Jamaica
30 day	0.5	360	Cherrapunji, India
90 day	0.3	648	Cherrapunji, India

Source: Adapted from Dominguez, 1996.

based on a desire to minimize the risk of residential flooding. In contrast, the design of storm sewers may be based on a 10-year precipitation event. For this example, failure of the storm sewers with attendant flooding of streets would be acceptable with a greater frequency than residential flooding.

The interactions between rainfall intensity, duration, and frequency are commonly presented graphically, as shown in Figure 6.1. These curves are defined empirically from rainfall records for any specific geographic region.

Example Problem 6.1

For the intensity/duration/frequency curve in Figure 6.1, determine the effect of recurrence interval (2, 5, 10, 15, 25, and 50-year event) on the rainfall intensity by identification of the design storm intensity (in/hour) for a 20-minute event.

Solution

The intensities are obtained from the graph as follows:

Recurrence Interval (yr)	Intensity (in/hr)
2	2.2
5	3.0
10	3.6
15	3.8
25	4.1
50	4.5

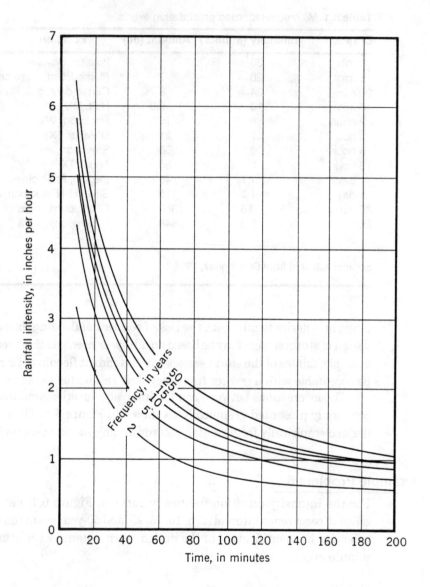

Figure 6.1

Rainfall intensity, duration, and frequency relationships. (*Adapted from ASCE/WPCF, 1969.*)

Example Problem 6.2

For the intensity/duration/frequency curve in Figure 6.1, determine the effect of duration (20, 40, 60, 120, and 200 minutes) on the rainfall intensity by identification of the design storm intensity (in/hour) for a 5-year recurrence interval event.

Solution

The intensities are obtained from the graph as follows:

Duration (min)	Intensity (in/hr)
20	3.0
40	2.0
60	1.5
120	1.0
200	0.7

6.1.3 Determination of Discharge Quantities

The determination of peak discharge for small watersheds can be established by a material balance approach. As an initial simplification, assume that 100% of the precipitation volume is recovered as runoff (output = input):

$$Q_I = i\,A \qquad (6.2)$$

where:

Q_I = instantaneous runoff, cfs

i = precipitation intensity, inches/hr

A = watershed area, acres

For customary units of expression, a conversion factor is needed:

$$Q_I\,(cfs) = i\,(inch/hr)\,A\,(acres)\,1.008\,(cfs\text{-}hr/acre\text{-}inch) \qquad (6.3)$$

However, the units conversion factor (1.008) is routinely dropped from calculations, so that the discharge (cfs) is normally calculated as the product of rainfall intensity (expressed in inches per hour) and the watershed area (expressed in acres).

The assumption that all precipitation is recovered as runoff is not accurate, as some precipitation will infiltrate into groundwater, evaporate, or be retained in depressions (or stormwater ponds). The fraction of precipitation that is recovered as discharge is defined as the runoff coefficient, C. The mass balance equation is modified to reflect the impact of this runoff coefficient, producing the **rational equation** for estimation of peak discharge values:

$$Q_P = C\,i\,A \qquad (6.4)$$

where:

Q_P = peak discharge (cfs)

C = runoff coefficient (dimensionless)

i = intensity (inches/hour)

A = watershed area (acres)

Values for the runoff coefficient are dependent on land use, soil type, vegetative cover, land slope, and soil moisture content. A range of values for the runoff coefficient are reported in Table 6.2 (Wanielista and Yousef, 1993). Use of the rational method may be appropriate for small watersheds that meet the following assumptions:

1. The rainfall intensity is constant for the length of time required to drain the watershed. This duration is referred to as the time of concentration.

2. The runoff coefficient remains constant throughout the time of concentration.

3. The watershed area does not change.

Wanielista and Yousef (1993) report that these assumptions are reasonable for watersheds with a time of concentration less than 20 minutes. More sophisticated methods are required for analysis of flows for larger watersheds.

Table 6.2 Runoff coefficients.

Land Use	C
Downtown business	0.70 to 0.95
Neighborhood business	0.50 to 0.70
Residential	
single family	0.30 to 0.50
multi-family, detached	0.40 to 0.60
multi-family, attached	0.60 to 0.75
apartment	0.50 to 0.70
Light industrial	0.50 to 0.80
Heavy industrial	0.60 to 0.90
Parks and cemeteries	0.10 to 0.25
Unimproved	0.10 to 0.30

Source: Adapted from Wanielista and Yousef, 1993.

Example Problem 6.3

Using the rational method, determine the peak discharge from an unimproved site with the following characteristics:

Area = 10 acres

Time of concentration = 40 minutes

Intensity/duration/frequency data in Figure 6.1

Use a 5-year design event for this analysis.

Solution

The runoff coefficient for an unimproved site is approximately 0.20 (see Table 6.2). The rainfall intensity for a 5-year storm with a duration of 40 minutes is 2.0 inches per hour (see example problem 6.2). The resulting peak discharge is calculated with the rational method (equation 6.4):

$$Q = C\,i\,A = (0.20)\,(2.0\text{ in/hr})\,(10\text{ acres}) = 4\text{ cfs}$$

Example Problem 6.4

Repeat example problem 6.3 for the post-development condition. Assume that the site has been developed into apartments. Site drainage improvements have reduced the time of concentration to 20 minutes.

Solution

The 5-year rainfall event for a 20-minute duration is 3.0 inches per hour (see example problem 6.2). The runoff coefficient for the developed site is estimated to be 0.60 (see Table 6.2). The corresponding peak discharge is calculated below:

$$Q = C\,i\,A = (0.60)\,(3.0\text{ in/hr})\,(10\text{ acres}) = 18\text{ cfs}$$

Comparison of the pre- and post-development peak discharge determined in example problems 6.3 and 6.4 suggests that the development of the site into an apartment complex may aggravate downstream flooding due to a substantial increase in peak discharge relative to the pre-development conditions. Mitigation measures, including stormwater detention facilities, may be required for storage of runoff to reduce peak discharge rates.

Example Problem 6.5

A stormwater detention pond is constructed with a bottom elevation of 50.00 ft and side slopes of 5 (horizontal) to 1 (vertical). The dimensions of the bottom of the pond are 70 feet (width) by 100 feet (length). Due to a high local groundwater table, the pond receives a continuous input of groundwater (Q_{gw}) at a rate of 1 cfs. The pond also receives surface runoff (Q), which is a function of precipitation. For a specific event following a prolonged dry period, the input runoff is defined as the following function of time:

$$Q = 0 \qquad \text{if } t < 0$$
$$Q = 5\,t \qquad \text{if } 0 \le t < 20$$
$$Q = 120 - t \qquad \text{if } 20 \le t < 120$$
$$Q = 0 \qquad \text{if } t \ge 120$$

where:

Q = surface runoff (cfs)

t = time (minutes)

The discharge from the pond is controlled by a 90-degree V-notch weir:

$$Q_{out} = K H^n$$

where:

Q_{out} = discharge (cfs)

K = weir coefficient = 2.5 $ft^{0.5}$/sec

n = 2.5

H = head on the weir (ft)

The elevation of the weir notch is set at 53.00 feet. Using a numerical simulation method, determine the maximum flow rate entering and leaving the pond in response to this storm event.

Solution

Development of a nonsteady-state mass balance for the pond yields the following equation:

$$\text{Accumulation} = \text{Inputs} - \text{Outputs} + \text{Generation}$$

$$\frac{d}{dt}(\rho V) = \rho Q_{gw} + \rho Q - \rho Q_{out} + 0$$

For a prismatoid, the volume (V) is defined as follows:

$$V = D (B_1 + 4 M + B_2)/6$$

where:

D = depth

B_1 = area of lower base (pond bottom)

M = area of mid-section (mid-depth)

B_2 = area of upper base (water surface)

For this rectangular pond:

$$B_1 = L W$$
$$B_2 = (L + 2 s D) (W + 2 s D)$$
$$M = (L + s D) (W + s D)$$

where:

L = bottom length (100 ft)

W = bottom width (70 ft)

D = depth (ft)

s = slope expressed horizontal to vertical (5:1)

Thus

$$V (ft^3) = D (L W + s W D + s L D + 4 s^2 D^2/3)$$

$$dV/dt = (dV/dD) (dD/dt) = (LW + 2 s W D + 2 s L D + 4 s^2 D^2) (dD/dt)$$

Recognizing that the density is constant and that $dD = dH$, the material balance equation is simplified as follows:

$$(LW + 2 s W D + 2 s L D + 4 s^2 D^2) (dH/dt) = Q_{gw} + Q - K H^{2.5}$$

The response of the pond to the input runoff is determined using the numerical simulation methods from chapter 3. Pseudocode is presented below:

Pseudocode

Define System Parameters
  L = 100
  W = 70
  s = 5
  K = 2.5
  n = 2.5
Define Initial Conditions
  t = 0
  Q_{gw} = 1
  $H = (Q_{gw} / K)^{(1/n)}$
  D = H + 3
  Q_{outmax} = 0
  Q_{max} = 0
Define Simulation Parameters
  dt = 1 (1 minute, must convert to seconds in calculation of dH)
  tend = 240 (4 hours, twice the storm duration)
  n = tend/dt
Enter Simulation Loop
  Begin Iterations
  Define Input Flow (Q)
  IF t < 0 THEN Q = 0
  IF t > 0 THEN Q = 5 t
  IF t > 20 THEN Q = 120 – t
  IF t > 120 THEN Q = 0
Solve For Differential Quantities
  $dH = (Q_{gw} + Q - K H^{2.5}) 60 dt / [L W + 2 D (s W + s L) + 4 s^2 D^2]$
  dD = dH
Increment Variables
  t = t + dt
  H = H + dH
  D = D + dD
Check for Maximum Flows
  IF Q > Q_{max} THEN Q_{max} = Q

IF $Q_{out} > Q_{outmax}$ THEN $Q_{outmax} = Q_{out}$
Continue Iterations
Print Output

Time	H	D	Q	dt	dH	dD	Q_{out}	Q_{max}	Q_{outmax}
min	ft	ft	cfs	min	ft	ft	cfs	cfs	cfs
0	0.693	3.693	0	1	0.000	0.000	1.000	0.000	1.000
1	0.693	3.693	5	1	0.020	0.020	1.000	5.000	1.000
2	0.714	3.714	10	1	0.041	0.041	1.076	10.000	1.076
3	0.764	3.754	15	1	0.060	0.060	1.235	15.000	1.235
4	0.814	3.814	20	1	0.078	0.078	1.495	20.000	1.495
5	0.892	3.892	25	1	0.096	0.096	1.881	25.000	1.881
10	1.513	4.513	50	1	0.158	0.158	7.039	50.000	7.039
15	2.351	5.351	75	1	0.173	0.173	21.189	75.000	21.189
20	3.192	5.192	100	1	0.156	0.156	45.498	100.00	45.498
25	3.776	6.776	95	1	0.069	0.069	65.558	100.00	69.282
30	4.022	7.022	90	1	0.025	0.025	79.525	100.00	81.103
35	4.096	7.096	85	1	0.003	0.003	84.886	100.00	84.886
36	4.099	7.099	84	1	0.000	0.000	85.030	100.00	85.030
37	4.099	7.099	83	1	-0.003	-0.003	85.026	100.00	85.030
38	4.096	7.096	82	1	-0.005	-0.005	84.894	100.00	85.030

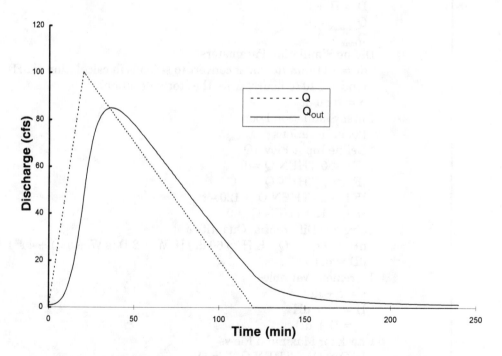

Stormwater pond response for example problem 6.5.

The simulation was completed to obtain maximum input and output flows of 100 cfs and 85.0 cfs, respectively. Partial output from the spreadsheet calculations is provided with a plot of the input and output hydrographs. It is observed that the pond has achieved marginal attenuation of the peak discharge. Dedication of more land area to the detention pond would achieve greater mitigation of peak flows.

6.1.4 Stormwater Quality Issues

Extensive financial resources have been committed to control point sources (municipal and industrial sources) of wastewater since passage of the Water Pollution Control Act Amendments of 1972. It is recognized that for many surface waters, further improvement in water quality will not be realized without a corresponding effort to control nonpoint pollution sources, defined as runoff from urban, residential, and agricultural lands. Pollutant contributions from point and nonpoint sources are reported in Table 6.3.

Water quality impairment associated with stormwater runoff includes oxygen depletion, nutrient enrichment and subsequent eutrophication, toxicity associated with metals and pesticides, contamination with pathogens, and sediment deposition. The concentration and pollutant mass loading associated with stormwater is extremely variable due to differences in land use. Water quality data for highway runoff are presented in Table 6.4.

Many of the pollutants in stormwater are associated with particulates. Phosphorus, in particular, is often adsorbed onto the surface of suspended solids. Provision of detention ponds may be beneficial to promote pollutant removal by sedimentation. Limited removal of soluble forms of the pollutants may also be achieved in wet detention stormwater ponds by a combination of chemical and biological transformations (Wanielista and Yousef, 1993).

Table 6.3 Relative point and nonpoint contributions of pollutants.

Pollutant	Point Source (%)	Nonpoint Source (%)
Chemical oxygen demand	30	70
Fecal coliform	10	90
Oil	30	70
Total phosphorus	34	66
Total Kjeldahl nitrogen	10	90
Metals		
Arsenic	95	5
Cadmium	84	16
Chromium	50	50
Copper	59	41
Iron	5	95
Lead	43	57
Mercury	98	2
Zinc	30	70

Source: Adapted from Wanielista and Yousef, 1993.

Table 6.4 Stormwater quality characterization for highway runoff.

Parameter	Average	Range
BOD_5, mg/L	24.0	2 to 133
Chemical oxygen demand, mg/L	15.0	5 to 1,058
Chloride, mg/L	386.0	5 to 13,300
Nitrite + nitrate nitrogen, mg/L	1.1	0.01 to 8.4
pH, standard units		6.5 to 8.1
Suspended solids, mg/L	261.0	4 to 1,156
Total Kjeldahl nitrogen, mg/L	3.0	0.1 to 14
Total organic carbon, mg/L	41.0	5 to 290
Total phosphate, mg/L	0.79	0.05 to 3.55
Total solids, mg/L	1147.0	145 to 21,640
Volatile suspended solids, mg/L	77.0	1 to 837
Metals		
Cadmium, mg/L	0.040	0.01 to 0.40
Chromium, mg/L	0.040	0.01 to 0.14
Copper, mg/L	0.103	0.01 to 0.88
Iron, mg/L	10.3	0.1 to 45
Lead, mg/L	0.96	0.02 to 13.1
Mercury, μg/L	3.22	0.13 to 67
Nickel, mg/L	9.92	0.1 to 49
Zinc, mg/L	0.41	0.01 to 3.4

Source: Adapted from Wanielista and Yousef, 1993.

6.1.5 Groundwater Issues

Protecting groundwater resources from contamination and depletion is a critical component of water resources management. Numerous instances of groundwater contamination have been documented throughout the United States, caused by inappropriate land disposal of toxic substances. Many of these sites now require extremely expensive remedial measures (see chapter 7 for more details). In the meantime, a potential water supply source is lost. Source protection is vital to the efficient management of groundwater resources due to the minimal rates of contaminant dispersion and/or purification that occur in groundwaters.

Water quality concerns also may result from excessive groundwater withdrawals. Removal of groundwater at a rate which exceeds recharge will result in lowered water table elevations. High population densities in coastal regions, and the associated demand for water, have often produced depression of the water table. Due to the proximity of saline waters, the reduction in water table has often been accompanied by salt water intrusion into the aquifers. This phenomenon has led to abandonment of many groundwater supplies in coastal areas, with development of surface water or inland groundwater resources at great expense. Astute management of groundwater resources, including maintenance of adequate recharge of stormwater and reclaimed water (wastewater effluent), can help prevent depletion.

6.2 Drinking Water Treatment

6.2.1 Drinking Water Standards

The evolution of drinking water treatment objectives throughout the developed world has followed a similar pattern. Efforts in the early twentieth century were directed to prevention of *acute* health effects, in particular the transmission of waterborne diseases, including various bacterial, viral, and protozoan infections. The widespread adoption of chlorination of public water supplies has largely been effective in the control of cholera, typhoid fever, and other diseases associated with tainted water. Subsequent objectives addressed the aesthetic qualities of the finished water, with a goal to produce a product with a superior taste, odor, and appearance. More recent treatment objectives relate to chronic health effects, principally increased incidence of cancer, associated with long-term (chronic) exposure to low concentrations of toxic substances present in potable water.

The preoccupation with preventing epidemics of waterborne disease is well-founded based on documentation of illness attributable to contaminated water supplies dating to the mid-nineteenth century (Okun, 1996). The pioneering work of Dr. John Snow to identify the cause of an outbreak of cholera in London in 1849 is widely recounted as the first documentation of disease transmission via public water supplies. This epidemiological study, which preceded the identification of a bacterial agent as the cause of cholera, resulted in the closure of a contaminated well and termination of the cholera outbreak.

Although public water supplies are generally accepted as free of infectious agents in the developed world, much of the population of the undeveloped world must rely on water supplies that are not adequately treated to prevent waterborne disease. It is reported that three million children, worldwide, under five years of age die from complications of diarrhea each year (Otterstetter and Craun, 1997). The majority of these deaths (in excess of 300 deaths per hour) are believed to result from exposure to contaminated water supplies.

The recent experience in the United States merits a brief discussion. In the past twenty-five years, a total of 740 outbreaks of infectious disease associated with water supplies have been reported in the United States. It is reasonable to assume that many more episodes have been undocumented (Otterstetter and Craun, 1997). A recent event (1993) in Milwaukee, Wisconsin, has received considerable scrutiny owing to the magnitude of the affected population. A total of 400,000 individuals became seriously ill, and the number of fatalities exceeded 100 due to exposure to a protozoan pathogen (*Cryptosporidium*) transmitted in the public water supply (Okun, 1996). The public health implications of these episodes in chlorinated public water supplies have spurred a renewed focus on the control of pathogens in water supplies.

Concurrent with the development of standards to address chronic health effects, it was discovered that application of chlorine for disinfection of water supplies may result in the formation of chlorinated organic compounds that are potential carcinogens. These compounds, referred to collectively as disinfection by-products, are formed by reaction of benign naturally occurring organic matter with chlorine. Regulation of trihalomethanes in the 1970s represented the initial recognition of the potential adverse effect of chlorination of potable waters. As more health effects information becomes available, it is reasonable to expect that regulation of additional disinfection by-products will occur.

Federal standards have been established by the Environmental Protection Agency pursuant to Public Law 93–523 (Safe Drinking Water Act of 1974). Numerical limits are classified as **primary standards** (health-related) and **secondary standards** (aesthetics). The primary standards are enforceable and are expressed as **maximum contaminant levels** (MCLs). In addition, maximum contaminant level goals (MCLGs) have been established for each health-related constituent at a level that is believed to produce no known adverse health effect. The MCLG for known or probable carcinogens has been set at zero. Identification of the enforceable limits (MCLs) is based on health effects, available treatment technology, and economic factors. The definition of MCLs and MCLGs is an ongoing activity, and changes in primary standards are imposed to reflect advances in understanding of health effects as well as advances in control technology. A partial listing of primary drinking water standards is provided in Table 6.5. The reader is cautioned that the numerical limits reported in Table 6.5 are subject to revision by the EPA, including addition of new compounds and alteration of numerical standards.

Action levels for public notification have been established for lead (0.015 mg/L) and copper (1.3 mg/L) concentrations measured at the consumer's tap. The presence of these metals is normally the result of corrosion of plumbing (lead solder) and piping materials (copper pipe) within the customer's property. Many utilities are currently engaged in programs to adjust product water quality to minimize corrosion of lead and copper. These efforts include control of pH, introduction of corrosion inhibitors (for example, zinc phosphates), and regulation of oxidant levels (dissolved oxygen and disinfectant residuals).

Nonenforceable secondary maximum contaminant levels (SMCLs) are summarized in Table 6.6 and address issues of palatability and appearance.

Development of the early standards was predicated on the use of uncontaminated natural waters as a drinking water source. These initial standards did not include all compounds that represented a potential health concern. Addition of MCLs for many synthetic organic compounds and metals which result from industrial activity (and associated compromise of environmental water quality) rather than natural occurrence has been necessary to account for the impact of waste disposal from an industrialized society. It is also recognized that reuse of treated wastewater (reclaimed water) as a potential water supply may dictate reconsideration of primary standards to protect public health due to the presence of compounds in reclaimed water that

Table 6.5 Primary drinking water standards.

Contaminant	MCLG mg/L	MCL mg/L
Pesticides and herbicides		
alachlor	zero	0.002
atrazine	0.003	0.003
chlordane	zero	0.002
dibromochloropropane	zero	0.0002
2,4-D	0.07	0.07
1,2-dichloropropane	zero	0.005
diquat	0.02	0.02
endrin	0.002	0.002
ethylene dibromide	zero	0.00005
heptachlor	zero	0.0004
heptachlor epoxide	zero	0.0002
lindane	0.0002	0.0002
methoxychlor	0.04	0.04
pentachlorophenol	zero	0.001
toxaphene	zero	0.003
2,4,5-TP	0.05	0.05
Disinfection by-products (interim standard)		
Trihalomethanes	zero	0.100
Chlorinated solvents		
carbon tetrachloride	zero	0.005
1,2-dichloroethane	zero	0.005
1,1-dichloroethylene	0.007	0.007
dichloromethane	zero	0.005
tetrachloroethylene	zero	0.005
1,1,1-trichloroethane	0.2	0.2
1,1,2-trichloroethane	0.003	0.005
trichloroethylene	zero	0.005
vinyl chloride	zero	0.002
Aromatic hydrocarbons		
benzene	zero	0.005
benzo(a)pyrene	zero	0.0002
ethylbenzene	0.7	0.7
styrene	0.1	0.1
toluene	1	1
xylenes (total)	10	10
Chlorinated synthetic organics		
chlorobenzene	0.1	0.1
o-dichlorobenzene	0.6	0.6
hexachlorobenzene	zero	0.001
PCBs	zero	0.0005
dioxin (2,3,7,8-TCDD)	zero	3×10^{-8}
Nonchlorinated synthetic organics		
dalapon	0.2	0.2
di(2-ethylhexyl)adipate	0.4	0.4
di(2-ethylhexyl)phthalate	zero	0.006

Metals		
antimony	zero	0.006
barium	2	2
beryllium	0.004	0.004
cadmium	0.005	0.005
chromium	0.1	0.1
copper	1.3	BAT
lead	zero	BAT
mercury	0.002	0.002
selenium	0.05	0.05
thallium	0.0005	0.002
Other inorganics		
arsenic (interim)		0.05
asbestos (million fibers/L)	7	7
cyanide	0.2	0.2
fluoride	4	4
nitrate nitrogen	10	10
nitrite nitrogen	1	1
Radioactivity (interim)		
gross alpha (pCi/L)	zero	15
radium 226 + 228 (pCi/L)	zero	5
beta/photon emitters (mrem/yr)	zero	4
Physical parameters		
turbidity (NTU)	N/A	BAT
Bacterial		
total coliform	zero	5% samples positive
Giardia lamblia	zero	BAT
Legionella	zero	BAT
heterotrophic plate count		BAT
viruses	zero	BAT

Note: BAT = best available technology

Source: Pontius, 1998.

would not be expected in natural waters. These factors will surely result in a continual evolution of federal drinking water standards.

6.2.2 Source Water Characteristics

The target concentrations for a finished water are established by the primary and secondary drinking water standards. These standards apply throughout the United States, therefore it would be expected that potable water standards would be uniform. Variations in raw water characteristics would dictate actual treatment requirements in any specific application. Potential water sources may be categorized based on common raw water characteristics. This classification may be extended to discuss several general water treatment facilities.

Table 6.6 Secondary drinking water standards.

Contaminant	SMCL mg/L
aluminum	0.05 to 0.2
chloride	250
color (cobalt platinum units, cpu)	15
copper	1
fluoride	2
foaming agents	0.5
iron	0.3
manganese	0.05
odor (taste and odor number)	3
pH	6.5 to 8.5
silver	0.10
sulfate	250
total dissolved solids	500
zinc	5

Source: Pontius, 1998.

Surface water sources (lakes, rivers, impoundments) may be described by high color associated with naturally occurring organic material, high turbidity (organic and inorganic particulate matter), and generally low concentrations of inorganic species (such as iron, calcium, magnesium). Surface water characteristics may exhibit pronounced seasonal variations associated with algal blooms, increased sediment load following precipitation events, and stratification of lakes and impoundments. These sources commonly display an elevated potential to form disinfection by-products due to the elevated organic concentration; thus it is sometimes necessary to remove organic compounds prior to chlorination to minimize formation of these by-products. Treatment objectives focus on removing turbidity and disinfecting without generating excessive disinfection by-products.

Groundwater sources often exhibit greater variability in composition than surface waters, although the water quality from an individual aquifer would be consistent. Groundwater characteristics at a single location may vary considerably with depth of wells, owing to different aquifer formations that would supply the water. Some groundwater sources require minimal treatment, perhaps only disinfection, in order to comply with drinking water standards. Other sources commonly require removal of iron, manganese, hydrogen sulfide, hardness (calcium and magnesium), and/or total dissolved solids. Removal of synthetic organic compounds (for example, chlorinated solvents) may be necessary in those cases where groundwater pollution from industrial sources has occurred. Agricultural activity may contaminate shallow aquifers with pesticides and/or nitrate.

Selection of a water source for a public water supply must consider both the quality of the source water and the reliable quantity that can be developed. Surface water sources often exhibit significant seasonal flow variations, so construction of a storage reservoir may be necessary to guarantee adequate supplies during extended low flow periods. The yield of groundwater supplies depends on the hydraulic characteristics of the specific aquifer formation. Recharge of groundwater sources must also be adequate to balance withdrawals; if not, mining of the water may result with an associated decline in water tables. It must be recognized that the water quantities that are available for development are finite for either surface or groundwater supplies.

6.2.3 Typical Treatment Facilities

6.2.3.1 *Rapid Sand Filtration of Surface Water.* Treatment of surface waters must, as a minimum, remove turbidity and provide for disinfection. The most common treatment plant configuration is shown in Figure 6.2. The process consists of a series of unit processes for turbidity removal: coagulation, flocculation, sedimentation, and filtration. Terminal disinfection also is required.

The particulates that are present in surface waters include inorganic (clay particles) and organic (algal biomass) components. These colloidal particles are typically very small in diameter, only slightly more dense than water, and negatively charged (surface charge). The settling velocity of particles is increased for larger diameter or greater density particulates. The presence of similar charges on the colloid surface prevents growth in particle size due to the electrostatic repulsive forces between particles. These conditions prevent efficient removal by gravity sedimentation and filtration without chemical treatment to alter the particle characteristics. A coagulant chemical, commonly alum ($Al_2(SO_4)_3 \cdot 14\ H_2O$), is added during a rapid mix process to destabilize the charge on the particle surface. After coagulation, collisions between the destabilized particles are promoted by gentle mixing (flocculation). These collisions result in larger diameter particles with an attendant increase in settling velocity. Subsequent sedimentation would then be effective in removing most suspended solids. Filtration is supplied as a final polishing step for turbidity removal and to enhance the efficiency of subsequent disinfection processes.

The process design of water treatment facilities is often based on empirical guidelines derived from experience with similar source waters (Ten States Standards, 1997a). Criteria for determining the size of process units are based on hydraulic residence time (HRT) and/or hydraulic loading rate (HLR), as defined in equations (6.5) and (6.6). Multiple units, in parallel, are specified for reliability and redundancy.

$$HRT = \frac{V}{Q} \tag{6.5}$$

$$HLR = \frac{Q}{A} \tag{6.6}$$

where:

HRT = hydraulic residence time, days

HLR = hydraulic loading rate, gal/day-ft^2

V = tank volume, gal

Q = flow rate, gal/day

A = tank surface area, ft^2

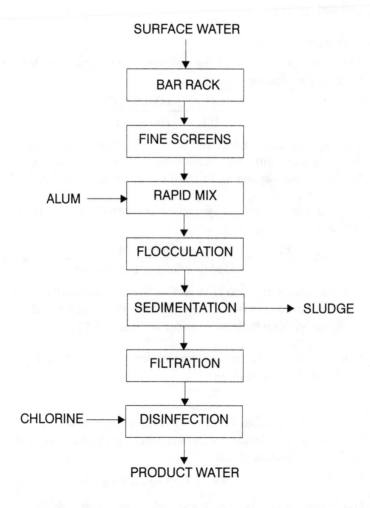

Figure 6.2
Schematic flow diagram of treatment plant for surface water. (*Reynolds and Richards, 1996.*)

Example Problem 6.6

Sedimentation facilities for a rapid sand filtration plant must be sized to satisfy the following criteria:

Hydraulic loading rate (maximum) = 1000 gallons/ft^2-day

Hydraulic residence time (minimum) = 2 hours

Flow rate = 4,000,000 gallons per day

Minimum of two sedimentation basins in parallel

Recommend dimensions for a system of circular sedimentation basins. Calculate the hydraulic loading rate and hydraulic residence time for the recommended system.

Solution

The total surface area of the sedimentation tanks is determined with the hydraulic loading rate constraint:

$$A = \frac{Q}{HLR} = \frac{4,000,000 \text{ gpd}}{1000 \text{ gpd/ft}^2} = 4000 \text{ ft}^2$$

Clarifier mechanisms are available in standard size increments of 5 ft. Therefore, select two units with a diameter of 55 ft. The total surface area is 4752 ft^2 and the corresponding hydraulic loading rate is 842 gpd/ft^2.

The depth is determined by application of the hydraulic residence time constraint:

$$\text{Depth} = \frac{Q}{A} HRT = \frac{4,000,000 \text{ gpd}}{(4752 \text{ ft}^2)(24 \text{ hr/day})(7.48 \text{ gal/ft}^3)}(2 \text{ hours}) = 9.38 \text{ ft}$$

Select a sidewater depth of 10 feet, with an additional freeboard allowance of two feet for a total tank depth of 12 feet. The actual liquid volume is 47,520 ft^3, which yields a hydraulic residence time of 2.13 hours.

Summary recommendation: two circular units with a diameter of 55 feet and a sidewater depth of 10 feet.

Disinfection is achieved most commonly by chlorination. Chlorine may be supplied as chlorine gas, which reacts with water to form hypochlorous acid and hydrochloric acid:

$$Cl_2 + H_2O \rightarrow HOCl + HCl \tag{6.7}$$

Hypochlorous acid is a weak acid, which dissociates to form a hydrogen ion and a hypochlorite ion (pK$_a$ = 7.5):

$$HOCl \leftrightarrow H^+ + OCl^- \tag{6.8}$$

The sum of the concentration of hypochlorous acid (HOCl) and hypochlorite ion (OCl^-) is referred to as the free chlorine residual.

Chlorine will react with ammonia, if present, to form several chloramines:

$$NH_4^+ + HOCl \rightarrow NH_2Cl + H^+ + H_2O \tag{6.9}$$

$$NH_2Cl + HOCl \rightarrow NHCl_2 + H_2O \tag{6.10}$$

$$NHCl_2 + HOCl \rightarrow NCl_3 + H_2O \tag{6.11}$$

The final product of this reaction is unstable and readily decomposes to form nitrogen gas which escapes from solution. The overall reaction for oxidation of ammonia to nitrogen gas is represented below:

$$2\,NH_3 + 3\,Cl_2 \rightarrow N_2 + 6\,H^+ + 6\,Cl^- \tag{6.12}$$

The sum of monochloramine (NH_2Cl) and dichloramine ($NHCl_2$) is referred to as the combined chlorine residual. The combined residual is not as strong an oxidant as the free residual, and thus is not as effective for disinfection. Consequently, longer reaction times are required to achieve disinfection with a combined residual than with a free chlorine residual. This disadvantage is partially offset by the relative stability of the combined residual in the distribution system. Combined residuals also have an advantage of minimal formation of disinfection by-products.

Example Problem 6.7

Balance the reaction for oxidation of ammonia to nitrogen gas with chlorine. Assume that the chlorine gas is reduced to chloride ion.

Solution

Balance the oxidation half-reaction:

$$2\,NH_3 \rightarrow N_2$$
$$2\,NH_3 \rightarrow N_2 + 6\,H^+$$
$$2\,NH_3 \rightarrow N_2 + 6\,H^+ + 6\,e^-$$

Balance the reduction half-reaction:

$$Cl_2 \rightarrow 2\,Cl^-$$
$$Cl_2 + 2\,e^- \rightarrow 2\,Cl^-$$

Multiply the reduction half-reaction by 3 (so the number of electrons is equal for both half-reactions) and add the half-reactions:

$$2\,NH_3 + 3\,Cl_2 \rightarrow N_2 + 6\,H^+ + 6\,Cl^-$$

Example Problem 6.8

Determine the quantity of chlorine gas (mg/L) required to react with 10 mg/L of NH_3–N (ammonia nitrogen).

Solution

The basis for this stoichiometry calculation is the quantity of ammonia (10 mg/L). Note that the ammonia is expressed as the quantity of nitrogen. This approach is customary for nitrogen species. Expression of all forms of nitrogen (ammonia, nitrate, nitrite, organic nitrogen, etc.) as just nitrogen simplifies calculations because the mass of nitrogen is conserved, even though the masses of the various nitrogen species are not.

$$Cl_2 = \frac{10 \text{ mg } NH_3\text{–N/L}}{14 \text{ mg N/mmole}} \frac{3 \text{ mmole } Cl_2}{2 \text{ mmole } NH_3\text{–N}} 71 \text{ mg } Cl_2/\text{mmole} = 76 \text{mg } Cl_2/L$$

Adding chlorine to waters that contain ammonia will result in the initial formation of a combined chlorine residual. After all of the ammonia has been oxidized to nitrogen gas, addition of chlorine will result in the formation of a free chlorine residual. Provision of sufficient chlorine dose to form a free residual is referred to as breakpoint chlorination, in which the breakpoint dose is defined as the chlorine dose that corresponds with the destruction of ammonia. In those cases where a combined residual is desired, a smaller chlorine dose will suffice. Breakpoint chlorination may be used to achieve ammonia removal from wastewaters, although other options are generally regarded as more cost effective.

Example Problem 6.9

Determine the required hydraulic residence time in an ideal plug flow reactor to achieve 99.99% pathogen destruction efficiency. Assume that a constant free chlorine residual of 0.2 mg/L is maintained during disinfection and the reaction kinetics are described by the following second-order rate expression:

$$-r = k \, C \, C_{Cl}$$

where:

        k = rate constant = 2 L/mg-min

        C = pathogen concentration (organisms per L)

        C_{Cl} = free chlorine residual concentration (mg/L)

        –r = disinfection rate (organisms per minute per L)

Solution

A steady-state mass balance around a differential volume of the reactor yields the following equation:

$$\text{Accumulation} = \text{Input} - \text{Output} + \text{Generation}$$

$$0 = Q\,C - Q\,(C + dC) + dV\,r$$

$$0 = -Q\,dC - dV\,k\,C\,C_{Cl}$$

Recognizing that C_{Cl} is constant for the problem conditions, the equation can be solved by separation of variables and integration:

$$\int_{0}^{V}\frac{dV}{Q} = \int_{C_0}^{C_e}\frac{-dC}{kCC_{Cl}} = \frac{-1}{kC_{Cl}}\int_{C_0}^{C_e}\frac{dC}{C}$$

$$\text{HRT} = \frac{V}{Q} = \frac{\ln(C_0/C_e)}{kC_{Cl}} = \frac{\ln\!\big(C_0/0.0001C_0\big)}{(2\ \text{L/mg-min})(0.2\ \text{mg/L})} = 23\,\text{min}$$

6.2.3.2 Lime-Soda Softening of Groundwater. Many groundwaters, particularly in the midwestern and western United States, contain substantial concentrations of calcium and magnesium. These ions contribute to hardness in the water. These compounds are not regulated by either the primary or secondary drinking water standards; however, removal of these compounds is commonly practiced in order to minimize scaling problems in the water distribution system, water heaters, and shower stalls. Scale formation represents the deposition of an insoluble salt of calcium or magnesium. Many industrial processes (for example, boiler operation) cannot tolerate any hardness due to problems associated with scale formation. Hardness also contributes to excessive consumption of detergents during laundry operations.

Neither calcium nor magnesium are believed to represent health concerns for the general population. Target concentrations for removal of hardness, also referred to as water softening, are therefore not defined by federal MCLs. Typical goals for water softening range from 80 to 120 mg/L of hardness, expressed as calcium carbonate equivalents (as $CaCO_3$). Conversion between mass and equivalent concentrations is illustrated in example problem 6.10. Classification of waters based on hardness is presented in Table 6.7.

Table 6.7 Characterization of water hardness.

Description	Hardness (mg/L as $CaCO_3$)
Soft	< 50
Moderately hard	50 to 150
Hard	150 to 300
Very hard	> 300

Example Problem 6.10

Analysis of a groundwater provided these results:

$$Ca = 88 \text{ mg/L} \qquad Mg = 50 \text{ mg/L}$$

Determine the water hardness in units of $CaCO_3$ equivalents.

Solution

Equivalent concentrations were introduced in chapter 2. It is likely that the principal application of equivalent concentrations in introductory chemistry courses focused on acid base reactions with concentrations expressed as normality. The approach for softening calculations is similar. Concentrations must be converted from mass to molar units by division by the molecular weight. Equivalent concentrations are then determined by multiplying by the number of equivalents per mole. Finally, equivalents as $CaCO_3$ are determined by multiplying by the equivalent weight of calcium carbonate, which is 50.

The molecular weight of $Ca = 40$ and there are two equivalents of charge per mole of calcium ion (Ca^{2+}). The molecular weight of $Mg = 24.3$ and there are two equivalents per mole (Mg^{2+}).

$$Ca = \frac{88 \text{ mg Ca/L}}{40 \text{ mg Ca/mmole}} \left(2 \frac{\text{meq}}{\text{mmole}}\right)\left(50 \frac{\text{mg CaCO}_3}{\text{meq}}\right) = 220 \text{ mg/ L as CaCO}_3$$

$$Mg = \frac{50 \text{ mg Mg/L}}{24.3 \text{ mg Mg/mmole}} \left(2 \frac{\text{meq}}{\text{mmole}}\right)\left(50 \frac{\text{mg CaCO}_3}{\text{meq}}\right) = 206 \text{ mg/L as CaCO}_3$$

The total hardness is the sum: $220 + 206 = 426$ mg/L as $CaCO_3$. This water would be classified as very hard per Table 6.7.

Treatment for removal of calcium and magnesium is commonly achieved by chemical precipitation of calcium as calcium carbonate and magnesium as magnesium hydroxide. A typical treatment plant configuration is presented in Figure 6.3. There are obvious similarities in the sequence of unit operations and processes with the rapid sand filtration facility. In the case of lime-soda softening, particulate material is produced during treatment by the following chemical precipitation reactions:

Removal of carbon dioxide:

$$CO_2 + Ca(OH)_2 \rightarrow CaCO_3(s) + H_2O \qquad (6.13)$$

Removal of calcium carbonate hardness:

$$Ca^{2+} + 2\,HCO_3^- + Ca(OH)_2 \rightarrow 2\,CaCO_3(s) + 2\,H_2O \qquad (6.14)$$

Removal of calcium noncarbonate hardness:

$$Ca^{2+} + SO_4^{2-} + Na_2CO_3 \rightarrow CaCO_3(s) + 2\,Na^+ + SO_4^{2-} \qquad (6.15)$$

Removal of magnesium carbonate hardness:

$$Mg^{2+} + 2\,HCO_3^- + 2\,Ca(OH)_2 \rightarrow 2\,CaCO_3(s) + Mg(OH)_2(s) + 2\,H_2O \quad (6.16)$$

Removal of magnesium noncarbonate hardness:

$$Mg^{2+} + SO_4^{2-} + Ca(OH)_2 + Na_2CO_3 \rightarrow$$
$$CaCO_3(s) + Mg(OH)_2(s) + 2\,Na^+ + SO_4^{2-} \qquad (6.17)$$

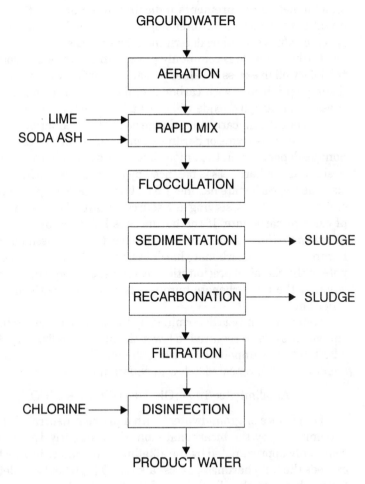

Figure 6.3

Schematic flow diagram of a lime-soda softening drinking water treatment plant for a groundwater source. (*Adapted from Reynolds and Richards, 1996.*)

Destruction of excess alkalinity:

$$2\ HCO_3^- + Ca(OH)_2 \rightarrow CaCO_3(s) + 2\ H_2O + CO_3^{2-} \qquad (6.18)$$

Recarbonation:

$$Ca^{2+} + CO_2 + 2\ OH^- \rightarrow CaCO_3(s) + H_2O \qquad (6.19)$$

Note that adding excess CO_2 can produce some bicarbonate ions, and this allows some calcium to remain in the water.

The chemicals that are required to achieve softening include lime $(Ca(OH)_2)$, soda or soda ash (Na_2CO_3), and carbon dioxide (CO_2). It is interesting to note that removal of calcium is achieved by adding lime, a compound which itself contains calcium. Removal of magnesium as the hydroxide requires a pH of about 11. This pH elevation requires lime in excess of the stoichiometric requirements indicated in equations (6.13) to (6.19). This quantity of excess lime is often approximated as 50 mg/L as $CaCO_3$, but site-specific values should be determined for each case.

Carbon dioxide is commonly present in groundwater sources in concentrations well in excess of values present in surface waters as a result of establishing equilibrium with carbonate minerals in the subsurface strata. The presence of carbon dioxide will exert a lime demand, with an associated production of calcium carbonate sludge as noted in equation (6.13).

The various forms of hardness may be classified as carbonate or noncarbonate, depending on the anion that is associated with the calcium or magnesium. If the hardness is balanced by bicarbonate (HCO_3^-), the hardness is classified as carbonate hardness. The bicarbonate ion participates in the precipitation reactions, serving as a source of the carbonate ion for precipitation of calcium carbonate. If the hardness is balanced by some ion other than bicarbonate (for example, sulfate), the hardness is classified as noncarbonate hardness. For noncarbonate hardness, the applicable anion does not participate in the chemical precipitation reactions. Consequently, it does not matter whether the noncarbonate hardness is associated with sulfate, chloride, or other anions.

Water alkalinity is determined by titration of a sample to determine the equivalents of acid required to lower the pH from 8.3 to 4.5 (Standard Methods, 1992). For unpolluted waters, the alkalinity was defined in chapter 2, equation (2.29), repeated below in equation (6.20):

$$\text{Alkalinity (eq/L)} = [OH^-] + [HCO_3^-] + 2\ [CO_3^{2-}] - [H^+] \qquad (6.20)$$

For surface or groundwaters with a pH near neutral (7.0), the alkalinity is dominated by the bicarbonate ion. Consequently, the bicarbonate ion is commonly approximated by the alkalinity. For waters in which the alkalinity exceeds the total hardness, excess alkalinity is present and defined as the difference between the alkalinity and the total hardness.

Waters may be classified into three groups based on the relative concentration of calcium, magnesium, and alkalinity. Determination of required chemical doses, and chemical precipitate sludge quantities, is facilitated by determining the quantities of the various forms of hardness reviewed in equations (6.13) to (6.19). Bar charts (example problems 6.11 to 6.13) are commonly used for this purpose.

In the first category, the alkalinity is less than the calcium concentration. For this type of water, some of the calcium is carbonate hardness (defined as the alkalinity) and some is noncarbonate hardness. All of the magnesium is considered to be noncarbonate hardness. This type of water is examined in example problem 6.11.

Example Problem 6.11

A groundwater was analyzed as follows:

CO_2 = 40 mg/L as $CaCO_3$

Ca^{2+} = 200 mg/L as $CaCO_3$

Mg^{2+} = 120 mg/L as $CaCO_3$

Alkalinity = 180 mg/L as $CaCO_3$

Chloride = 140 mg/L as $CaCO_3$

Determine the basis (reactant quantities) for stoichiometric calculations for equations (6.13) to (6.19) for this groundwater.

Solution

A bar chart may be constructed to facilitate determination of the desired quantities (all quantities in mg/L as $CaCO_3$).

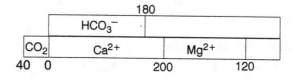

The results are summarized below:

Compound	Equation	Basis, mg/L as $CaCO_3$
Carbon dioxide	6.13	40
Calcium carbonate	6.14	180
Calcium noncarbonate	6.15	20
Magnesium carbonate	6.16	0
Magnesium noncarbonate	6.17	120
Excess alkalinity	6.18	0
Recarbonation	6.19	50
Excess lime		50

Water in the second category has an alkalinity that exceeds the calcium but which is less than the total hardness. For this type of water, all of the calcium hardness is carbonate hardness, but the magnesium hardness is part carbonate and part noncarbonate hardness. An example of this type of water is reviewed in example problem 6.12.

Example Problem 6.12

A groundwater was analyzed as follows:

CO_2 = 40 mg/L as $CaCO_3$

Ca^{2+} = 200 mg/L as $CaCO_3$

Mg^{2+} = 120 mg/L as $CaCO_3$

Alkalinity = 280 mg/L as $CaCO_3$

Chloride = 40 mg/L as $CaCO_3$

Determine the basis for stoichiometric calculations for equations (6.13) to (6.19) for this groundwater.

Solution

A bar chart may be constructed to facilitate determination of the desired quantities, again in mg/L as $CaCO_3$.

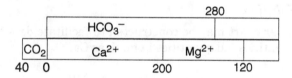

The results are summarized as follows:

Compound	Equation	Basis, mg/L as $CaCO_3$
Carbon dioxide	6.13	40
Calcium carbonate	6.14	200
Calcium noncarbonate	6.15	0
Magnesium carbonate	6.16	80
Magnesium noncarbonate	6.17	40
Excess alkalinity	6.18	0
Recarbonation	6.19	50
Excess lime		50

The final category of water has an alkalinity that exceeds the total hardness. For this example, all of the hardness is carbonate hardness. In addition, the excess alkalinity is not zero. It should be noted that addition of sufficient lime to destroy the excess alkalinity is necessary before pH elevation will be achieved for precipitation of magnesium. An example of this type of water is reviewed in example problem 6.13.

Example Problem 6.13

A groundwater was analyzed as follows:

CO_2 = 40 mg/L as $CaCO_3$
Ca^{2+} = 200 mg/L as $CaCO_3$
Mg^{2+} = 120 mg/L as $CaCO_3$
Na^+ = 60 mg/L as $CaCO_3$
Alkalinity = 380 mg/L as $CaCO_3$

Determine the basis for stoichiometric calculations for equations (6.13) to (6.19) for this groundwater.

Solution

A bar chart may be constructed to facilitate determination of the desired quantities.

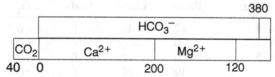

The results are summarized below:

Compound	Equation	Basis, mg/L as $CaCO_3$
Carbon dioxide	6.13	40
Calcium carbonate	6.14	200
Calcium noncarbonate	6.15	0
Magnesium carbonate	6.16	120
Magnesium noncarbonate	6.17	0
Excess alkalinity	6.18	60
Recarbonation	6.19	50
Excess lime		50

Recarbonation is practiced to stabilize the water. Addition of carbon dioxide will achieve precipitation of excess soluble calcium and/or produce $Ca(HCO_3)_2$ prior to distribution of the finished water. Stabilizing the water at the treatment plant will reduce scaling problems associated with supersaturated solutions of calcium carbonate in the distribution system. Recarbonation also will lower the pH from the elevated values required to achieve magnesium precipitation.

Example Problem 6.14

A groundwater was analyzed as follows:

CO_2 = 40 mg/L as $CaCO_3$

Ca^{2+} = 200 mg/L as $CaCO_3$

Mg^{2+} = 120 mg/L as $CaCO_3$

Alkalinity = 240 mg/L as $CaCO_3$

SO_4^{2-} = 80 mg/L as $CaCO_3$

Determine chemical requirements for softening and sludge quantities produced. Express all answers as concentration of the chemical, not as $CaCO_3$ equivalents.

Solution

Chemical quantities (as $CaCO_3$) are determined for equations (6.13) to (6.19):

					Sludge Production	
Equation	Basis	Lime	Soda	CO_2	$CaCO_3$	$Mg(OH)_2$
6.13	40	40	0	0	40	0
6.14	200	200	0	0	400	0
6.15	0	0	0	0	0	0
6.16	40	80	0	0	80	40
6.17	80	80	80	0	80	80
6.18	0	0	0	0	0	0
6.19	50	0	0	50	50	0
Excess Lime	50	50				
Total		450	80	50	650	120

Conversion of the chemical doses and sludge quantities to mg/L as the chemical is summarized below:

Chemical	mg/L as $CaCO_3$	Equivalent Weight	mg/L as chemical
$Ca(OH)_2$	450	37	333
Na_2CO_3	80	53	85
CO_2	50	22	22
$CaCO_3$ sludge	650	50	650
$Mg(OH)_2$ sludge	120	29	70
		Total Sludge	720

6.2.3.3 *Aeration for Removal of Hydrogen Sulfide from Groundwater.*
Many compounds that are present in groundwater sources are volatile, that is, these compounds may be transferred from water to the atmosphere by aeration (or air stripping). This strategy is employed to remove certain industrial contaminants (for example, hydrocarbons), although controls must be implemented to assure that solving a groundwater pollution problem is not achieved at the expense of contaminating atmospheric resources. A far more common application of air stripping in potable water treatment is for removing naturally occurring dissolved gases, principally hydrogen sulfide (H_2S), from groundwater (recall example problem 2.12).

Hydrogen sulfide is a common contaminant in many groundwaters. It produces an unpleasant taste and odor with a characteristic smell of rotten eggs. Disinfection, typically with chlorine, is practiced after aeration. For those water sources which do not require more extensive treatment, aeration/chlorination represents a very economical treatment process in comparison to rapid sand filtration or lime-soda softening.

6.3   Wastewater Treatment*

6.3.1 Receiving Water Standards
Federal goals for wastewater treatment were outlined in the Water Pollution Control Act Amendments of 1972 (PL 92-500) as follows:

1. To eliminate the discharge of pollutants.
2. Wherever possible, to have water quality suitable for sustaining fish, shellfish, and wildlife and for recreational purposes.
3. To prohibit the discharge of toxic pollutants.

Water quality criteria have been established to protect beneficial uses of surface waters and may be classified as follows (Dietz, 1996):

1. Maintenance of adequate dissolved oxygen to support desirable aquatic life forms. A minimum dissolved oxygen of 5 mg/L is commonly accepted for support of sport fish species.
2. Reduction of plant and algal nutrient levels to avoid cultural eutrophication problems. Eutrophication is more likely to be a concern for discharge to lakes and reservoirs than for discharge to free-flowing rivers.
3. Maintenance of concentrations of toxic substances at values which do not pose a threat to aquatic species and/or other potential uses (for example, potable water supply). Toxic substances are more commonly associated with industrial sources and are typically regulated for each industrial category. Whole effluent toxicity bioassay testing may be required to verify the absence of toxic agents in effluents.

---

*Portions of the discussion in section 6.3 are taken from W. Christopher King, Environmental Engineering P.E. Examination Guide and Handbook, American Academy of Environmental Engineers, 1996. Used with permission.

4. Elimination of pathogens to control transmission of waterborne diseases. Disinfection of wastewaters which pose a risk of disease transmission is prescribed prior to discharge.

5. Maintenance of suitable aesthetic qualities to foster recreational use of the surface water resources.

6.3.2 Typical Treatment Standards

6.3.2.1 *National Pollutant Discharge Elimination System.*
Section 402 of PL 92-500 established procedures for issuance of discharge permits for municipal and industrial wastewaters (Nemerow and Dasgupta, 1991; Dietz, 1996). These permits identify maximum allowable concentrations of pollutants that may be present in a facility discharge. Monitoring and reporting requirements also are prescribed in the permits.

6.3.2.2 *Municipal Discharge Standards.* The level of treatment
that must be provided for municipal wastewaters will vary as necessary to maintain receiving water quality standards. In those cases where dissolved oxygen concentrations can be maintained above criteria (typically 5 mg/L), secondary treatment would be adequate, as defined in Table 6.8. Secondary treatment imposes a maximum effluent BOD_5 and suspended solids concentration of 30 mg/L. These standards can be achieved by many biological treatment processes. Disinfection would typically also be required. Dechlorination may be practiced prior to discharge to mitigate any toxic effects associated with the disinfectant residual.

Table 6.8 Secondary treatment standards.

Parameter	Maximum Concentration mg/L
BOD_5	30
Suspended solids	30

In those cases where discharge of a secondary effluent would not satisfy receiving water quality criteria, greater removal of oxygen-demanding materials would be required. Actual numerical limits would be site specific, depending on receiving water characteristics. Removal of additional BOD_5 and reduced nitrogen compounds (TKN) may be specified to maintain adequate dissolved oxygen levels.

For wastewater discharge into lakes or impoundments, removal of nutrients (nitrogen and phosphorus) may be prescribed to minimize algal growth potential. Specific numerical limits will be site specific. One possible set of effluent standards which is common in Florida for a nutrient removal system is noted in Table 6.9. Various combinations of biological processes (nitrification/denitrification, biological phosphorus uptake) and chemical processes

Table 6.9 Effluent standards for nutrient removal.

Parameter	Maximum Concentration mg/L
BOD$_5$	5
Suspended solids	5
Total nitrogen	3
Total phosphorus	1

(phosphorus precipitation) can be employed to achieve these limits. Terminal filtration, disinfection, and dechlorination also would typically be provided.

6.3.2.3 *Industrial Wastewater Discharge Standards.* Industrial facilities with direct surface water discharge are subject to specific discharge requirements established in a NPDES permit issued to the industry (U.S. EPA, 1985). In addition to regulation of conventional pollutants (pH, BOD$_5$, SS, oil and grease), categorical industry standards have been established for each industry to define Best Available Technology (BAT) for reduction of toxic compounds. Pretreatment standards for toxic substances also are established for industries that discharge to a Publicly Owned Treatment Works (POTW). The federal standards represent minimum pretreatment requirements; individual municipalities may establish pretreatment standards that are more restrictive than the federal mandates. The specific effluent standards faced by industrial dischargers are quite variable in light of the different wastewater characteristics and local sewer use ordinances (Dietz, 1996).

6.3.3 Characterization of Wastewaters

Wastewater treatment objectives may be quite varied for either municipal or industrial wastewater applications due to various effluent standards and/or a wide range of raw wastewater characteristics. Untreated municipal wastewaters exhibit consistency in composition, so that generalizations may be offered regarding quantity and quality. Industrial wastewaters are extremely variable in composition so that site-specific sampling is necessary for characterization. A review of the literature (for example, Nemerow and Dasgupta, 1991) may be useful for approximation of industrial wastewater characteristics.

6.3.3.1 *Characterization of Municipal Wastewaters.* Typical domestic wastewater concentrations are reported in Table 6.10. Caution must be exercised, however, owing to the variability in flow rates that results from differing water use, infiltration, and inflow. Average per capita flows are commonly estimated at 100 gallons per day (Ten States Standards, 1997b), including an allowance for normal infiltration/inflow. Modification of this value is often appropriate to account for local infiltration/inflow conditions and water consumption.

Table 6.10 Composition of untreated domestic wastewater.

Parameter	Weak	Typical	Strong
Alkalinity (as $CaCO_3$)	50	100	200
Ammonia (as N)	12	25	50
BOD_5	110	220	400
COD	250	500	1000
Fixed Suspended Solids	20	55	75
Grease	50	100	150
Inorganic Phosphorus (as P)	3	5	10
Nitrate (as N)	0	0	0
Nitrite (as N)	0	0	0
Organic Nitrogen (as N)	8	15	35
Organic Phosphorus (as P)	1	3	5
Total Dissolved Solids	250	500	850
Total Nitrogen (as N)	20	40	85
Total Phosphorus (as P)	4	8	15
Total Suspended Solids	100	220	350
TOC	80	160	290
Volatile Suspended Solids	80	165	275

Note: All units in mg/L.
Source: Adapted from Metcalf and Eddy, 1991.

6.3.3.2 *Industrial Waste Characterization.* Current regulatory emphasis (Pollution Prevention Act of 1990) has been placed on modification of manufacturing processes to alter the characteristics of industrial wastewaters, thereby reducing the quantity (either volume or mass) and toxicity of materials which require treatment. Examples of waste minimization, or pollution prevention, are reviewed in case studies in chapter 8. This strategy has been widely practiced for many years to minimize costs associated with end-of-pipe wastewater treatment by minimizing water use and recovering by-products from waste streams. These pollution prevention or waste minimization efforts are outlined in the following hierarchy (U.S. EPA, 1988):

1. Source Reduction. Reduce the amount of waste at the source, through changes in industrial processes.

2. Recycling. Reuse and recycle wastes for the original or some other purpose, such as materials recovery or energy production.

3. Incineration/Treatment. Destroy, detoxify, and neutralize wastes into less harmful substances.

4. Secure Land Disposal. Deposit wastes on land using volume reduction, encapsulation, leachate containment, monitoring, and controlled air and surface/subsurface waste releases.

Similar priorities are appropriate for reducing the volume of solid and hazardous wastes.

6.3.4 Biological Treatment

6.3.4.1 Overview. Biological wastewater treatment processes have been used for municipal and many industrial wastewater treatment applications. Biological processes have been demonstrated to achieve BOD removal, suspended solids removal, ammonia removal, nitrogen removal (nitrification/denitrification), phosphorus removal, and removal of many synthetic organic compounds from municipal and industrial wastewaters. The robust potential of biological processes is summarized in a recent review (Zitomer and Speece, 1993).

The design of biological treatment systems may be accomplished with empirical guidelines developed from experience. Heuristic guidelines are well established to support design of municipal treatment facilities due to the similarity in domestic wastewater characteristics (Ten States Standards, 1997b). Due to substantial variability in industrial wastewater characteristics and to a scarcity of historical data, the design of many industrial wastewater treatment facilities must pursue a more fundamental approach based on study of reaction kinetics for the specific wastewater. The kinetic results are integrated into a design approach using mass balances and reactor engineering principles. This fundamental approach is presented in chapter 3 and is valid for all applications. The fundamental approach is preferred whenever site-specific kinetic data are available to support the analysis.

6.3.4.2 Kinetics. The design of biological wastewater treatment facilities must address reaction rates for removal of substrate and production of biomass. In this context, substrate may be represented by any measure of organic material which is oxidized by the biomass to supply energy (for example, BOD_5, or any specific organic compound). For nitrification processes, the substrate is ammonia (measured as TKN). Knowledge of the rate of substrate utilization is necessary to evaluate BOD_5 or TKN removal efficiency relative to effluent standards.

Biological processes for wastewater treatment achieve substrate removal by both converting substrate to biomass and by oxidizing substrate (see chapter 4). The assimilative removal results in the production of suspended solids (biomass). The quantity of suspended solids affects the design of many unit operations and processes, including sludge-handling facilities. Knowledge of the rate of biomass production is therefore necessary to estimate residuals (sludge) quantities and complete facility design.

Various empirical models have found widespread use for analysis of kinetics. Selected models were reviewed in chapter 4. Relevant terminology also was introduced in that chapter. Four common models are presented for specific substrate utilization rate in equations (6.21) to (6.24), representing zero-order, first-order, variable-order (Monod), and inhibitory kinetics, respectively:

$$q = \frac{-r_s}{X} = k_0 \qquad (6.21)$$

$$q = \frac{-r_s}{X} = k_1 S \qquad (6.22)$$

$$q = \frac{-r_s}{X} = \frac{kS}{K_s + S} \qquad (6.23)$$

$$q = \frac{-r_s}{X} = \frac{kS}{K_s + S + \dfrac{S^2}{K_I}} \qquad (6.24)$$

The variable-order Monod model in equation (6.23) is widely used due to flexibility to describe both zero- and first-order kinetics. The inhibitory model is appropriate for specific organic compounds where the rate of biological activity is reduced at excessive substrate concentrations. The inhibitory model degenerates to give the Monod model in the limit as the inhibition constant approaches infinity. Various synthetic organic compounds are reported to exhibit inhibition at high concentrations (Grady, 1990).

The net rate of production of biomass is related to the rate of substrate utilization by the reaction stoichiometry. Recall that the specific growth rate (μ) was defined in chapter 4 as the net growth rate of biomass (growth minus decay). The mass ratio of cells produced to substrate removed is referred to as the observed yield (Y_{obs}):

$$\mu = \frac{r_x}{X} = Y_{obs} \frac{-r_s}{X} \qquad (6.25)$$

where: μ=specific growth rate, days^{-1}

The observed yield varies with the system operation and is therefore not constant. The observed yield may be expressed as a function of the system specific growth rate and the maximum yield (Y_{max}), as reported in equation (6.26). The maximum yield (or true yield) is independent of system operation and may be determined as a constant that is characteristic of the wastewater. Specification of growth kinetics also requires knowledge of the endogenous decay coefficient (k_e).

$$Y_{obs} = \frac{Y_{max}}{1 + \dfrac{k_e}{\mu}} \qquad (6.26)$$

Equations (6.25) and (6.26) may be combined to yield equation (6.27):

$$\mu = Y_{max} q - k_e \qquad (6.27)$$

6.3.4.3 Identification of Growth Rate or Organic Loading as Key Variables for Design of Biological Treatment Systems.
The preceding kinetic equations (substrate removal and biomass production) can be coupled with mass balance equations and solved simultaneously to

establish a relationship between the specific growth rate and the substrate concentration. For first-order substrate removal kinetics (equation 6.22):

$$S = \frac{\mu + k_e}{Y_{max}\, k_1} \qquad (6.28)$$

A similar derivation for variable-order kinetics (equation 6.23) yields:

$$S = \frac{K_s\,(\mu + k_e)}{Y_{max}\, k - \mu - k_e} \qquad (6.29)$$

It is evident from these equations that the effluent quality (represented by the substrate concentration, S) is determined by the experienced specific growth rate (μ). As the growth rate (μ) is decreased, the removal of substrate is improved (Lawrence and McCarty, 1970).

Through development of material balances for biomass, the specific growth rate is easily related to conventional design parameters for suspended growth systems. For complete-mix suspended growth systems without solids recycle (such as aerated lagoons, aerobic digestion, or anaerobic digestion) at steady state, the growth rate is related to the reactor hydraulic residence time (see Figure 6.4).

$$\text{Accumulation} = \text{Input} - \text{Output} + \text{Generation} \qquad (6.30)$$

$$0 = Q\, X_0 - Q\, X + V\, r_x \qquad (6.31)$$

$$0 = 0 - Q\, X + V\, r_x \qquad (6.32)$$

$$\mu = \frac{r_x}{X} = \frac{Q}{V} = \frac{1}{HRT} \qquad (6.33)$$

where:

Q = influent flow rate, L/day

X_0 = influent biomass concentration (assumed negligible)

V = reactor volume, L

HRT = hydraulic residence time, days

X = biomass concentration in the reactor, mg/L

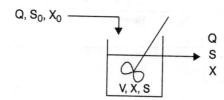

Figure 6.4
Complete mix reactor without solids recycle.

For complete-mix suspended growth systems with solids recycle (such as activated sludge, Figure 6.5) at steady state, the growth rate is related to the solids residence time.

$$\text{Accumulation} = \text{Input} - \text{Output} + \text{Generation} \qquad (6.34)$$

$$0 = Q\,X_0 - Q_e\,X_e - Q_w\,X_w + V\,r_x \qquad (6.35)$$

$$\mu = \frac{r_x}{X} = \frac{Q_e\,X_e + Q_w\,X_w}{V\,X} = \frac{1}{SRT} \qquad (6.36)$$

where:

Q = influent flow rate, L/day

X_0 = influent biomass concentration (assumed negligible)

Q_e = effluent flow rate, L/day

X_e = effluent biomass concentration, mg/L

Q_w = waste sludge flow rate, L/day

X_w = waste sludge biomass concentration, mg/L

V = reactor volume, L

SRT = solids residence time, days

It is noted that the solids residence time is defined as the total inventory of biomass (VX) in the reactor divided by the rate at which biomass is removed in the effluent and waste sludge. Thus, the SRT represents the average time that biomass remains in the system. The solids residence time is also commonly referred to as the sludge age or the mean cell residence time.

The organic loading may be used as an alternative to the specific growth rate for design of biological treatment systems. The use of organic loading is attractive for attached growth systems (trickling filters and rotating biological contactors) due to difficulties in measurement of biomass concentration

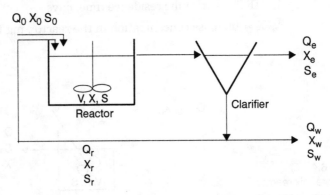

Figure 6.5
Complete mix reactor with solids recycle.

in these systems. The **organic loading** is defined as the rate of supply of substrate per unit biomass:

$$\text{Organic Loading} = \frac{\text{Mass of Organics per Unit Time}}{\text{Mass of Biomass in Reactor}} \tag{6.37}$$

$$\text{Organic Loading} = \frac{Q\,S_0}{V\,X} \tag{6.38}$$

where:

Q = influent flow rate, L/day

S_0 = influent substrate concentration, mg/L

V = reactor volume, L

X = reactor biomass concentration, mg/L

In those systems where direct measurement of biomass is difficult, it is often assumed that the biomass is proportional to the reactor volume or the media surface area. For these estimates of biomass, the organic loading is calculated as follows:

$$\text{Organic Loading} = \frac{\text{Mass of Organics per Unit Time}}{\text{Reactor Volume}} \tag{6.39}$$

$$\text{Organic Loading} = \frac{\text{Mass of Organics per Unit Time}}{\text{Reactor Surface Area}} \tag{6.40}$$

For all systems that operate at a high efficiency, the effluent substrate concentration (S) is much less than the influent (S_0); thus the organic loading is approximately equal to the specific substrate utilization rate:

$$\text{Organic Loading} \cong \frac{Q\,S_0}{V\,X} = \frac{Q(S_0 - S)}{V\,X} = \frac{-r_s}{X} = q \tag{6.41}$$

As noted previously in equation (6.27), the specific substrate utilization rate can not be specified independently of the specific growth rate:

$$\mu = Y_{max}\,q - k_e \tag{6.27}$$

Thus, the organic loading is also dependent on the growth rate. Therefore, specification of either the specific growth rate or the organic loading rate would determine the substrate removal efficiency for a biological wastewater treatment process. Empirical design guidelines have been developed using indicators of growth rate (for example, solids residence time) and organic loading (for example, pound BOD_5 per 1000 ft^3-day).

6.3.4.4 Typical Kinetic Values.

Substrate utilization kinetic data are provided in Table 6.11 for a variety of industrial wastewaters. A typical first-order rate constant for domestic wastewater is included for comparison. It is evident from these data that many industrial wastewaters do not sustain the same magnitude of biological reaction rates as a municipal wastewater. Biological treatment may be satisfactory for these industrial wastes but the required growth rate could be very different than that for municipal applications to achieve the same effluent criteria.

For municipal wastewaters, the maximum yield (Y_{max}) and endogenous decay coefficients have been estimated as 0.60 mg cells per mg BOD_5 and 0.06 days^{-1}, respectively (Metcalf and Eddy, 1991).

Table 6.11 First-order reaction rate constant for selected wastewaters.

Source	k_1, $\dfrac{L}{\text{mg VSS-day}}$
Brewery	0.0053
Domestic wastewater	0.0294
Organic chemical	0.0015
Pharmaceutical	0.0094
Phenolic	0.0022
Pulp and paper	0.0100
Refinery	0.0162
Rendering	0.0360
Textile	0.0036
Vegetable oil	0.0074

Source: Grady and Lim, 1980.

Example Problem 6.15

Determine the required hydraulic residence time for aerated lagoon treatment of a municipal wastewater to achieve compliance with an effluent BOD_5 standard of 20 mg/L. Use appropriate kinetic values from Table 6.11. Assume that an aerated lagoon may be described as a complete mix reactor without solids recycle.

Solution

As developed in equation (6.33) from a material balance for biomass, the HRT equals the reciprocal of the growth rate for a complete mix, no recycle reactor. The required growth rate for the specified kinetics is calculated as follows:

$$\mu = Y_{max}\,(q) - k_e = Y_{max}\,k\,S - k_e$$
$$\mu = (0.60 \text{ mg VSS/mg BOD}_5)\,(0.0294 \text{ L/mg VSS-day})$$
$$(20 \text{ mg BOD}_5/\text{L}) - 0.06 \text{ days}^{-1}$$
$$\mu = 0.293 \text{ days}^{-1}$$

The required hydraulic residence time is therefore 3.42 days.

Example Problem 6.16

Determine the required hydraulic residence time for aerated lagoon treatment of a pulp and paper wastewater to achieve compliance with an effluent BOD_5 standard of 20 mg/L. Use appropriate kinetic values from Table 6.11. Assume that an aerated lagoon may be described as a complete mix reactor without solids recycle. For purposes of this calculation, assume $Y_{max} = 0.4$ mg VSS/mg BOD_5 and $k_e = 0.06$ per day.

Solution

As developed in equation (6.33) from a material balance for biomass, the HRT equals the reciprocal of the growth rate for a complete mix, no recycle reactor. The required growth rate for the specified kinetics is calculated as follows:

$$\mu = Y_{max} (q) - k_e = Y_{max} \, k \, S - k_e$$

$$\mu = (0.40 \text{ mg VSS/mg BOD}_5) \, (0.0100 \text{ L/mg VSS-day})$$
$$(20 \text{ mg BOD}_5/\text{L}) - 0.06 \text{ days}^{-1}$$

$$\mu = 0.020 \text{ days}^{-1}$$

The required hydraulic residence time is therefore 50 days. The different kinetics between the municipal and pulp and paper wastewaters result in a much different design for the two systems.

6.3.5 Activated Sludge Treatment

The most popular biological process for municipal or industrial wastewater treatment is the activated sludge process. A typical facility for treatment of municipal wastewater is diagrammed in Figure 6.6. The activated sludge portion of the treatment plant includes a biological reactor (the aeration basin), equipment for supply of oxygen to the biological reactor, sedimentation tanks, return activated sludge-pumping facilities, and waste sludge-pumping facilities. Biomass is produced by aerobic growth in the aeration basin. The biomass is separated from the effluent in the sedimentation tanks, and the majority of this biomass is returned to the aeration basin to maintain a large inventory of microorganisms. A portion of the solids is removed from the plant as waste sludge. The aeration equipment must supply adequate oxygen to maintain aerobic conditions and to achieve mixing in the reactor.

The treatment facility includes components for removal of grit, rags, and other debris which might damage equipment or clog pipes. Processing of residuals (sludge) is required to reduce odors and minimize volumes for final disposal. Digestion is provided for biological conversion of cell mass into gaseous products. The digestion process reduces the organic content of the sludge, simultaneously destroying pathogens and minimizing potential for creation of nuisance odors. Components are provided for solids-liquid separation, including thickening and dewatering. The volume reduction achieved by the solids-liquid separation reduces costs for other sludge-handling operations, including transport costs for final disposal. Residuals are commonly disposed of into landfills or on agricultural lands. Effluent disinfection, typically with chlorine, is provided prior to discharge. In some cases, dechlorination is required prior to discharge to surface waters to mitigate toxicity to aquatic organisms associated with the residual chlorine.

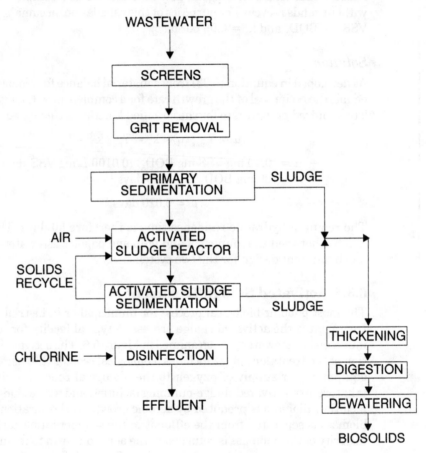

Figure 6.6
Activated sludge treatment facility. (*Reynolds and Richards, 1996.*)

Example Problem 6.17

A food-processing wastewater contains 1000 mg/L of glucose ($C_6H_{12}O_6$). The flow rate is 2 million gallons per day (MGD). Biological treatment with the activated sludge process achieves removal of 90% of the glucose according to the following reaction:

$$25\ C_6H_{12}O_6 + 3\ NH_4^+ + 3\ HCO_3^- + 135\ O_2$$

$$\rightarrow 3\ C_5H_7O_2N + 138\ CO_2 + 147\ H_2O$$

Determine the quantity of biomass ($C_5H_7O_2N$) that is produced in pounds per day. Determine the quantity of oxygen that must be supplied by the aeration equipment in pounds per day. Determine the observed yield for this biological process in units of pounds VSS / pound glucose.

Solution

The basis for this stoichiometric calculation is the mass of glucose removed:

$$\text{Mass glucose} = (1000 \text{ mg/L}) (2 \text{ MGD}) \left(\frac{8.34 \text{ lbs/MGAL}}{\text{mg/L}} \right) (0.90 \text{ removal})$$

$$= 15{,}012 \text{ lbs glucose/day}$$

Stoichiometric quantities of cell mass and oxygen are calculated as follows:

$$\text{Cell mass} = \frac{15{,}012 \text{ lbs glucose/day}}{180 \text{ lb/lb-mole}} \frac{3 \text{ lb-mole cells}}{25 \text{ lb-mole glucose}} (113 \text{ lb cells/lb-mole})$$

$$= 1131 \text{ lbs biomass/day}$$

$$\begin{array}{l} \text{Oxygen=} \\ \text{mass} \end{array} \frac{15{,}012 \text{ lbs glucose/day}}{180 \text{ lb/lb-mole}} \frac{135 \text{ lb-mole oxygen}}{25 \text{ lb-mole glucose}} (32 \text{ lb oxygen/lb-mole})$$

$$= 14{,}412 \text{ lbs oxygen/day}$$

The observed yield is the ratio of rate of biomass production to rate of substrate removal:

$$Y_{obs} = \frac{1131 \text{ lbs VSS/day}}{15{,}012 \text{ lbs glucose/day}} = 0.075 \text{ lb VSS/lb glucose}$$

It is noted that a source of nitrogen must be available to support cell synthesis. If the wastewater does not contain sufficient ammonia, it may be necessary to add a source of nitrogen to achieve successful biological treatment. Municipal wastewaters are rarely deficient in nutrients; however, certain industrial wastewaters, particularly high-carbohydrate food processing wastewaters, may be deficient in nitrogen and/or phosphorus.

6.3.5.1 *Design Equation Development from Material Balances.*
Development of the necessary set of equations for characterization of a complete-mix activated sludge facility requires defining equations for conservation of the mass of substrate around the entire system, for conservation of the mass of biomass around the secondary clarifier, and for conservation of the mass of biomass around the entire system.

Knowledge of appropriate kinetic relationships for substrate removal and biomass production is also necessary. Development of key design equations is presented for the activated sludge process represented in Figure 6.7.

A steady-state substrate mass balance around the treatment facility yields the following equation:

$$\text{Accumulation} = \text{Input} - \text{Output} + \text{Generation} \qquad (6.42)$$

$$0 = Q\,S_0 - Q_e\,S_e - Q_w\,S_w + V r_s \qquad (6.43)$$

where:

Q = influent wastewater flow rate, L/day

S_0 = substrate concentration in the influent to the activated sludge process, mg/L

Q_e = effluent flow rate, L/day

S_e = effluent substrate concentration, mg/L

Q_w = waste sludge flow rate, L/day

S_w = waste sludge substrate concentration, mg/L

V = aeration basin volume, L

$-r_s$ = substrate removal rate, mg S/L-day

If the reaction in the clarifier is assumed to be negligible, then the substrate concentration in the reactor (S) is equal to the concentration in the effluent (S_e) and the waste sludge (S_w). Furthermore, the sum of the effluent

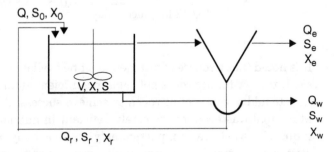

Figure 6.7(a)
Waste sludge
from aeration basin.

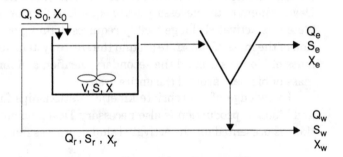

Figure 6.7(b)
Waste sludge
from clarifier.

flow rate (Q_e) and the waste sludge flow rate (Q_w) is equal to the influent flow rate (Q). Thus:

$$q = \frac{-r_s}{X} = \frac{Q(S_0 - S)}{VX} \tag{6.44}$$

The aeration basin biomass concentration (X) may be determined by simultaneous solution of equations (6.44), (6.27), and (6.36) as reported in equation (6.45):

$$\mu = Y_{max}\, q - k_e \tag{6.27}$$

$$\mu = \frac{1}{SRT} \tag{6.36}$$

$$X = \frac{Y_{max}}{1 + k_e\, SRT} \frac{Q}{V}(S_0 - S)\, SRT \tag{6.45}$$

For first-order kinetics, the relationship between the solids residence time and the effluent soluble substrate concentration is determined as follows:

$$q = k_1 S \tag{6.22}$$

$$S = \frac{1 + k_e\, SRT}{Y_{max}\, k_1\, SRT} \tag{6.46}$$

$$SRT = \frac{1}{S\, Y_{max}\, k_1 - k_e} \tag{6.47}$$

For variable-order kinetics, a similar relationship between the solids residence time and the effluent soluble substrate concentration is obtained:

$$S = \frac{K_s\,(1 + k_e\, SRT)}{Y_{max}\, k\, SRT - (1 + k_e\, SRT)} \tag{6.48}$$

$$SRT = \frac{K_s + S}{(Y_{max}\, k - k_e)\, S - k_e\, K_s} \tag{6.49}$$

A steady-state biomass balance around the activated sludge treatment facility produces the following equations:

$$\text{Accumulation} = \text{Input} - \text{Output} + \text{Generation} \tag{6.50}$$

$$0 = Q X_0 - Q_e X_e - Q_w X_w + V r_x \tag{6.51}$$

$$0 = 0 - Q_e X_e - Q_w X_w + V r_x \tag{6.52}$$

$$\frac{r_x}{X} = \frac{Q_e X_e + Q_w X_w}{VX} = \frac{1}{SRT} \tag{6.53}$$

where:

Q = influent flow rate, L/day

X_0 = influent biomass concentration (assumed negligible)

Q_e = effluent flow rate, L/day

X_e = effluent biomass concentration, mg/L

Q_w = waste sludge flow rate, L/day

X_w = waste sludge biomass concentration, mg/L

V = reactor volume, L

SRT = solids residence time, days

Equation (6.53) can be solved to define the necessary waste sludge flow rate:

$$Q_w = \frac{V\,X}{SRT\,X_w} - \frac{Q_e\,X_e}{X_w} \tag{6.54}$$

For the special case where sludge is wasted directly from the aeration basin ($X_w = X$) and negligible effluent suspended solids ($X_e = 0$), determination of the necessary waste sludge flow rate is simplified because the SRT may be maintained at the desired value without consideration of the clarifier thickening performance (represented by X_w) or the reactor biomass concentration (X):

$$Q_w = \frac{V}{SRT} \tag{6.55}$$

A steady-state biomass balance around the final clarifier provides a relationship between the recycle flow rate and underflow concentration for systems which practice sludge wasting from the clarifier underflow:

$$\text{Accumulation} = \text{Input} - \text{Output} + \text{Generation} \tag{6.56}$$

$$0 = (Q + Q_r)\,X - Q_e\,X_e - Q_w\,X_w - Q_r\,X_r + 0 \tag{6.57}$$

where:

Q = influent flow rate, L/day

Q_r = recycle flow rate, L/day

Q_e = effluent flow rate, L/day

X_e = effluent biomass concentration, mg/L

Q_w = waste sludge flow rate, L/day

X_w = waste sludge biomass concentration, mg/L

X_r = recycle (or underflow) biomass concentration, mg/L

If the mass of biomass removed in the effluent and waste sludge streams is negligible in comparison to the mass of biomass in the recycled sludge, equation (6.57) is simplified:

$$X_r = \frac{(Q + Q_r)\,X}{Q_r} \tag{6.58}$$

$$Q_r = \frac{X\,Q}{X_r - X} \qquad (6.59)$$

Example Problem 6.18

Determine the required solids residence time for activated sludge treatment of a municipal wastewater to achieve compliance with an effluent BOD$_5$ standard of 20 mg/L. Use appropriate kinetic values from Table 6.11.

Solution

As developed in equation (6.36) from a material balance for biomass, the SRT is equal to the reciprocal of the growth rate for a complete-mix activated sludge system. The required SRT is calculated using equation (6.47):

$$SRT = \frac{1}{S\,Y_{max}\,k_1 - k_e}$$

$$SRT = \frac{1}{\left(20\,\dfrac{mg\ BOD_5}{L}\right)\left(0.6\,\dfrac{mg\ VSS}{mg\ BOD_5}\right)\left(0.0294\,\dfrac{L}{mg\ VSS\text{–}day}\right) - 0.06\,\dfrac{1}{day}}$$

$$= 3.42\ days$$

Increase of the SRT to 5 days would be desirable to improve the settling characteristics of the biomass.

6.3.5.2 *Empirical Design Approach.* General design criteria for selected activated sludge modifications are reported in Table 6.12 for municipal wastewater applications. A minimum solids residence time of five days is normally selected to avoid adverse settling characteristics associated with high growth rates (Bisogni and Lawrence, 1971). Higher values of solids residence time (15+ days) may be required to achieve compliance with specific effluent standards.

Table 6.12 Typical design guidelines for municipal activated sludge systems.

SRT (days)	5 to 15
Organic Loading $\left(\dfrac{lb\ BOD_5}{lb\ VSS\text{-}d}\right)$	0.2 to 0.6
Organic Loading $\left(\dfrac{lb\ BOD_5}{10^3\ ft^3\text{-}d}\right)$	10 to 120
MLSS (mg SS/L)	2500 to 4000
HRT (hours)	3 to 8
Q_r/Q (%)	25 to 125
Clarifier Hydraulic Loading (gpd/ft^2)	400 to 800

Source: Adapted from Metcalf and Eddy, 1991.

6.3.6 Disinfection

The most common unit process to achieve disinfection is chlorination. Ten State Standards (1997b) recommends a minimum contact period of 15 minutes at peak hourly flow. Contact basins should be designed to minimize short-circuiting and longitudinal dispersion. Reactors that approach plug-flow conditions are preferred. Actual requirements are often dictated by state regulatory agencies or conditions of discharge permits, which may specify a range of acceptable chlorine residuals and hydraulic residence times.

Dechlorination, if required, is often achieved by adding sulfur dioxide:

$$SO_2 + HOCl + H_2O \rightarrow Cl^- + SO_4^{2-} + 3\,H^+ \tag{6.60}$$

Example Problem 6.19

Balance the oxidation reduction reaction for dechlorination with sulfur dioxide. Assume that hypochlorous acid is reduced to chloride ion, and sulfur dioxide is oxidized to sulfate.

Solution

Balance the half-reaction for reduction of HOCl:

$$HOCl \rightarrow Cl^-$$

$$HOCl \rightarrow Cl^- + H_2O$$

$$HOCl + H^+ \rightarrow Cl^- + H_2O$$

$$HOCl + H^+ + 2\,e^- \rightarrow Cl^- + H_2O$$

Balance the half-reaction for oxidation of SO_2:

$$SO_2 \rightarrow SO_4^{2-}$$

$$SO_2 + 2\,H_2O \rightarrow SO_4^{2-}$$

$$SO_2 + 2\,H_2O \rightarrow SO_4^{2-} + 4\,H^+$$

$$SO_2 + 2\,H_2O \rightarrow SO_4^{2-} + 4\,H^+ + 2\,e^-$$

Add the half-reactions, canceling the electrons:

$$SO_2 + HOCl + H_2O \rightarrow Cl^- + SO_4^{2-} + 3\,H^+$$

Alternative disinfection processes exist, including treatment with alternate chlorine compounds (for example, chlorine dioxide, sodium or calcium hypochlorite, or bromine chloride), ozonation, and UV radiation.

6.3.7 Sludge Digestion

Production of potable water supplies and treatment of municipal and industrial wastewaters often involve conversion of various dissolved species into sludges that can be separated from the liquid phase by physical processes (for example, sedimentation, filtration, or dewatering). These sludges, also known as residuals, require further processing prior to disposal in landfills or on agricultural lands. The residuals produced during potable water treatment are typically low in organic content, as these solids are often produced by chemical precipitation processes. In contrast, the residuals associated with wastewater treatment are rich in organic materials. Stabilization of these sludges is necessary to reduce the organic content and minimize odors that would compromise options for disposal. Biological processes are commonly used to achieve stabilization, including both aerobic and anaerobic digestion processes. In addition to reduction in odor potential, benefits of digestion include reduction in sludge mass, sludge volume, and pathogen content.

Aerobic sludge digestion is achieved in reactors that are similar to the aeration basins of the activated sludge process. The process is aerobic, and oxygen must be supplied with aeration equipment to satisfy biological requirements and to achieve mixing. Reactor design is based on the hydraulic residence time, typical values range from 20 to 40 days (depending on temperature). The process is capable of reducing the volatile suspended solids content (organic content) of the sludge by 40 to 55%. The stoichiometry of digestion of waste activated sludge ($C_5H_7O_2N$) is illustrated in equation (6.61):

$$C_5H_7O_2N + 7 O_2 \rightarrow 5 CO_2 + 3 H_2O + H^+ + NO_3^- \qquad (6.61)$$

Due to economic factors, aerobic digestion is often favored for small treatment facilities. The process is simple to operate and relatively insensitive to toxic materials (for example, heavy metals) that are present in the sludge.

Anaerobic digestion proceeds in an enclosed reactor to exclude oxygen. The reaction products include methane gas (CH_4), which is recovered and used as a fuel to heat the digester. Operating temperatures range from 95 °F to 120 °F. The stabilization reactions proceed at a greater rate at the elevated temperatures, thus reducing the necessary size of reactors. Greater pathogen removals are also achieved at these elevated temperatures. The required volume of anaerobic digesters is determined by the hydraulic residence time, with similar values as reported previously for aerobic digestion. The anticipated reduction in volatile suspended solids is also comparable to that reported for aerobic digestion. The process stoichiometry for anaerobic digestion is noted in equation (6.62):

$$38 C_5H_7O_2N + 152 H_2O \rightarrow 38 HCO_3^- + 38 NH_4^+ + 57 CO_2 + 95 CH_4 \quad (6.62)$$

The anaerobic digestion process requires greater skill to operate than aerobic digestion. Careful monitoring of the system alkalinity, pH, temperature, and gas production is necessary to achieve stable operation. The bacteria that produce methane are extremely sensitive to environmental factors (temperature and pH) and toxic agents (including heavy metals, ammonia, and sulfide). In spite of these complications, anaerobic digestion processes are commonly selected for large municipal treatment facilities due to favorable

economics. In particular, power costs are reduced, relative to aerobic digestion, because anaerobic processes do not require a supply of oxygen.

End-of-Chapter Problems

6.1   A precipitation event produced 2.0 inches of rainfall in two hours. Using
the rainfall intensity/duration/frequency chart in Figure 6.1, determine
the annual probability of occurrence of this event.

6.2   A precipitation event produced 3.0 inches of rainfall in three hours. Using
the rainfall intensity/duration/frequency chart in Figure 6.1, determine
the annual probability of occurrence of this event.

6.3   The rainfall intensities in problems 6.1 and 6.2 are identical (1.0 inch per
hour). Would the recurrence interval be the same for each event? Why or
why not?

6.4   A 20-acre site is undeveloped. Using the rainfall data from Figure 6.1, estimate the peak discharge for a 25-year storm if the time of concentration
for the watershed is 30 minutes. State all assumptions.

6.5   Assume that the 20-acre site in problem 6.4 has been developed as single
family residential. Estimate the peak discharge for a 25-year storm.
Assume that the drainage improvements which occur during site development have decreased the time of concentration for the watershed from 30
to 20 minutes.

6.6   Assume that the 20-acre site in problem 6.4 has been developed as single
family residential. Estimate the peak discharge for a 25-year storm.
Assume that stormwater detention ponds were included in the site development which have increased the time of concentration for the watershed
from 30 to 60 minutes.

6.7   Based on answers to problems 6.4, 6.5, and 6.6 above, discuss the advantages and disadvantages of stormwater detention ponds for mitigation of
peak post-development flows.

6.8   Repeat example problem 6.5 using a larger stormwater pond. Assume that
the dimensions of the pond bottom are 200 feet (width) by 350 feet
(length). All other system parameters remain the same. Determine the
peak discharge from the pond in response to the specified surface runoff
input.

6.9   Repeat example problem 6.5 assuming the output discharge is controlled
by a rectangular weir with a length of 10 feet:

$$Q_{out} = C\,L\,H^n$$

where:

$$Q_{out} = \text{discharge (cfs)}$$
$$C = \text{weir coefficient} = 3.33 \text{ ft }^{0.5}/\text{sec}$$
$$L = \text{weir length (ft)}$$
$$n = 1.5$$

6.10 Compare the water quality of highway runoff (Table 6.4) with untreated (Table 6.10) and treated (Tables 6.8 and 6.9) municipal wastewater. Discuss adverse environmental impacts associated with release of highway runoff to surface waters.

6.11 It has been demonstrated that HOCl molecules are more effective for disinfection than OCl^- ions. Assume you put a certain amount of HOCl into water and adjust the pH independently with HCl. Which pH would produce better disinfection: pH 7 or pH 8? Prove your answer with calculations. The K_a for HOCl is 3.2×10^{-8}.

6.12 Using a spreadsheet for calculations and plotting, determine the molar concentrations of HOCl and OCl^- for pH values between 6 and 8 (increment by 0.1 pH units). Assume the total free chlorine residual is 10^{-4} molar. Plot the results on a log scale (log concentration versus pH).

6.13 A potential groundwater supply for a small community was analyzed as follows:

$$CO_2 = 10 \text{ mg/L}$$
$$Fe = 1 \text{ mg/L}$$
$$H_2S = 0.5 \text{ mg/L}$$
$$Ca = 40 \text{ mg/L}$$
$$Mg = 10 \text{ mg/L}$$
$$\text{Alkalinity} = 150 \text{ mg/L as } CaCO_3$$
$$pH = 7.1$$

Compare the raw quality data with primary and secondary water standards, and determine treatment objectives for development of this source as a potable supply.

6.14 A groundwater was analyzed as follows:

$$CO_2 = 30 \text{ mg/L as } CaCO_3$$
$$Ca = 180 \text{ mg/L as } CaCO_3$$
$$Mg = 120 \text{ mg/L as } CaCO_3$$
$$Na = 80 \text{ mg/L as } CaCO_3$$
$$\text{Alkalinity} = 360 \text{ mg/L as } CaCO_3$$
$$SO_4^{2-} = 10 \text{ mg/L as } CaCO_3$$
$$Cl^- = 10 \text{ mg/L as } CaCO_3$$
$$pH = 7.2$$

Would softening be recommended for this water? Determine the following quantities (as mg/L of $CaCO_3$): total hardness, calcium carbonate hardness, calcium noncarbonate hardness, magnesium carbonate hardness, magnesium noncarbonate hardness, and excess alkalinity.

6.15 Describe the purpose of adding excess lime during lime-soda softening.

6.16 Describe the purpose of recarbonation after lime-soda softening.

6.17 A groundwater was analyzed as follows:

CO_2 = 35 mg/L
Ca = 200 mg/L
Mg = 60 mg/L
Na = 35 mg/L
Alkalinity = 400 mg/L as $CaCO_3$
pH = 7.0

Convert the stated mass concentrations into mg/L as $CaCO_3$. Determine the total hardness of this water. Would softening be recommended for this water?

6.18 A groundwater was analyzed as follows:

CO_2 = 35 mg/L
Ca = 200 mg/L
Mg = 60 mg/L
Na = 35 mg/L
Alkalinity = 400 mg/L as $CaCO_3$
pH = 7.0

For a water production rate of 3 million gallons per day (MGD), determine chemical requirements (in lbs per day) and sludge production rates (in lbs per day) for lime-soda softening of this water.

6.19 A groundwater was analyzed as follows:

CO_2 = 35 mg/L
Ca = 100 mg/L
Mg = 60 mg/L
Na = 35 mg/L
Alkalinity = 300 mg/L as $CaCO_3$
pH = 7.0

For a water production rate of 1 million gallons per day (MGD), determine chemical requirements (in lbs per day) and sludge production rates (in lbs per day) for lime-soda softening of this water.

6.20 Rapid mixing facilities in a rapid sand filtration plant must be designed to achieve a minimum hydraulic residence time of 30 seconds. For a water production rate of 500,000 gallons per day, determine the dimensions of rapid mixing tanks. There should be a minimum of two tanks in parallel for reliability. Calculate the actual residence time for the recommended facilities. Use a square tank configuration with the depth equal to 1.25 times the width.

6.21 Flocculation basins in a rapid sand filtration plant must be designed to achieve a minimum hydraulic residence time of 45 minutes. For a water production rate of 500,000 gallons per day, determine the dimensions of flocculation basins. There should be a minimum of two basins in parallel for reliability. Calculate the actual residence time for the recommended facilities. Use a rectangular tank configuration with a width equal to approximately one third of the tank length. The width and the depth should be approximately equal.

6.22 Sedimentation tanks in a rapid sand filtration plant must be designed to achieve a minimum hydraulic residence time of 3 hours and a maximum hydraulic loading rate of 750 gal/day-ft^2. For a water production rate of 500,000 gallons per day, determine the dimensions of sedimentation tanks. There should be a minimum of two tanks in parallel for reliability. Calculate the actual residence time and hydraulic loading rate for the recommended facilities. Use a rectangular tank configuration with a width equal to approximately one tenth of the length.

6.23 Sand filters in a rapid sand filtration plant must be designed to achieve a maximum hydraulic loading rate of 5 gal/min-ft^2. For a water production rate of 500,000 gallons per day, determine the dimensions of filtration facilities. There should be a minimum of four filters in parallel for reliability. Calculate the actual hydraulic loading rate for the recommended facilities. Use a square tank configuration with a filter media (sand) depth of 24 inches. Allow a head space of 5 feet above the sand media.

6.24 For the specifications of example problem 6.9, determine the required hydraulic residence time for a complete-mix reactor.

6.25 For the specifications of example problem 6.9, determine the required total hydraulic residence time for a dispersed plug flow reactor that can be described as three equal-volume complete-mix reactors in series.

6.26 For the specifications of example problem 6.9, determine the required total hydraulic residence time for a dispersed plug flow reactor that can be described as five equal-volume complete-mix reactors in series.

6.27 Using the typical characteristics for a municipal wastewater from Table 6.10 and assuming a per capita wastewater generation rate of 100 gallons per day, determine the per capita generation rate of BOD$_5$ and suspended solids. Express the answers in lbs per capita-day.

6.28 Using the kinetic information from example problem 6.15 for a municipal wastewater, determine the effluent BOD$_5$ concentration for a complete-mix activated sludge facility operated at a five-day solids residence time.

For a wastewater flow rate of 2 million gallons per day (MGD) with characteristics of a typical wastewater from Table 6.10, determine the rate of residuals generation (lbs/day).

6.29 Using the kinetic information from example problem 6.15 for a municipal wastewater, determine the effluent BOD_5 concentration for a complete-mix activated sludge facility operated at a fifteen-day solids residence time. For a wastewater flow rate of 2 million gallons per day (MGD) with characteristics of a typical wastewater from Table 6.10, determine the rate of residuals generation (lbs/day).

6.30 Determine the required hydraulic residence time for an aerated lagoon for removal of 95% of the BOD_5 from an industrial wastewater. The kinetics were determined as follows:

> Variable order kinetics (equation 6.23)
> $k = 1.2$ mg BOD_5/mg VSS-day
> $K_s = 90$ mg BOD_5/L
> $Y_{max} = 0.5$ mg VSS/mg BOD_5
> $k_e = 0.04$ day^{-1}
> The influent wastewater BOD_5 is 800 mg/L.

6.31 A municipal complete-mix activated sludge facility has the following characteristics:

> Flow = 3 million gallons per day (MGD)
> Influent $BOD_5 = 250$ mg/L
> Aeration basin volume = 100,000 ft^3
> Two circular clarifiers (diameter = 60 feet, depth = 12 feet)
> Return activated sludge flow rate = 1 MGD per clarifier
> Mixed liquor suspended solids concentration = 3500 mg/L
> Waste sludge flow rate (from aeration basin) = 100,000 gal/day
> Effluent suspended solids concentration is negligible

Determine the following parameters: organic loading rate, aeration basin hydraulic residence time, percent recycle, clarifier hydraulic loading rate, and solids residence time. Compare these values with the empirical guidelines presented in Table 6.12 for a municipal activated sludge system.

6.32 A municipal complete-mix activated sludge facility has the following characteristics:

> Flow = 4 million gallons per day (MGD)
> Influent $BOD_5 = 200$ mg/L
> Aeration basin volume = 100,000 ft^3
> Two circular clarifiers (diameter = 70 feet, depth = 12 feet)
> Return activated sludge flow rate = 1.5 MGD per clarifier
> Mixed liquor suspended solids concentration = 3500 mg/L

Effluent suspended solids concentration is negligible

If sludge is wasted from the aeration basin, determine the daily volume of sludge (gallons per day) that must be removed to maintain a solids residence time of 5 days.

6.33 A municipal complete-mix activated sludge facility has the following characteristics:

Flow = 4 million gallons per day (MGD)

Influent BOD_5 = 200 mg/L

Aeration basin volume = 100,000 ft^3

Two circular clarifiers (diameter = 70 feet, depth = 12 feet)

Return activated sludge flow rate = 1.5 MGD per clarifier

Mixed liquor suspended solids concentration = 3500 mg/L

Effluent suspended solids concentration is negligible

If sludge is wasted from the return activated sludge pipeline, determine the daily volume of sludge (gallons per day) that must be removed to maintain a solids residence time of 5 days.

6.34 For the plant configuration in problem 6.32 (waste sludge from aeration basin), how would the required waste sludge volume be affected if the return activated sludge flow rate were increased to 2 MGD per clarifier? For the plant configuration in problem 6.33 (waste sludge from return activated sludge pipeline), how would the required waste sludge volume be affected if the return activated sludge flow rate were increased to 2 MGD per clarifier? Which configuration for sludge wasting is easier to operate? Why?

6.35 For a municipal wastewater that receives treatment with the complete-mix activated sludge process, plot the effluent BOD_5 as a function of solids residence time using values of SRT which range from 1 to 20 days. The influent BOD_5 concentration is 220 mg/L. Assume the following wastewater characteristics:

Variable order kinetics (equation 6.23)

k = 5 mg BOD_5 /mg VSS-day

K_s = 60 mg BOD_5 /L

Y_{max} = 0.6 mg VSS/mg BOD_5

k_e = 0.06 day^{-1}

6.36 For a municipal wastewater that receives treatment with the complete-mix activated sludge process, recommend an operating value for the solids residence time to achieve compliance with secondary treatment standards. The influent BOD_5 concentration is 220 mg/L. Assume the following wastewater characteristics:

Variable order kinetics (equation 6.23)

k = 5 mg BOD_5 /mg VSS-day

$$K_s = 60 \text{ mg BOD}_5 \text{/L}$$
$$Y_{max} = 0.6 \text{ mg VSS/mg BOD}_5$$
$$k_e = 0.06 \text{ day}^{-1}$$

6.37 A sludge digester must be designed to achieve a minimum hydraulic residence time of 20 days. The input sludge mass is 5000 lbs (dry weight) per day and the suspended solids concentration is 10,000 mg/L. Determine the required digester volume.

6.38 For the conditions in problem 6.37, it has been proposed to pretreat (thicken) the sludge prior to digestion. It is expected that thickening will achieve an increase in the input sludge solids content to 3% solids. Determine the required digester volume to achieve a 20-day hydraulic residence time with a thickened sludge input.

6.39 Treatment of hard-chrome-plating wastewaters includes precipitation of chromium as $Cr(OH)_3$. The chromium is present in a hexavalent form (H_2CrO_4) that must be reduced to the trivalent form prior to precipitation. Sulfur dioxide is commonly used as the reducing agent with production of sulfate. Balance the oxidation reduction reaction for reduction of hexavalent chromium with sulfur dioxide.

6.40 For the biological growth reaction in example problem 6.17, determine nitrogen supplementation requirements (express the answer in pounds of ammonia per day). The raw wastewater contains 500 mg/L of glucose and 2 mg/L of TKN. The treatment process achieves 98% removal of glucose. The wastewater flow rate is 1 million gallons per day (MGD).

References

ASCE/WPCF. 1969. *Design and Construction of Sanitary and Storm Sewers*. New York, NY and Washington, DC: American Society of Civil Engineers and Water Pollution Control Federation.

Bisogni, J. J., and A. W. Lawrence. 1971. "Relationships Between Biological Solids Retention Time and Settling Characteristics of Activated Sludge." *Water Research* 5:753.

Dietz, John D. 1996. "Wastewater Treatment." In *Environmental Engineering P.E. Examination Guide & Handbook*, edited by W. Christopher King. Annapolis, MD: American Academy of Environmental Engineers.

Dominguez, R. 1996. "Hydraulics and Hydrology." In *Environmental Engineering P.E. Examination Guide & Handbook*, edited by W. Christopher King. Annapolis, MD: American Academy of Environmental Engineers.

Grady, C. P. L. 1990. "Biodegradation of Toxic Organics: Status and Potential." *Journal of Environmental Engineering Div ASCE* 116:805.

Grady, C. P. L., and H. C. Lim. 1980. *Biological Wastewater Treatment*. New York: Marcel Dekker.

Lawrence, A. W., and P. L. McCarty. 1970. "A Unified Basis for Biological Treatment Design and Operation." *Journal of Sanitation Engineering Div ASCE* 96:757.

Metcalf and Eddy, Inc. 1991. *Wastewater Engineering*. 3rd ed. New York: McGraw-Hill.

Nemerow, N. L., and A. Dasgupta. 1991. *Industrial and Hazardous Waste Treatment*. New York: Van Nostrand Reinhold.

Okun, D. A. 1996. "From Cholera to Cancer to Cryptosporidiosis." *Journal of Environmental Engineering Div ASCE* 122:453.

Otterstetter, H., and G. Craun. 1997. "Disinfection in the Americas: A Necessity." *Journal of American Water Works Assoc* 89(9):8.

Pontius, F. W. 1998. "New Horizons in Federal Regulation." *Journal of American Water Works Assoc* 90(3):38.

Reynolds, T. R., and P. A. Richards. 1996. *Unit Operations and Processes in Environmental Engineering.* 2nd ed. PWS Publishing.

Standard Methods. 1992. *Standard Methods for the Examination of Water and Wastewater.* 18th ed. Washington, DC: APHA, AWWA, WEF.

Ten States Standards. 1997a. *Recommended Standards for Water Works.* Albany, NY: Health Education Service.

Ten States Standards. 1997b. *Recommended Standards for Wastewater Facilities.* Albany, NY: Health Education Service.

U.S. Environmental Protection Agency. 1985. *Environmental Regulations and Technology: The Electroplating Industry.* EPA/625/10-85/001.

U.S. Environmental Protection Agency. 1988. *Waste Minimization Opportunity Assessment Manual.* EPA/625/7-88/003.

Wanielista, M. P., and Y. A. Yousef. 1993. *Stormwater Management.* New York: John Wiley & Sons.

Zitomer, D. H., and R. E. Speece. 1993. "Sequential Environments for Enhanced Biotransformation of Aqueous Contaminants." *Environmental Science & Technology* 27:226.

Chapter 7

Management of Land-Disposed Waste

For many centuries, the only standard for waste disposal was "out of sight, out of mind." Public concern led to pollution-control regulations during the early 1970s that focused on the control of emissions to the air and water. However, solid or semisolid residuals resulting from the treatment of any kind of pollution ultimately were land-disposed. These residuals often were solid, hazardous, and/or radioactive wastes. Historically, these residuals were buried in the closest piece of vacant land. The land was believed to have infinite capacity for disposal, and therefore would be resilient to the effects of pollution. Over time, it has become obvious that the land environment does not have the ability to completely assimilate all types of waste. Furthermore, some wastes are particularly resistant to physical or biological decomposition, and their disposal has resulted in a blighted landscape (see Figure 7.1). In many cases in the past, hazardous waste disposal to the land resulted in serious human health problems and ecological damage, which prompted legislation in the 1970s and 1980s governing the management and disposal of solid, hazardous, and radioactive wastes.

7.1   Overview of Municipal Solid Waste

7.1.1 Sources, Composition, and Quantity

Nonhazardous solid wastes are primarily generated by people during community activities. Domestic and commercial wastes are broadly called **municipal solid waste** (MSW); nonhazardous solid wastes also include wastes generated by nondomestic sources such as industrial wastes, municipal combustion ash, agricultural waste, construction and demolition debris, and oil, gas, and mining waste. MSW includes durable goods such as appliances, furniture, tires, and electronics; nondurable goods such as news-

Figure 7.1
Discarded tire pile. (*Jose Azel/Aurora.*)

papers, clothing, paper towels, and office paper; containers and packaging; food wastes; yard wastes; and miscellaneous inorganic material such as soil, ashes, glass, metals, and stone.

Every two years, the U.S. EPA contracts for a study of the composition and quantity of MSW generated in the United States. The study is based on a materials balance approach that looks at the production and expected life cycles of products and attempts to estimate the composition and quantity of MSW from these and other factors (Franklin and Associates, 1998). This approach considers waste generation to be the end result of a product's life cycle. The most recent results of this study show that Americans generated over 209.7 million tons of MSW in 1996, averaging 4.3 pounds per person per day. Considering that the United States has only 5% of the world's population, this country produces a disproportionate amount of the world's solid waste (approximately 40%).

Waste quantity varies considerably both temporally and geographically. Historically, per capita generation in the United States has been found to be strongly correlated with the gross national product and has increased at a rate of 2 to 4% per year from 1960 through the early 1990s. It also has been observed that per capita generation rates vary according to climate, age of population, season, use of garbage grinders, size of household, per capita income, public attitude, local legislation, and frequency of collection. More recently the rate of growth in MSW generation has slowed to approximately 1 to 2% per year as a result of public education regarding waste reduction, county and city recycling programs, and the diversion of yard wastes to composting operations. The per capita waste generation rate has actually declined from 4.4 lb/person/day in 1993 to 4.3 lb/person/day in 1996 (Franklin and Associates, 1998). However, coupled with the projected increase in population, the annual MSW production in the United States is expected to surpass 225 million tons within the next ten years.

Figure 7.2 represents the composition of MSW on a weight basis (U.S. average). The composition of waste can reveal a great deal about the characteristics of a community, and varies with location, season, economic conditions, and demographics. As can be seen in Figure 7.2, MSW in the United States is predominantly paper and plastics, which is typical of highly developed countries (due to the pervasive use of packaging and containers, and our intensive use of computers).

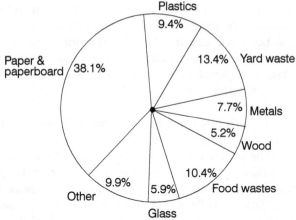

Figure 7.2
MSW materials generated in the U.S., by weight in 1996. (*Franklin and Associates, 1998.*)

Yard waste became a major component of MSW in the 1970s because of the decline in burning of debris as well as the increase in the number of suburban households. However, since many states recently banned yard wastes from their landfills (to save space), the percent of yard waste in MSW has dropped significantly. Plastics have been increasingly used as containers and primary components of many products over the past twenty years, while glass and metal use has decreased. Food waste, although still a significant fraction of MSW, has declined in quantity as the use of prepared food and garbage grinders has increased.

7.1.2 Regulation of MSW

Significant regulation of land-based waste disposal did not occur until the Resource Conservation and Recovery Act (RCRA) was passed in 1976, which mandated development of requirements for the management of solid and hazardous wastes. In the two decades since RCRA was promulgated, U.S. EPA policy has shifted from an emphasis on pollution control to pollution prevention.

RCRA defines a **solid** waste as any solid, liquid, semisolid, or contained gaseous discarded material that is no longer useful for its intended purpose or is an unintended or unusable by-product. RCRA then distinguishes between **nonhazardous** and **hazardous** solid wastes and regulates them in very different ways. Nonhazardous wastes are regulated under Subtitle D of RCRA, which provides the legislative mandate for environmental control of the transportation, processing, and disposal of these wastes.

7.1.3 Waste Management Practices

Current waste management practices involve a hierarchy that includes source reduction, recycling, incineration, and landfilling. This practice is often referred to as **integrated waste management**. Source reduction is probably the most important management element because it eliminates or minimizes the production of waste. A waste that is not created will never harm the environment. Source reduction involves downsizing and light-weighting of containers, elimination of single-use items (pens, razor blades, batteries, paper or plastic cups), and use of concentrated material (juice, detergent, softeners).

7.1.3.1 Collection and Transport.

In most cities and counties, collection and transport of MSW, recyclables, or yard waste represent more than half of the solid waste management budget. Collection is either accomplished by a public agency or, more frequently, by private companies that receive franchises giving exclusive license to collect wastes within a certain area. Waste, placed in either stationary or hauled containers, is collected in vehicles with capacities ranging from 10 to 40 yd^3 (7.6 to 31 cubic meters). Stationary containers are generally used at low waste production sites (such as residential or commercial areas). Hauled containers are used in high waste production locations such as construction and demolition sites with relatively short haul distances to disposal sites. Frequently, waste is compacted in trucks such as the one shown in Figure 7.3 to maximize transport efficiency. Source separated recyclables, such as newspapers, glass bottles, tin cans, and aluminum cans, are generally placed in separate bins on the collection vehicle without compaction.

Figure 7.3
A compactor truck unloading collected municipal solid waste.

Example Problem 7.1

Compare the time required to collect waste generated by a large shopping center using either a 30 yd³ (23 cubic meters) mechanically loaded compactor vehicle or a hoist truck which hauls each container separately to the landfill. The waste is stored in 12 containers, 10 yd³ (7.6 m³) each. Use the following information:

Compaction ratio:	2.5
Average time to travel to the landfill:	30 min
Average time at the landfill:	10 min
Time to pick up and unload container (compactor):	7 min
Time to load container on hoist truck:	10 min
Time to drive between containers:	2 min

Solution

(a) Hoist Truck

Time per trip = time to load container + time to landfill + time at landfill + time to return from landfill

$$= (10\,\text{min} + 30\,\text{min} + 10\,\text{min} + 30\,\text{min})\,\text{per trip}$$
$$= 80\,\text{min/round trip}$$

Number of trips (hoist truck)	$= 12$
Total time	$= 960\,\text{min}$

(b) Compactor Truck

Number of trips (compactor)	= total volume waste/vehicle capacity
Vehicle capacity	= volume x compaction ratio
	$= 30 \times 2.5$
	$= 75\,\text{yd}^3/\text{trip}$
Number of trips	$= 120\,\text{yd}^3/(75\,\text{yd}^3/\text{trip})$
	= 1.6 trips (round to 2 trips)
Total Time	= no. of containers x unload time + drive between containers x (number of containers – 1) + number of whole trips x (haul time + at-site time + return time)
	= 12 containers (7 min/container) + 2 min/drive (11 drives between containers) + 2 trips (30 + 10 + 30) min/trip
	= 246 min

The compactor requires approximately one-fourth of the time the hoist truck requires. However, capital and operating costs for the large compactor vehicle are much greater than for the hoist truck and must be considered in an economic analysis of collection systems.

It is not uncommon today to use transfer stations as intermediate stops for wastes prior to disposal at landfills or incinerators located far away from the waste generators. In some cases, the waste may be transferred to a rail car or barge for transport to disposal sites located several states away. Advantages of transfer stations include cost savings associated with the use of larger volume long-haul trucks and smaller crews for the longer trip to the disposal site, shorter turnaround for collection vehicles, opportunity for inspection of loads for hazardous wastes, and ease of locating disposal sites in less populated areas.

7.1.3.2 Recycling. In 1993, approximately 22 percent of U.S. MSW was recycled, although many communities have achieved much higher recycling rates. Materials most successfully recycled include aluminum, glass, high density polyethylene (HDPE) and polyethylene terephthalate (PET) plastics, ferrous and nonferrous metals, paper, cardboard, lead-acid batteries, waste oil, and yard waste. In some places, local markets also have developed for recovered components from aseptic packaging, tires, construction wastes, polystyrene, vinyl, polyvinyl chloride (PVC) plastics, and household batteries. Materials recovered from the waste stream either are used directly, used as raw materials for remanufacturing, or used as feed sources for chemical or biological waste conversion (i.e., composting or anaerobic digestion). Most of the materials are separated and collected at the source of generation. Drop-off sites and buy-back centers (in bottle bill states) also provide a significant quantity of recovered material.

Example Problem 7.2

Determine the percentage of generated waste that is recycled given the following information. The waste generated is 30% paper, 12% cardboard, 18% yard waste, 8% glass, 5% plastic, 3% aluminum, 9% ferrous metal, and 15% miscellaneous. Of that, 50% of the paper, 30% of cardboard, 40% of glass, 20% of plastic, 50% of aluminum, and 10% of ferrous metal is recycled. Determine the final composition of the waste after recyclables are removed.

Solution

The results of calculations necessary to solve this problem are provided in the table below.

Waste component	Generated waste composition, lbs/100 lbs	Recycling efficiency, %	Weight disposed, lbs/100 lbs generated[1]	Composition after recycling, % by weight[2]
Paper	30	50	15	20
Cardboard	12	30	8.4	11.2
Yard waste	18	0	18	24
Glass	8	40	4.8	6.4
Plastic	5	20	4	5.3
Aluminum	3	50	1.5	2
Ferrous metal	9	10	8.1	10.8
Miscellaneous	15	0	15	20
Total	100	–	74.8	99.7[3]

[1]Column 4, the mass of each component remaining after removing the recycled material, is generated by multiplying Column 2 by [1 - (Column 3/100)].
[2]To determine the composition of the waste disposed (Column 5), divide Column 4 by 74.8 and express as %.
[3]Column does not add up to 100 due to round-off error.

The percentage of waste recycled is determined by subtracting the total of Column 4 from 100: Percent Recycled = 100 – 74.8

$$= 25.2\%$$

Sometimes MSW is transported to material recovery facilities (MRFs) for processing. Source separated materials are handled at "clean" MRFs while mixed wastes are separated at "dirty" MRFs. Separation of mixed wastes is done both manually as well as mechanically through shredding and screening, magnetic separation of ferrous metals, optical separation of glass, electromagnetic separation of aluminum, and air classification to separate light organic materials. A typical process flow diagram for processing mixed waste is provided in Figure 7.4. Materials are further processed in order to improve the quality and usefulness for recycling through cleaning, separating glass by color, baling, crushing, and packaging.

The cost of collecting and processing recyclable materials is quite high (as high as $300/ton in some communities) but is offset by revenues from the sale of recyclables and by avoided cost of disposal. Revenue from the sale of recyclables is extremely variable, subject to supply and demand forces and international market factors. For example, in 1991, Russian bauxite (raw material for aluminum) was sold at extremely low prices to get hard currency to finance the new governments formed following the breakup of the former Soviet Union. Virgin raw material costs therefore fell sharply and the demand for recovered aluminum declined. In the early 1990s, markets for plastics and old newspaper strengthened significantly, following many years of extremely low prices. This increase was attributed to the growing mill capacity for recycled raw materials. More recently, recycled newsprint prices fell again, due to the low cost of virgin materials.

Recycling is an important element of successful integrated waste management and has been encouraged through state and federal regulations. Expansion of markets for recycled materials may require the following:

- public education to encourage purchase of recycled materials
- elimination of government subsidies of virgin raw materials
- mandatory recycled content of products
- accurate labeling of products
- established standards/specification for recovered materials to ensure quality
- restricted production of materials which cannot be easily recycled

7.1.3.3 MSW Disposal. Disposal options for MSW include composting, landfilling, or incineration (with subsequent landfilling of ash). Composting may be thought of as recycling since it involves the beneficial reuse of waste. Composting employs aerobic biological processes to convert the waste into a humic material that can be used as a soil amendment. The nutrient content is generally too low to use compost as a fertilizer without chemical additives or cocomposting with wastewater sludge. Incineration and landfilling are the major disposal options for municipal solid waste and are discussed in detail in sections 7.4 and 7.5.

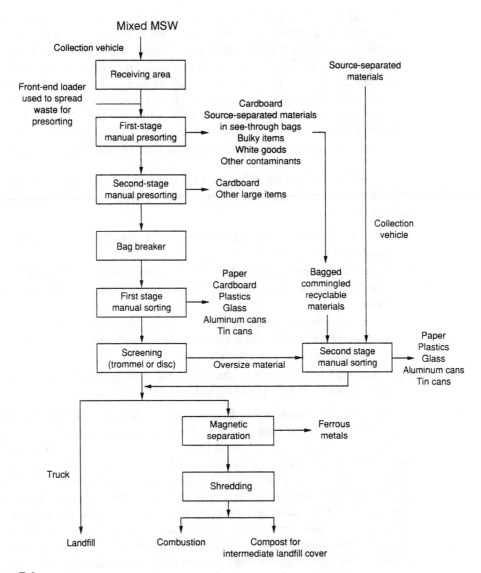

Figure 7.4
Flow diagram for the recovery of waste materials from mixed MSW. (*After Tchobanoglous, 1993.*)

7.2   Overview of Hazardous Waste

7.2.1 Sources, Composition, and Quantity

The primary legislation regulating hazardous waste in the United States, the Resource Conservation and Recovery Act, defines hazardous waste as "... a solid waste, or combination of solid wastes, which may, because of its quantity, concentration, or physical, chemical or infectious characteristics ... pose a substantial present or potential hazard . . . when improperly treated, stored, transported, or disposed of, or otherwise mismanaged." This definition is structured to provide the U.S. EPA broad authority to identify and regulate hazardous wastes. The statute addresses all hazardous waste activities regardless of whether historical evidence exists for poor handling of hazardous wastes. Additionally, RCRA does not consider cost as a basis for failing to institute regulatory controls. In other words, RCRA regulations must be followed regardless of economic impact (Fortuna and Lennett, 1987). Failure to do so can result in criminal charges against high-level facility managers and owners.

Based upon RCRA definition, a solid waste is designated as a hazardous waste if it falls within one of the four following categories (and is not specifically excluded):

1. EPA may list a waste as hazardous based upon the presence of specific hazardous constituents.
2. Wastes not specifically listed may be identified as hazardous on the basis that they exhibit one or more of the four characteristics of a hazardous waste.
3. If a mixture of a listed hazardous waste or characteristic waste and any other material occurs, the waste is hazardous provided the mixture still exhibits hazardous characteristics.
4. If the waste is derived from the treatment, storage, or disposal of a listed waste, then the waste is hazardous. (Wentz, 1989)

A **listed** hazardous waste is one that has been specifically identified under RCRA as a hazardous waste. Presently, there are over 700 listed wastes, some of which are provided in Table 7.1. A **characteristic** waste is not a listed waste but exhibits one or more hazardous characteristics defined as corrosivity, reactivity, ignitability, and/or toxicity (as summarized in Table 7.2). **Toxicity** is defined as the ability to leach certain constituents by the Toxic Characteristic Leaching Procedure (TCLP), a laboratory procedure. A characteristic waste can be rendered nonhazardous by eliminating the characteristic that caused it to be hazardous. However, a listed waste and any waste derived from its treatment continues to be regulated as a hazardous waste even if the treated stream no longer exhibits hazardous characteristics.

RCRA now divides generators of hazardous waste into three categories: large quantity generators, small quantity generators, and conditionally exempt small quantity generators. Large quantity generators (LQGs) produce more than 2200 lbs (1000 kg) of hazardous waste each month. LQGs are required to test their waste to determine whether the waste is a hazardous waste. If the waste is hazardous and will be shipped off-site, the generator is required to acquire an EPA identification number, start a manifest (a set of forms that allows each waste shipment to be tracked) for the waste, containerize and label the waste, and notify the receiving facility that the waste is being shipped. Examples of labels for a hazardous waste shipment are shown

Table 7.1 Examples of listed hazardous wastes.

Hazardous Waste	Basis for Listing
Spent halogenated solvent used in degreasing	Toxicity
Spent nonhalogenated solvents	Ignitability
Wastewater treatment sludge from the production of chrome yellow and orange pigments	Toxicity
Stripping still tails from the production of methyl ethyl pyridines	Toxicity
2,6-dichlorophenol waste from the production of 2,4-dichlorophenol	Toxicity
Ammonia still lime sludge from coking operations	Toxicity
Cyanogen chlorine	Acute Toxicity
Thiophenol	Acute Toxicity
Emission control dust/sludge from secondary lead smelting	Toxicity

Table 7.2 U.S. hazardous characteristics.*

Hazardous Characteristic	Definition
Corrosivity	Waste that is highly acidic or alkaline (pH $\leq$ 2 or $\geq$ 12.5)
Ignitability	Waste that is easily ignited and poses a fire hazard during routine management
Reactivity	Waste that is capable of potentially harmful, sudden reactions
Toxicity	Waste capable of leaching any of eight heavy metals (As, Ba, Cd, Cr, Pb, Hg, Se, Ag) and/or six specific pesticides (endrin, lindane, methoxychlor, toxaphene, 2,4-D, and 2,4,5-TP) into slightly acidic water

*The U.S. EPA has specified laboratory procedures for each characteristic.

in Figure 7.5. The generator may store hazardous waste on the production site for up to 90 days without acquiring a TSD (treatment, storage, disposal) permit. The 90-day time clock starts from the time the first bit of waste enters the storage/shipment container.

In 1993, there were 22,615 LQGs in the United States who collectively produced 235 million tons of RCRA hazardous waste (U.S. EPA, 1995a). These numbers represent a decrease of 80 large generators and a reduction of 70 million tons of hazardous waste since 1991. During the 1993 operating year, 60% of the LQGs generated more than 12 million tons of hazardous

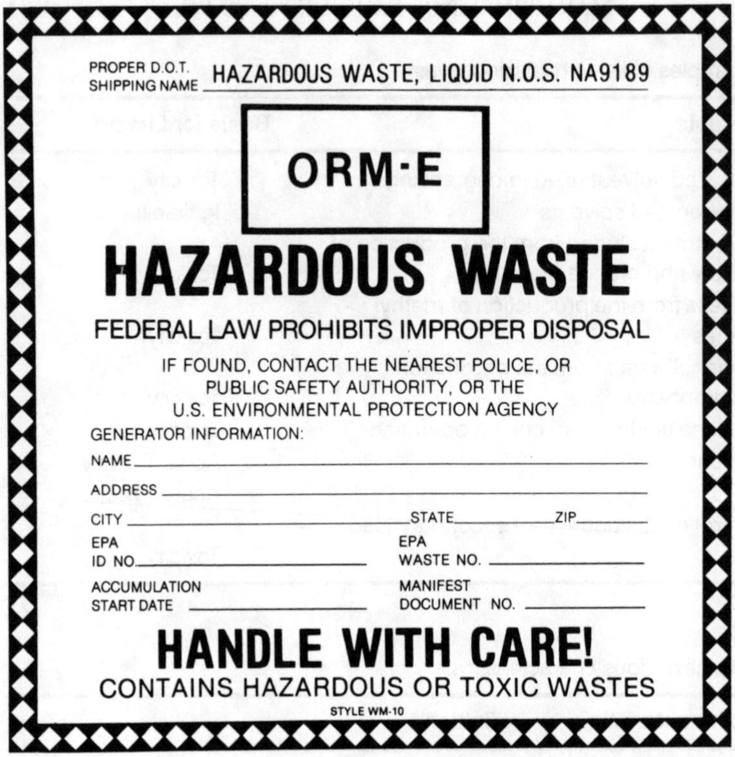
PROPER D.O.T.
SHIPPING NAME HAZARDOUS WASTE, LIQUID N.O.S. NA9189

ORM-E

HAZARDOUS WASTE

FEDERAL LAW PROHIBITS IMPROPER DISPOSAL

IF FOUND, CONTACT THE NEAREST POLICE, OR
PUBLIC SAFETY AUTHORITY, OR THE
U.S. ENVIRONMENTAL PROTECTION AGENCY

GENERATOR INFORMATION:
NAME ______
ADDRESS ______
CITY ______ STATE ______ ZIP ______
EPA ID NO. ______ EPA WASTE NO. ______
ACCUMULATION START DATE ______ MANIFEST DOCUMENT NO. ______

HANDLE WITH CARE!

CONTAINS HAZARDOUS OR TOXIC WASTES

STYLE WM-10

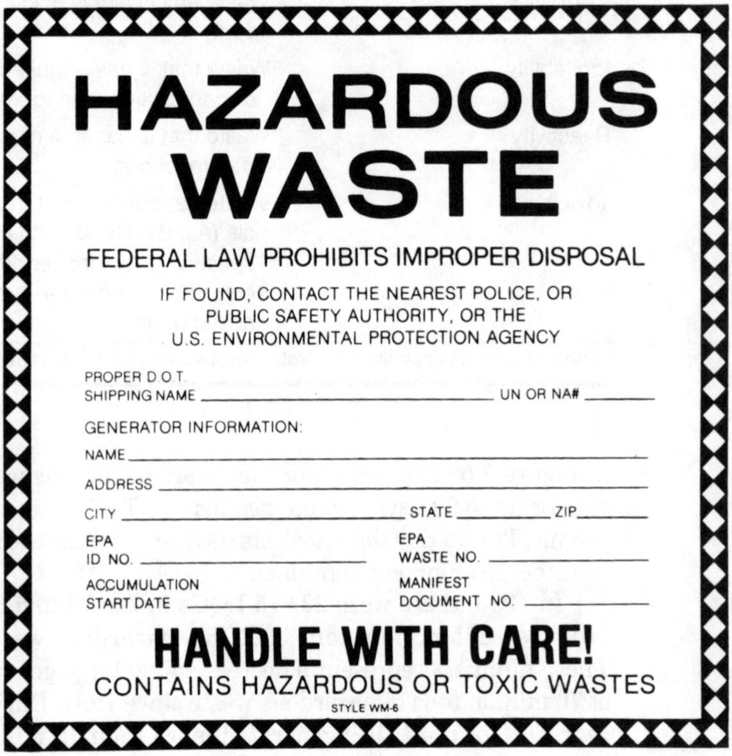
HAZARDOUS WASTE

FEDERAL LAW PROHIBITS IMPROPER DISPOSAL

IF FOUND, CONTACT THE NEAREST POLICE, OR
PUBLIC SAFETY AUTHORITY, OR THE
U.S. ENVIRONMENTAL PROTECTION AGENCY

PROPER D.O.T.
SHIPPING NAME ______ UN OR NA# ______

GENERATOR INFORMATION:
NAME ______
ADDRESS ______
CITY ______ STATE ______ ZIP ______
EPA ID NO. ______ EPA WASTE NO. ______
ACCUMULATION START DATE ______ MANIFEST DOCUMENT NO. ______

HANDLE WITH CARE!

CONTAINS HAZARDOUS OR TOXIC WASTES

STYLE WM-6

Figure 7.5
Hazardous waste labels–EPA.

waste, while 33% generated between 1 and 12 million tons, and 7% generated less than 1 million tons. In 1993, the states generating the most hazardous waste were Texas, New York, Louisiana, and Michigan. These four states accounted for 70 percent of the national total for large quantity generators.

To be classified as a small quantity generator (SQG) under RCRA, a generator would produce between 220 and 2200 lbs (100 and 1000 kg) of hazardous waste per month. SQGs basically have the same RCRA requirements as LQGs except for some minor differences in storage time limits. SQGs can store hazardous waste on-site for up to 180 days as long as the total weight does not exceed 13,200 lbs (6000 kg). SQGs that ship wastes off-site are still required to start a manifest and to ensure a signed copy is received from the TSD facility within 35 days. SQGs are not required to report their generation statistics to the EPA on a biannual basis (as are LQGs) and therefore their impact on total hazardous waste generation is not as easily documented. It has been noted that, historically, SQGs produce only 1% of hazardous waste in the United States (U.S. EPA, 1987), yet their number exceeds 200,000 (1995a).

Conditionally exempt small quantity generators (CESQGs) produce less than 220 lbs (100 kg) of hazardous waste each month. These generators are not subject to RCRA Subtitle C requirements, but they are required to properly manage their solid waste under RCRA Subtitle D (nonhazardous waste regulations). Because of the potential liability associated with improper disposal of hazardous waste, CESQGs often dispose of their hazardous waste at a RCRA permitted TSD facility. In general, they typically store waste on-site for longer periods of time than larger quantity generators, until they accumulate enough volume to warrant a trip to a disposal site.

7.2.2 Regulation of Hazardous Waste

Subtitle C of RCRA specifically addresses hazardous waste management. When RCRA was promulgated in 1976, it provided the first legislative action which gave the U.S. government authority to regulate the disposal of hazardous waste. RCRA is a "cradle-to-grave" legislation, which means it addresses every phase of hazardous waste management, from creation to ultimate disposal. Subtitle C is divided into six major components (Harris et al., 1987):

- identification and listing of hazardous wastes
- hazardous waste tracking from "cradle to grave"
- standards for hazardous waste generators; transporters; and treatment, storage, and disposal (TSD) facilities
- issuing of permits for TSD facilities
- authorization for states to administer the hazardous waste program
- provisions relating to enforcement

7.2.3 Waste Management Practices

7.2.3.1 Waste Minimization. Source reduction efforts that pertain to hazardous waste include good housekeeping practices by hazardous-waste generators (spill prevention, preventive maintenance, emergency procedures), elimination or replacement of hazardous raw materials such as solvents or plating baths with nontoxic materials, and waste segregation. However, even under the most stringent operating procedures, the production of some amounts of hazardous waste is inevitable in many manufacturing processes, necessitating proper management and disposal.

Because of complex regulations, recycling of hazardous waste is presently done on a very small scale in the United States, accounting for only 1.6% of generated hazardous waste. Recycling is accomplished through solvent recovery, fuel blending, metals recovery, acid regeneration, waste oil recovery, and nonsolvent organic recovery (U.S. EPA, 1995a). Waste minimization, or pollution prevention, however, is a high priority of the EPA as well as hazardous waste generators. Voluntary programs for pollution prevention are encouraged by RCRA, the Pollution Prevention Act of 1990, and by concerns of companies about their potential liability during storage, treatment, and disposal of hazardous wastes under CERCLA, the Comprehensive Emergency Response, Compensation, and Liability Act.

7.2.3.2 Treatment, Storage, and Disposal of Hazardous Wastes. The Resource Conservation and Recovery Act was amended in 1984 to mandate treatment of hazardous waste prior to land disposal. Hazardous waste TSD facilities are regulated and permitted under RCRA Subtitle C and have their own operating requirements for recordkeeping, monitoring and inspection, contingency plans, personnel training, and financial responsibility. Additionally, groundwater monitoring is required of the TSD owner/operator in order to detect and evaluate any migration of contaminants from the property. In 1993, some 3,792 TSD facilities were subject to regulation under RCRA (U.S. EPA, 1995a). These facilities managed 210 million tons of hazardous waste. There was a decline in the number of permitted facilities between 1991 and 1993, and a consequential 84 million-ton decrease in the amount of waste managed at TSDs. More recent data indicate that the number of treatment facilities has continued to decline since 1993 due to public opposition to facility construction, consolidation of existing facilities, phase out of certain chlorinated solvents, and waste minimization initiatives (Woods, 1995).

In 1987, 96% of all RCRA hazardous waste was managed at the site of generation, while only 4% was sent off-site for commercial treatment and disposal (U.S. EPA, 1987). On-site treatment, storage, and disposal is generally practiced at larger facilities that generate sufficient quantities to justify the expense and necessary space for storage and disposal. Smaller firms, and those in crowded urban locations, are more likely to transport their waste off-site where it is managed by a commercial firm or a publicly owned and operated facility. RCRA regulations apply to both on-site and off-site facilities.

Treatment of hazardous wastes involves physical, chemical, and biological processes. Physical processes are used to concentrate contaminants or effect a phase change so the hazardous constituent can be more conveniently processed or disposed. These processes include activated carbon adsorption, stripping, sedimentation, dewatering, and distillation. Chemical treatment actually alters the hazardous constituents through chemical reaction, usually rendering the waste nonhazardous. Chemical processes include oxidation and reduction, neutralization, precipitation, and solidification. Biological treatment is usually applied for the destruction of organic hazardous constituents using activated sludge processes, trickling filters, stabilization ponds, and anaerobic digestion.

Disposal of hazardous wastes generally involves thermal treatment or land application. These processes are described in detail in sections 7.4 and 7.5.

7.3   Overview of Radioactive Waste

7.3.1 Sources, Composition, and Quantity

Exposure to radiation is a natural part of our daily lives. There are three types of radioactive emissions: alpha particles, beta particles, and gamma radiation. Alpha particles are essentially helium nuclei; they are slow moving and easily stopped. They can only travel a short distance in air, and are only dangerous if ingested. Beta particles are essentially free electrons; they are more energetic and can move farther and penetrate deeper than alpha particles. Consequently, this type of radiation is more dangerous and requires shielding by concrete, lead, steel, or water to protect exposed individuals. Gamma radiation is highly penetrating and must be heavily shielded using dense material such as concrete or lead. Sources of natural radiation include cosmic rays, radon gas, sunlight, soil, and rocks. Certain devices also are sources of radiation—color televisions, smoke detectors, and X-ray equipment (medical and security). Approximately 80% of the radiation dose we receive each day comes from natural sources and approximately 20% comes from artificial sources.

Nuclear technology is another potential source of artificial radiation. Such technology contributes significantly to our high quality of life, through power generation, agricultural and industrial research, disinfection, and medical diagnosis. The level of risk from these sources remains low.

For many decades, our national security was ensured through the secretive and complex production of nuclear weapons. This effort required tremendous technological innovation with large budgets but limited time. Thousands of large sites, with tremendous quantities of equipment and dangerous materials, and huge volumes of water (for cooling and shielding) were necessary to achieve worldwide defensive supremacy. More recently, since the pressure of weapons manufacturing has lifted due to the end of the Cold War, it has become apparent that this mammoth effort has resulted in severe environmental consequences related to both hazardous and radioactive contamination of land, water, and equipment. The clean-up of these wastes (as well as wastes generated during processing of radioactive materials for energy and industrial related uses) is a challenge on par with any defense effort in history. The United States spent an estimated $300 billion manufacturing nuclear weapons up until 1991, when the Cold War ended; while the Department of Energy (DOE) has estimated that it will take 75 years and $227 billion to clean up nuclear wastes at U.S. weapons production sites.

Radioactive wastes are managed under separate regulations according to their level of radioactivity. High-level waste (HLW) consists of materials generated during the processing of spent nuclear fuel, wastes from commercial utilities, and defense-related wastes. HLWs are highly radioactive, short-lived fission products and long-lived isotopes. Spent fuel (SF) from nuclear power plants is not actually considered waste but requires careful management because of its high level of radioactivity and potential security threat. A third category, transuranic (TRU) waste, contains elements that are heavier than uranium and are primarily generated by the U.S. defense program. Low-level waste (LLW) includes materials that do not fall in the above categories and, as the name implies, contain materials with low radionuclide content. LLW does not require shielding during management.

To date, approximately three million cubic meters of LLW have been disposed in the United States (U.S. DOE, 1996). LLW is generated by nuclear power plants, universities, hospitals, uranium mining activities, and manufacturers of smoke detectors, watches, and medical devices. LLW consists of rags, contaminated clothing, waste from decontamination and decommissioning, construction debris, filters, and scrap metal. These wastes are short-lived alpha and beta particle emitters and their radiation will generally decline to safe levels within approximately 200 to 300 years.

TRU wastes were created during the design, development, testing, and production of nuclear weapons. Clean-up of weapon-manufacturing sites and decommissioning of weapons will actually produce more TRU wastes in the future. Most TRU wastes consist of material used by employees at national defense plants such as laboratory gloves, shoe covers, plastic bags, tools, dried sludges and other laboratory and production facility waste, but also include contaminated soils and material contaminated during accidents. Approximately 100,000 cubic meters of TRU wastes are being stored across the United States (U.S. DOE, 1996). TRU elements require a very long time to decay. For example, plutonium-239 has a half life of 24,000 years. TRU wastes may emit either alpha, beta, or gamma radiation.

Approximately 30,000 metric tons of spent nuclear fuel (SF) are stored in casks approved by the Nuclear Regulatory Commission (NRC) at some 100 commercial nuclear power reactors across the United States (U.S. DOE, 1996). SF is generated during the routine maintenance of nuclear power plants. Approximately 10 cubic meters of SF are generated per year per reactor. By the year 2000, an estimated 40,000 metric tons of spent fuel and an estimated 8,000 metric tons of defense HLW will have accumulated. HLWs are highly radioactive, heat generating, and often long-lived, and must be placed into special containers prior to transport. Some 30 million liters of HLW are stored in 243 underground tanks in Washington, South Carolina, and Idaho (U.S. DOE, 1996).

7.3.2 Regulation of Radioactive Waste

The United States has struggled since the 1950s with the issues surrounding regulation of radioactive waste disposal. It was in the 1950s that the National Academy of Science and the Atomic Energy Commission first recommended disposal of highly radioactive wastes (HLW and TRU) in deep underground repositories. During the next several decades a search for appropriate sites was made. The Nuclear Waste Policy Act of 1982 (amended in 1987) specified a detailed approach for HLW and SF disposal and assigned the U.S. Department of Energy the responsibility for the transport, storage, and geologic disposal of HLW. The Nuclear Regulatory Commission regulates the management of HLW.

The Low-Level Radioactive Waste Policy Act (LLRWPA) of 1980 (amended in 1985) provided for the formation of regional compacts to manage LLW. These compacts were to be formed by 1986, however the 1985 amendments of the LLRWPA extended the deadline to 1992. Currently there are nine regional compacts encompassing 41 states that provide for the disposal of LLW generated within their boundaries.

7.3.3 Waste Management Practices

7.3.3.1 Waste Minimization. As with solid and hazardous wastes, minimization of radioactive waste is attempted wherever possible. Minimization is accomplished through recycling, separation of radioactive waste from nonradioactive waste, and compaction.

7.3.3.2 Low-Level Radioactive Waste Management. To facilitate the management of the high volume of waste, LLW has been divided into four categories: A, B, C, and greater-than-C waste. Class A waste is the least radioactive and consists primarily of paper, clothing, and rags. The majority (97%) of LLW is Class A. Class B and C wastes are more radioactive than A and include used nuclear power plant filters and solidified liquid wastes. Disposal of Class A, B, and C wastes are the responsibility of the states and generally require a facility design life of 300 years. Greater-than-C wastes include radionuclides which do not decay to safe levels within 100 years. The disposal of these wastes is the responsibility of the Department of Energy.

Transporting of LLW generated during nondefense-related activities is regulated by the Department of Transportation and the NRC. LLWs are packaged in drums, steel boxes, or steel casks, depending on the radioactivity of the waste. As a result of the LLRWPA, a number of active licensed disposal facilities have been constructed and issued permits by the nine regional compacts. Three facilities are located in Barnwell, South Carolina; Hanford, Washington; and Clive, Utah. Sites in Ward Valley, California; Boyd County, Nebraska; Hudspeth County, Texas; and Wake County, North Carolina are under construction. A site at Beatty, Nevada ceased operation on January 1, 1993.

Generally the waste is placed in engineered disposal trenches for shallow land burial (see Figure 7.6). Clay and/or geomembrane bottom liners are provided. Waste is placed in high-integrity containers and concrete overpacks prior to burial. The waste is covered with an engineered cap consisting of layers of low permeability soil and geomembranes. Defense-related LLWs are stabilized through evaporation, encapsulation with polyethylene, vitrification (converted into a glass-like material under high temperature), or mixing with concrete prior to disposal in landfills or concrete vaults.

7.3.3.3 *Transuranic Waste Management.* TRU wastes have been accumulating since the 1940s with the advent of nuclear weapons. Currently, TRU wastes are being temporarily stored either on concrete or asphalt pads under weather resistant structures or in earth-covered berms at ten nuclear defense sites located in California, Colorado, Idaho, Illinois, Nevada, New Mexico, Ohio, Tennessee, South Carolina, and Washington. Waste storage is designed for easy retrieval of containers when final disposal is possible. Disposal of TRU wastes in deep geologic repositories like salt beds was first suggested in the 1950s. Salt is an acceptable medium because its presence testifies to the lack of circulating groundwater. In addition, it is easy to mine and fractures tend to heal in a short time.

In order to demonstrate the feasibility of deep disposal of TRU waste, the DOE has designed and constructed a Waste Isolation Pilot Plant (WIPP) 26 miles east of Carlsbad, New Mexico, in 225-million-year-old bedded salt formations. The WIPP facility is composed of surface buildings; four vertical shafts (for transport of waste, workers, air, and salt); and horizontal underground storage rooms, alcoves, and tunnels. The design provided for nearly 16 kilometers of shafts and tunnels 650 meters below the surface. Construction was completed in 1989; however, testing of radioactive waste continues at existing national laboratories, not at the WIPP facility. Once the repository geology is better understood, radioactive waste will be transported and stored at the WIPP facility. The EPA ruled in May 1998 that the WIPP can safely meet environmental standards for isolation of TRU waste for 10,000 years. The WIPP facility will be operated for the next 20 to 40 years. Ultimately, the site will be closed and the above-ground area will be returned to its pre-construction state.

7.3.3.4 *Spent Fuel and High-Level Waste Management.* There are three options for the management of spent nuclear fuel: store indefinitely

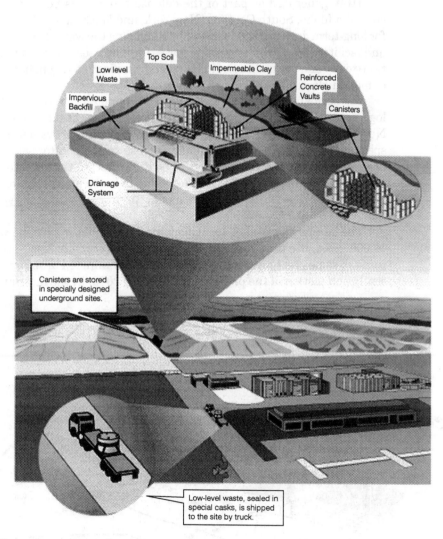

Figure 7.6
Diagram of a low-level waste disposal site.

at the point of generation, reprocess for its fuel value, or isolate the waste in the ocean, polar ice sheets, outer space, or in deep underground repositories. Reprocessing of the SF to chemically recover uranium and plutonium is being done commercially in several countries. However, reprocessing is currently more expensive than mineral mining. In addition, a highly radioactive liquid by-product remains after reprocessing which must be vitrified prior to disposal. SF is currently stored at the reactor site either in pools of water or above-ground casks.

HLW generated as part of the defense program is being temporarily stored in Idaho, South Carolina, New York, and Washington. In preparation for long-term disposal, the DOE will be stabilizing the waste to reduce volume and facilitate transport. Several processes will be used, including lime addition (to dry and neutralize liquid HLW) and vitrification through the addition of molten glass.

For permanent disposal of HLW and SF, remote isolation in a deep geologic repository was recommended as the preferred management option. The Nuclear Waste Policy Act directed the DOE to develop a system to manage spent fuel and HLW (commercial and defense related). After selecting nine possible sites, the NWPA amendment of 1987 directed the DOE to evaluate only Yucca Mountain, Nevada, for suitability to host a monitored, retrievable temporary storage site while a permanent repository is built at the same location (see Figure 7.7).

The site for disposal of spent fuel and HLW must isolate the waste for 10,000 years, the time required for the waste to return to levels that would pose the same risk or less to the public as that of unmined uranium ore. Currently, completion of the project is targeted for 2010. In the interim, the site

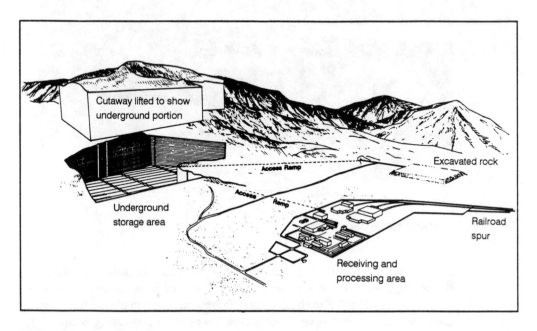

Figure 7.7
High-level nuclear waste storage area conditionally planned for Yucca Mountain, Nevada, will require vast and remote acreage to hold radioactive waste for the next 10,000 years.

is being well characterized to provide information necessary to design the repository and waste packages (for example, the ability of the rock to contain the tremendous amount of heat generated during waste decay). Site characterization costs are estimated at over $6 billion. Another important aspect of the project is the development of a transportation system to move waste to Yucca Mountain safely. The transportation, storage, and geologic deposition of the waste will be regulated by the NRC. The Yucca Mountain facility will be operated by the DOE and funded by taxes on nuclear power use.

7.3.3.5 Contaminated Sites Remediation. Weapons manufacturing has left a legacy of contaminated sites across the United States that will require hundreds of billions of dollars to clean. Waste disposal practices during the 1940s through the 1970s (which were considered acceptable at the time) have actually created extensive problems. These sites present many challenges, and contain wastes that are unique within the environmental remediation industry, including radioactive hazards and huge volumes of contaminated wastes and soils, as well as a large number of contaminated structures such as reactors, evaporating ponds, and manufacturing facilities. For example, the Hanford Site in southeastern Washington contains more than 60% of the nation's HLW generated from the defense program. These wastes are stored in tanks, or were buried in trenches or dumped on the ground. It is estimated that it may take more than 30 years and $50 billion to clean up the billions of cubic meters of contaminated soil and tens of millions of liters of HLW. Similar situations are found at the Savannah River Plant in South Carolina and the Fernald Site in Ohio, as well as other locations. Remediation will require innovative techniques to reduce the waste volume and provide long-term stability.

7.4   Thermal Destruction of Waste

Thermal destruction of waste involves the high-temperature oxidation of combustible matter, typically with the addition of excess air to ensure nearly 100% combustion. Approximately 16% of MSW in the United States is combusted in 148 incinerators (121 of which recover energy). Less than 2% of the hazardous waste generated in this country is treated thermally in commercial incinerators, on-site incineration units, or used as fuel in boilers or industrial furnaces (U.S. EPA, 1995a). Based on a recent census, there were 211 hazardous waste incinerators (HWIs) and 130 boilers/industrial furnaces (BIFs) in the country (1995b). Incinerators are regulated both under RCRA (Subtitles C and D) and the Clean Air Act. A variety of devices are currently available for thermal destruction of waste, including mass-fired and refuse derived fuel (RDF) MSW incinerators, liquid injection incinerators, rotary kiln incinerators, fluidized bed incinerators, infrared incinerators, wet air oxidation, supercritical water oxidation, and vitrification. Figure 7.8 provides a conceptual flow diagram of the incineration process. Figure 7.9 provides a schematic diagram of an integrated hazardous waste incineration facility.

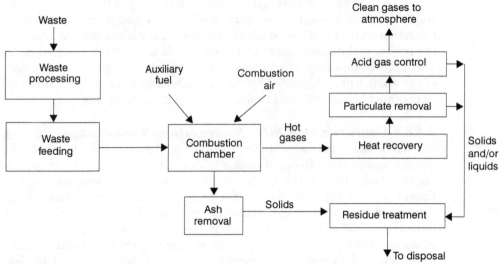

Figure 7.8
Conceptual flow diagram of the incineration process.

The waste incinerator is designed to provide optimum temperature, time of combustion, and mixing of air and waste (turbulence) within the furnace. The waste feeder is designed according to waste characteristics such as form (gas, solid, or liquid), quantity, viscosity, corrosivity, and particle size. Waste is moved through the primary combustion unit using moving or inclined grates, a rotating kiln, spray nozzles, and/or air injection. Movement is designed to control the time and turbulence of waste exposure to air. Temperature as well as gaseous product residence time is regulated by the strategic introduction of combustion air to meet stoichiometric oxygen requirements using forced air or induced draft blowers. Excess air is often added to provide additional oxygen to ensure complete combustion, and to help control temperature and residence time. The refractory-lined furnace is sized to provide adequate volume to retain waste and combustion products for a sufficient amount of time to complete all of the combustion reactions. Typically, the gaseous by-products from the primary combustion unit are directed to a secondary combustion chamber (afterburner) to continue the combustion process. When a low Btu waste is burned it may be necessary to provide supplemental fuel to support the combustion process. In many units, energy recovery is possible for subsequent steam or electric power generation. Energy recovery can be accomplished using closely spaced steel tubes along the primary combustion unit walls and circulating water through them. Alternatively, energy can be recovered using boiler tubes suspended within the furnace with water flowing countercurrent to the direction of hot gas flow. The following series of problems explores the energy and mass balance associated with a MSW incinerator.

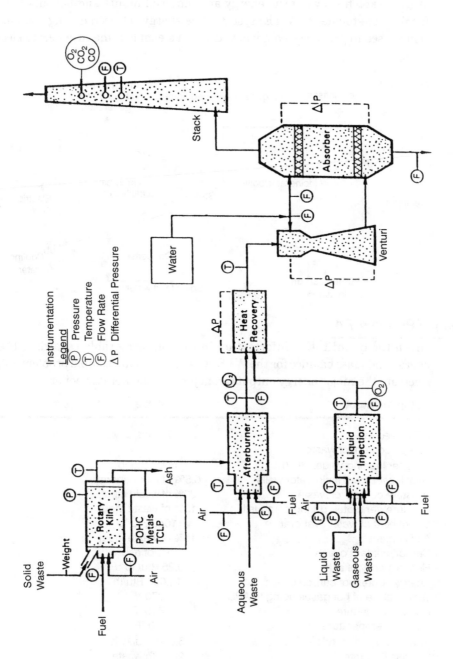

Figure 7.9

An integrated hazardous waste incineration facility. (*Adapted from Oppelt, 1987.*)

Example Problem 7.3

Draw a sketch showing the energy and material inputs and outputs for a MSW waste-to-energy (WTE) system. The energy released during combustion is used to produce steam which drives a steam turbine to generate electricity.

Solution

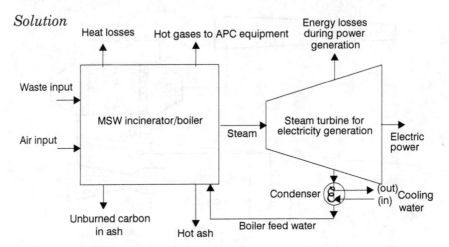

Example Problem 7.4

Given the quantitative information in the following table regarding the energy and mass balance for the WTE system shown in example problem 7.3, calculate the major energy outputs from just the incinerator/boiler.

Input/Output	Value
MSW feed rate	100 tons/day
Energy content of waste (higher heating value, HHV)	5000 Btu/lb
Heat losses from incinerator walls	0.5% of input energy
Moisture content of waste	20% of waste input
Moisture generated during combustion	15% of waste input
Latent heat of water evaporation	1040 Btu/lb
Ash content	15% of waste input
Carbon content of ash	5% of ash
Heat capacity of ash	0.25 Btu/ lb-°F
Energy content of carbon in ash	14,000 Btu/lb
Temperature of hot gases going to APC	370 °F
Grate temperature	850 °F
Ambient temperature	70 °F
Power conversion rate[1]	6,000 Btu/kwh
Exit gas flow rate	20 lb/lb waste
Heat capacity of air and exit gases	0.25 Btu/ lb-°F

[1] Includes turbine and generator mechanical and electrical losses, and heat lost to cooling water.

Solution

1. Calculate energy content (HHV) of incoming waste stream.

    5,000 Btu/lb x 100 tons/d x 2000 lb/ton = 1×10^9 Btu/d

2. Calculate heat losses from furnace walls.

    $0.005 \times 1 \times 10^9$ Btu/d = 5×10^6 Btu/d

3. Calculate heat "loss" to evaporate moisture. Moisture is derived from water in the waste initially (20% of waste) and water is generated as a result of oxidation of waste (15% of waste). Both must be subtracted from the HHV of the waste.

    (0.2 + 0.15) lb water /lb waste x 100 tons/d x 2000 lb/ton = 70,000 lb/d

    70,000 lb/d x 1040 Btu/lb = 7.3×10^7 Btu/d

4. Calculate heat associated with ash leaving incinerator.

    0.15 lb ash/lb waste x 100 tons/d x 2000 lb/ton = 30,000 lb/d

    30,000 lb/d x (850 – 70)°F x 0.25 Btu/°F-lb = 5.9×10^6 Btu/d

5. Calculate energy losses due to the fact that waste is not completely burned and carbon remains in the ash (if the carbon does not combust, it will not release its heating value).

    0.05 lb carbon/lb ash x 30,000 lb/d x 14,000 Btu/lb = 2.1×10^7 Btu/d

6. Calculate heat lost in hot gases.

    20 lb/lb waste x 100 tpd x 2000 lb/t x (370–70) °F x 0.25 Btu/lb-°F
    = 3.0×10^8 Btu/d

7. Calculate heat contained in steam to turbine/generator.

    Heat in steam = Input – Summation of losses =

    $1 \times 10^9 - (5 \times 10^6 + 7.3 \times 10^7 + 5.9 \times 10^6 + 2.1 \times 10^7 + 3.0 \times 10^8) = 6.0 \times 10^8$ Btu/d

Example Problem 7.5

Calculate the energy efficiency of the waste incinerator/boiler described in example problem 7.4.

Solution

$$Efficiency = Useful\ energy/input\ energy$$
$$= (6.0 \times 10^8 / 1.0 \times 10^9) \times 100$$
$$= 60\ \%$$

Example Problem 7.6

Calculate the electric power generation from the steam turbine described in example problem 7.4.

Solution

The energy available for power generation is 6.0 x 10⁸ Btu/d. Calculate the electric power generated.

$$[(6.0 \times 10^8 \text{ Btu/d})/(6{,}000 \text{ Btu/kwh})]1 \text{ day}/24 \text{ hr} = 4170 \text{ kw } (4.2 \text{ MW})$$

Example Problem 7.7

Calculate the overall energy efficiency of power generation from MSW for the WTE system described in example problem 7.4.

Solution

$$\text{\% Efficiency} = (\text{electricity out/energy in}) \times 100$$
$$= 100 \times (4170 \text{ kw} \times 3412 \text{ Btu/kwh} \times 24 \text{ hr/d})/$$
$$(1 \times 10^9 \text{ Btu/d})$$
$$= 34\%$$

Stack emissions from incinerators are stringently regulated under the Clean Air Act. Pollutants of concern include particulate matter (PM-2.5), acid gases (primarily HCl), nitrogen oxides, heavy metals, products of incomplete combustion (such as carbon monoxide and certain organic compounds such as chlorinated dioxins and furans), and principal organic hazardous constituents in the waste feed. These pollutants are controlled using good combustion practice and air pollution control devices. Good combustion practice requires using a good amount of excess air, and providing sufficient time, temperature, and turbulence conditions, with proper monitoring of operational conditions and emission levels. In addition, elimination of certain elements from the waste stream may also minimize the production of undesirable emissions (such as mercury in fluorescent bulbs, thermometers, etc.). Air pollution control devices typically used to control emissions from waste incinerators include wet and dry scrubbers coupled with electrostatic precipitators in older units and fabric filters in newer units. In addition, innovative processes such as catalytic and noncatalytic reduction of nitrogen oxides, and powdered lime and carbon injection followed by fabric filtration for dioxin and mercury control are being used with greater frequency. Many of these air pollution control processes were described in chapter 5.

7.5   Land Disposal of Waste

Of the hazardous wastes generated in 1993, 12.6% were managed by off-site land disposal, primarily using landfills, deep underground injection wells, surface impoundments, and land treatment units (U.S. EPA, 1995a). The remaining hazardous wastes were either incinerated (2%) or treated on-site. Approximately 55.5% of MSW was placed in some 2000 landfills in 1996 (Franklin and Associates, 1998). Land disposal of waste is regulated under Subtitles C and D of RCRA, although many states have adopted more stringent requirements. As noted in section 7.3, all radioactive wastes are placed in repositories either near the surface (LLW) or in deep geologic sites (TRU and HLW). This section will emphasize landfill waste management because of its predominant use in land disposal of hazardous and nonhazardous waste. A **landfill** is an engineered method for land disposal of solid or hazardous wastes in a manner that protects the environment.

Landfill siting requires extensive evaluation of potential locations with consideration of distance from waste generation, climate, surface and groundwater hydrology, topography and geology, seismic activity, distance from airports, and neighborhood acceptance. The size of the site is a function of the amount of waste to be disposed over the life of the landfill and the area required for support facilities such as maintenance and administration buildings, access roads, and ancillary waste management operations (composting, handling of tires, recyclables, handling of bulky items, household hazardous waste collection, construction and demolition debris disposal, etc.). Example problem 7.8 illustrates the basic sizing calculations.

Example Problem 7.8

Determine the area required to provide ten years of landfill disposal capacity for waste generated by a stable population of 250,000. Use the following information:

Per capita daily generation rate:	4.4 lb/cap/d
In place density:	950 lb/yd^3
Average landfill depth:	40 ft

Solution

Total mass of waste = population x per capita generation x landfill life

= 250,000 x 4.4 lb/cap/d x 10 yrs x 365 d/yr

= 4 x 10^9 lb

Volume required = mass of waste/density of waste

= 4 x 10^9 lb/950 lb/yd^3

= 4,226,000 yd^3

= 114,111,000 ft^3

$$\begin{aligned}
\text{Area required} &= \text{volume/depth} \\
&= 114{,}111{,}000 \text{ ft}^3 / \ 40 \text{ ft} \\
&= 2{,}853{,}000 \text{ ft}^2 \\
&= 66 \text{ acres}
\end{aligned}$$

The actual site would be larger than the above calculated landfill area (by as much as 50% or so) to accommodate auxiliary facilities (buildings, roads, equipment, maintenance, storage, etc.), depending on the extent of waste management practiced at the site.

Within a landfill, biological, chemical, and physical processes occur which promote the degradation of wastes and result in the production of **leachate** (polluted water emanating from the base of the landfill) and gases. Thus, the landfill design and construction must include elements that permit control of landfill leachate and gas.

The major design components of a landfill include the liner, leachate collection and management system, gas management facilities, and the final cap. A schematic of a typical landfill is provided in Figure 7.10. The liner system is required to prevent migration of leachate from the landfill and to facilitate removal of leachate. It generally consists of multiple layers composed of natural material (clay or silt) and/or geomembranes. Landfills may be designed with single, composite, or double liners depending on the applicable local, state, or federal regulations. A single liner provides one clay or geomembrane layer. A composite liner consists of two layers, the bottom is a clay material and the top layer is a geomembrane. The two layers of a composite

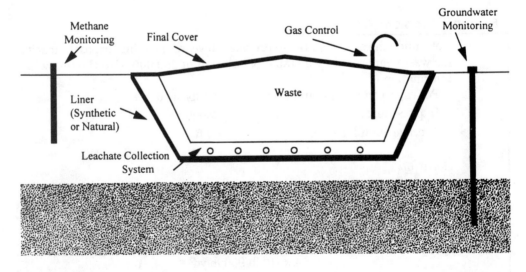

Figure 7.10
Schematic diagram of a typical landfill.

liner are in intimate contact to maximize moisture restriction. A double liner may be either two single liners or two composite liners (or even one of each) separated by a leak detection system—a series of pipes placed between the liners to collect and monitor any water that leaks through the top liner. Clearly, the more layers that are included, the more protective the liner system will be; however, costs will increase dramatically.

Leachate is directed to low points at the bottom of the landfill through the use of an efficient drainage layer composed of sand, gravel, or a geosynthetic net. Perforated pipes are placed at low points to collect leachate and are sloped to allow the moisture to move out of the landfill. Regulations usually restrict the leachate (free liquid) depth on the liner to 30 cm or less.

With the introduction of landfill liners, leachate collection and treatment has become necessary. The return of leachate to the landfill (recirculation) is practiced at an increasing number of biologically active landfills because of advantages associated with enhanced waste stabilization and gas production, improved leachate quality, on-site leachate storage and treatment, and reduced volume due to evaporation. Leachate treatment needs depend upon the final disposition of the leachate. Treatment is often difficult because of high organic strength, irregular production rates and composition, variation in biodegradability, and low phosphorous content (if biological treatment is considered). Generally, where on-site treatment and discharge is used, several unit processes are required to address the range of contaminants present. For example, a leachate treatment facility constructed at the Al Turi Landfill in Orange County, New York, utilizes polymer coagulation, flocculation, and sedimentation followed by anaerobic biological treatment, two-stage aerobic biological treatment, and filtration prior to discharge to the Wallkill River (King and Mureebe, 1992). This was a significant and complex wastewater treatment process that was built just to treat landfill leachate!

Final disposal of leachate may be accomplished through (1) codisposal at a publicly owned treatment works (POTW), (2) on-site treatment followed by direct discharge to a receiving body of water, (3) deep well injection, (4) land application, or (5) natural or mechanical evaporation. Pretreatment of leachate may be necessary to address specific contaminants that could create problems at the POTW providing final treatment. High lime treatment has been practiced at the Alachua County Southwest Landfill in Gainesville, Florida, to ensure low concentrations of heavy metals in the leachate sent to the county POTW.

At an active landfill, daily deliveries of waste are placed in lifts or layers on top of the liner and leachate collection system to depths of 60 meters or greater. Typical waste placement methods used for landfill construction are shown in Figure 7.11. Waste is covered at the end of each working day with soil or alternative daily cover such as textiles, geomembrane, carpet, foam, or other proprietary materials. The landfill sides are sloped to facilitate maintenance and to increase slope stability. Once the landfill reaches design height, a final cap is placed to minimize infiltration of rainwater, minimize dispersal of wastes, accommodate settling, and facilitate long-term maintenance. The cap may consist (from top to bottom) of vegetation and supporting soil, a filter and drainage layer, a hydraulic barrier, foundation for the hydraulic barrier, and a gas control layer.

Because of the prevailing anaerobic conditions within a biologically active landfill, these sites produce large quantities of gas composed of methane, carbon dioxide, water, and various trace components such as ammonia, hydrogen sulfide, and nonmethane volatile organic carbon compounds (VOCs). Tables 7.3 and 7.4 provide typical composition data for MSW landfill gas. Landfill gas is generally controlled by installing vertical or horizontal wells within the landfill. These wells are either vented to the atmosphere (if

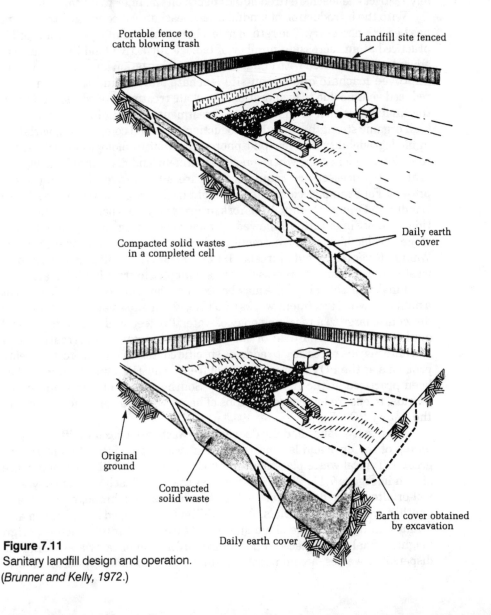

Figure 7.11
Sanitary landfill design and operation.
(*Brunner and Kelly, 1972.*)

gas migration control is the primary intent of the system) or connected to a central blower system which pulls gas to a flare or treatment process.

The gas can pose an environmental threat because methane is a potent greenhouse gas and many of the VOCs are odorous and/or toxic. However, the

Table 7.3 Typical constituents of MSW landfill gas.

Component	% by volume (dry)
Methane	45-60
Carbon dioxide	40-60
Nitrogen	2-5
Oxygen	0.1-1.0
Ammonia	0.1-1.0
Hydrogen	0-0.2

Source: Tchobanoglous, Theisen, and Vigil, 1993.

Table 7.4 Summary of trace organic compounds in Florida landfill gases.

Compound	Concentration in ppb	
	Average	Range
Benzene	240	BDL-470
Chlorobenzene	247	BDL-1978
Chloroethane	161	BDL-796
Cis-1,2-Dichloroethylene	76	BDL-379
Dichlorodifluoromethane	858	BDL-1760
1,1-Dichloroethane	241	BDL-1113
Dichlorotetrafluoroethane	229	BDL-544
Ethylbenzene	5559	2304-7835
M,P-Xylene	8872	3918-12,675
Methylene Chloride	94	BDL-749
O-Xylene	2872	1244-4378
P-Dichlorobenzene	372	BDL-699
Styrene	449	BDL-963
Tetrachloroethylene	131	BDL-561
Toluene	6313	2443-12,215
Trichloroethylene	19	BDL-149
Trichlorofluoromethane	20	BDL-160
Vinyl Chloride	333	BDL-1448
1,2,4-Trimethylbenzene	1328	244-2239
1,3,5-Trimethylbenzene	573	BDL-1140
1,1,1-Trichloroethane	268	BDL-1137
4-Ethyltoluene	1293	265-2646

Note: BDL = Below Detection Limit
Source: Reinhart, Cooper, and Ruiz, 1994.

gas has a high energy content and can be captured and burned for power, steam, or heat generation. Treatment of landfill gas is required prior to its beneficial use. Treatment may be limited to condensation of water and some of the organic acids, or may include removal of sulfide, particulates, heavy metals, VOCs, and carbon dioxide. Some of the more innovative uses of landfill gas include power generation using fuel cells, vehicle fuel (compressed or liquid natural gas), and methanol production.

7.6   Site Remediation

The American industrial revolution produced innovations that vastly enhanced the standard of living of the country's inhabitants. However, the revolution occurred with general disregard for the environmental impacts of these industrial and technological advances. During this time, waste disposal was almost completely unregulated, consequently industry focused on production levels without regard to waste generation. These attitudes towards waste generation prevailed for decades, and the environment eventually began to show signs of mistreatment and neglect.

Several events in the 1950s, 1960s, and 1970s opened the eyes of the American public to the problems associated with historical methods of waste disposal. One of the more publicized hazardous waste contamination problems occurred at Love Canal near Niagara Falls, New York. Improper disposal of hazardous waste at this site caused the evacuation of an entire neighborhood in an effort to reduce health risks from exposure to highly toxic chemicals. This site and its clean-up are discussed in more detail in chapter 8.

7.6.1 CERCLA

Environmental horror stories such as Love Canal were highly publicized throughout the United States and served as a catalyst for congressional reform of pollution control and site remediation legislation. During the 1970s, Congress imposed regulations for the disposal and handling of hazardous wastes and materials. Currently, the most important regulations are the Resource Conservation and Recovery Act and the Comprehensive Emergency Response, Compensation, and Liability Act (CERCLA). RCRA has been described in some detail in previous sections. CERCLA addresses the government monitored clean-up of past hazardous waste sites that are contaminating the environment, and defines the liability of parties responsible for the contamination.

CERCLA, or "Superfund" as it is sometimes called, was signed into law in 1980, with major amendments in 1986 (Superfund Amendments and Reauthorization Act, SARA). The primary purpose of CERCLA is the clean-up of problems associated with the past mismanagement of hazardous wastes. The United States is the only country with legislation of this type. RCRA had no provisions to deal with abandoned hazardous waste sites, and the EPA had no authority to respond to spills, leaks, and explosions until CERCLA was passed.

The heart of CERCLA is the Superfund, which, in essence, is a pool of money generated by special taxes to ensure that funds are available for clean-up efforts. CERCLA granted to EPA the authority to take any necessary short-term and emergency steps to address hazards to human health and the environment that were instigated by burning, leakage, or explosion of hazardous substances; imminent contamination of food chains; or pollution of a drinking-water source (Wentz, 1989). Under CERCLA, the EPA can also undertake long-term actions (duration greater than six months) at a complex hazardous waste site. CERCLA funding comes from taxes on petroleum and 42 listed chemicals, and SARA funding comes from taxes on petroleum and chemicals, corporate environmental tax, general appropriations, parties responsible for site clean-up, and Superfund penalties (Wentz, 1989).

7.6.2 Remedial Action

To be eligible for Superfund dollars, and hence clean-up under CERCLA legislation, a contaminated site must first appear on the National Priority List (NPL). Sites are placed on the NPL following an evaluation through the Hazardous Ranking System (HRS). HRS is a model that assesses the relative risk to the public and the environment from hazardous substances in groundwater, surface water, air, and soil. There are potentially tens of thousands of Superfund sites throughout the United States (Wentz, 1989), but as of mid-1995 only 1231 sites actually appeared on the NPL (U.S. EPA, 1995c). As of the end of 1997, a total of 498 NPL sites had been remediated.

Long-term remedial actions undertaken by the EPA at Superfund sites attempt to find a permanent solution to threats posed by hazardous substances at a location. Any and all wastes that are derived from Superfund clean-up efforts must be treated, stored, or disposed of in accordance with RCRA regulations.

The intent of a remedial response is to define the nature and extent of the contamination, and identify potential remedial alternatives and the best remedial design alternative. There are four steps outlined in the National Contingency Plan which must be completed for a Superfund remedial action:

- site evaluation and scoping of response actions
- initial evaluation and screening of technologies
- detailed evaluation of technologies
- selection of remedy

Following the selection of a remedial technology, the remedial design and subsequent remedial action begins. As of April 1995, the average cost under Superfund was $1.35 million for a remedial investigation/feasibility study, $1.26 million for a remedial design, and $22.5 million for a remedial action. The average operations and maintenance costs for the remedial action, spread out over a 30-year period and expressed in 1994 U.S. dollars, is $5.6 million per year (U.S. EPA, 1995d).

7.6.3 Liability

CERCLA is unique in its definition of liable parties. CERCLA imposes liability on potentially responsible parties (PRPs) for their involvement in the improper disposal of hazardous substances. PRPs may be past owners or operators of the site, generators of the hazardous substances that have polluted the site, and/or transporters who brought the hazardous wastes to the site. PRPs are retroactively liable for hazardous waste problems under CERCLA. In other words, even if waste was disposed in accordance with regulatory requirements in effect at the time of the disposal, and the contamination shows up years later, the PRPs are still liable for the clean-up of the waste. The only other country that imposes retroactive liability is The Netherlands; however, authorities there must demonstrate that the polluter was aware that his/her actions could endanger public health or the environment (American Council, 1992).

CERCLA provides only three limited defenses to liability. These statutory defenses include an act of God, an act of war, or acts of omissions of contractually unrelated third parties where the defendant(s) exercised due care and took appropriate actions in response to the threatened or actual release. United States courts have consistently upheld these provisions as the only defenses available to CERCLA defendants (Freilich, 1992). In addition, the nation's courts have recognized that PRPs are "jointly and severally liable" for clean-up of polluted sites, which means that each PRP is potentially liable for all of the clean-up costs. This ruling often results in those companies with the "deepest pockets" paying the most.

7.6.4 Remediation Technologies

Almost 60% of the remediation technologies applied to Superfund sites involve incineration or solidification/stabilization (the addition of chemicals to immobilize the hazardous contaminants either in place or in a land disposal site). Over 40% of the remaining technologies used are considered innovative. These innovative technologies may be categorized as **ex situ**, involving the treatment of wastes after removal from the site, or **in situ**, meaning that the waste is treated in place. Ex situ treatments include biological treatment of excavated sludges and sediments; pumping of groundwater followed by physical, chemical or biological treatment of the groundwater; soil washing; thermal desorption of contaminants from soils and sediments; and solvent extraction of contaminants from soils and sediments. Biological treatment of soils and groundwater can also be practiced in situ by adding essential nutrients and/or oxygen to the subsurface. Treatment of soils and groundwater is also possible in situ through the introduction of heat, steam, chemicals, or microwaves to the subsurface.

Soil vapor extraction (SVE) and sparging of air into the subsurface are frequently applied to volatile and/or biodegradable subsurface contaminants such as fuels or chlorinated solvents. Soil vapor extraction is being used more and more frequently to remediate fuel leaks and/or spills at gasoline stations. SVE involves the creation of a vacuum in the subsurface through the use of extraction wells, as shown in Figure 7.12. The creation of the vacuum causes movement of air through the contaminated site. Many components of gasoline are quite volatile (gasoline is actually composed of hundreds of compounds) and the vapors are removed by the moving air.

Another possible in situ remediation approach is intrinsic attenuation, where natural phenomena such as dilution, biodegradation, sorption, and

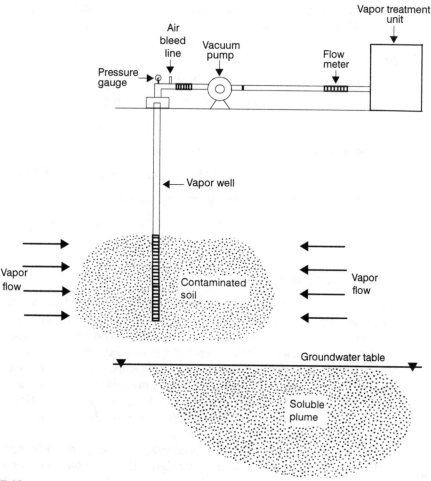

Figure 7.12
Schematic of a typical vapor extraction operation.

advection are permitted to reduce contaminant concentration to levels that do not pose significant risk to human or ecological health.

Contaminated site remediation is presently an exciting and dynamic area of environmental engineering. Ideally, these technologies will become obsolete in the not too distant future, as sites are successfully remediated and pollution prevention serves to eliminate the possibility of contaminated sites in the future.

Example Problem 7.9

Approximately 500 gal of gasoline contaminate the soil underneath a gasoline station. The SVE system air flow is designed at 100 cfm. Calculate the amount of time required to remove 80% of the gasoline if the average "concentration" of gasoline in the extracted air is 1500 mg/m^3. The density of liquid gasoline is 6.6 lb/gal.

Solution

1. Calculate the mass of gasoline in the spill.

$$500 \text{ gal} \times 6.6 \text{ lb/gal} = 3300 \text{ lb}$$
$$= 1.5 \times 10^6 \text{ g}$$

2. Calculate the rate of mass removal.

$$100 \text{ ft}^3/\text{min} \times 0.0283 \text{ m}^3/\text{ft}^3 \times 1500 \text{ mg/m}^3 \times 0.001 \text{ g/mg} = 4.25 \text{ g/min}$$

3. Calculate the time required to remove 1.5×10^6 g.

$$1.5 \times 10^6 \text{ g}/4.25 \text{ g/min} = 353,000 \text{ min (245 days)}$$

7.7 Risk Assessment

Risk assessment is a scientific decision-making process which assists scientists, managers, and regulators in quantifying real or potential harm to human health and the environment. The process can be used to assess accidental and workplace chemical exposure, to evaluate chemical tolerance levels in food, to identify the need and extent of remediation at hazardous waste release sites, to compare various treatment process technologies, and to conduct comparative risk management. The discussion in this section emphasizes the role of risk assessment in setting standards in protecting human health and the environment.

Risk is the probability that a specific negative outcome will occur. Safety is the complement of risk, or the probability that an adverse effect will not occur. Risk is expressed numerically. For example, if 2000 out of every 100,000 motorcyclists die from motorcycling accidents over a lifetime, then the individual lifetime risk of accidental death to motorcyclists is 2000/100,000 or 0.02.

Risk to human health comes from voluntary personal activities, accidents, natural disasters, and chemical exposure. Generally, risks taken during voluntary personal activities (such as driving, smoking, overeating, parachuting, etc.) are freely accepted and generally perceived by the individual to be lower than they really are. Consequently, people are willing to tolerate relatively high risk levels from these activities. Involuntary (externally imposed) risks (such as living near industries, landfills, or incinerators; eating contaminated food; or exposure to radiation) are not willingly accepted and are perceived to be higher than they really are. Intolerance of these imposed risks results in public demand for stringent environmental controls. Table 7.5 provides actual risk levels for a variety of exposures. As can be seen in Table 7.5, actual risk levels for involuntary exposures are much lower than for voluntary exposures, but are typically perceived as being objectionably high by most people.

7.7.1 Risk Assessment Process

Interest in risk assessment began in the 1960s when the process was applied to the weighing of benefits versus side effects of pharmaceutical drugs. The risk assessment process was first applied to hazardous waste site evaluations in the 1980s when the risk assessment process was defined by the National Academy of Science and its use was adopted by the U.S. Environmental Protection Agency in guidelines published in 1989. Since that time risk assessment has evolved to become a sophisticated tool in hazardous waste management. Although the process has many limitations and uncertainties (see section 7.7.3), it has become an accepted and pervasive tool in setting standards and clean-up goals and will play an increasingly important role in the future.

Table 7.5 Annual risk of death for common activities.

Activity	Annual Risk
Smoking 10 cigarettes/day	1.25×10^{-3}
Motor vehicle accident	2×10^{-4}
Manufacturing work accident	8×10^{-5}
Pedestrian hit by automobile	4×10^{-5}
Drinking two beers/day (cirrhosis only)	4×10^{-5}
Person in a room with a smoker	1×10^{-5}
Eating peanut butter (4 teaspoons/d)	8×10^{-6}
Drinking water with EPA limit of trichloroethene	2×10^{-9}

Sources: Rowe, 1977:290, 354, 356; Wilson and Crouch, 1987: 267–270.

Risk is a product of two elements—toxicity and exposure. Reduce or increase either component and risk decreases or increases as well. The risk assessment process, described in Figure 7.13, involves four steps that are necessary to evaluate the degree of chemical toxicity and to quantify potential or actual exposure:

- hazard identification
- dose-response assessment
- exposure assessment
- risk characterization

Hazard identification (or toxicity assessment) determines whether exposure to a chemical, physical, or biological agent can cause an increase in the incidence of an adverse effect. A hazard does not create risk by itself, but its presence is a necessary condition for a health or safety risk. Hazard identification involves establishing the physical, metabolic, and chemical properties of the agent; potential routes of exposure; toxicological effects; results of animal studies; and site characteristics.

The **dose-response assessment** evaluates the relationship between the level of exposure and the extent of injury. Responses are generally divided between carcinogenic and noncarcinogenic effects. Responses are typically evaluated by exposing animals to various doses of agents, however more recent procedures have utilized Salmonella cells, specific organ cells, and animal eggs or embryos. Responses vary from death to such problems as tumors,

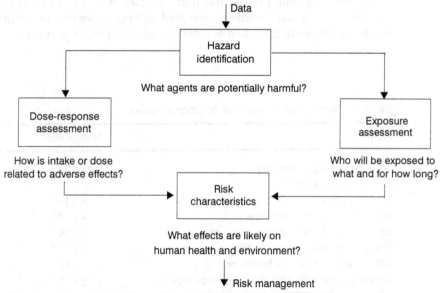

Figure 7.13
Multi-step risk assessment process.

skin irritation, respiratory effects, genetic mutation, and fetal development problems. Dose is generally expressed as the mass intake of the chemical normalized to the body weight of the exposed individual and the time period of exposure. The dose-response relationship is illustrated as a plot of response (fraction of exposed population) versus dose. Typical response curves for carcinogenic and noncarcinogenic agents are presented in Figure 7.14 (a and b). Note that for carcinogenic agents, it is assumed that a threshold does not exist; that is, a cancer risk is present at all doses. For noncarcinogenic agents, it is assumed that a dose below the "no observable adverse effect level" (NOAEL) does not cause harmful effects.

A cancer slope factor (CSF) is determined from the dose-response curve as shown in Figure 7.14a (see also Table 7.6). The CSF is developed by assuming a linear relationship between dose and response at low dosages and is an upper boundary estimate of risk. For noncarcinogens, a reference dose (RfD) is determined from the noncarcinogenic dose-response curve using the NOAEL. To calculate the RfD, the NOAEL is divided by safety factors (typically multiples of ten) to acknowledge the uncertainty associated with the extrapolation of response data from animals, the variability in animal and human response, and extrapolation from relatively high doses to low environmental doses, among other factors. Typical RfDs and CSFs for many chemicals are maintained in various databases for use during risk analysis.

The **exposure assessment** calculates the actual or potential dose that an exposed individual receives and delineates the affected population by identifying possible exposure paths. The process measures or estimates the intensity, frequency, and duration of human or ecological exposure to the agent or agents. The exposure assessment answers questions regarding the release routes from the site; attenuation along the transport pathways; concentrations in the air, drinking water, food, soil, and dust; body burdens in the pop-

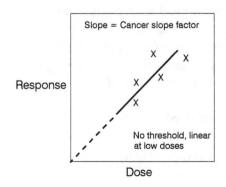

Figure 7.14a
Carcinogenic dose-response curve.

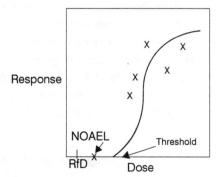

Figure 7.14b
Noncarcinogenic dose-response curve.

Table 7.6   Toxicity data for selected potential carcinogens.

Chemical	Cancer Slope Factor oral route $(mg/kg/day)^{-1}$	Cancer Slope Factor inhalation route $(mg/kg/day)^{-1}$
Arsenic	1.75	50
Benzene	2.9×10^{-2}	2.9×10^{-2}
Benzo(a)pyrene	11.5	6.11
Cadmium	—	6.1
Carbon tetrachloride	0.13	—
Chloroform	6.1×10^{-3}	8.1×10^{-2}
Chromium VI	—	41
DDT	0.34	—
1,1-Dichloroethylene	0.58	1.16
Dieldrin	30	—
Formaldehyde	—	4.5 E-2
Heptachlor	3.4	—
Hexachloroethane	1.4×10^{-2}	—
Methylene chloride	7.5×10^{-3}	1.4×10^{-2}
Nickel and compounds	—	1.19
Polychlorinated biphenyls (PCBs)	7.7	—
2,3,7,8-TCDD (dioxin)	1.56×10^{-5}	—
Tetrachloroethylene	5.1×10^{-2}	$1.0–3.3 \times 10^{-3}$
1,1,1-Trichloroethane (1,1,1-TC)	—	—
Trichloroethylene (TCE)	1.1×10^{-2}	1.3×10^{-2}
2,4,6-Trichlorophenol	1.10E-02	1.10E-02
Toxaphene	1.1	1.1
Vinyl chloride	2.3	0.295

Source: U.S. EPA IRIS database, 1989.

ulation; and work, play, and consumption habits of the population. Exposure routes include ingestion, inhalation, and dermal absorption. Multi-media computer models are available to predict the fate of chemical or biological agents in the environment and determine exposure concentrations. To quantify the rate of uptake by an individual, field measurements are often made where possible. If data are unavailable, typical exposure and uptake rates are frequently used as detailed in Table 7.7.

From exposure data, a daily dose is calculated as shown in equation (7.1).

$$\text{Daily Dose (mg/kg-day)} = \frac{(C)(I)(EF)(ED)(AF)}{(AT)(BW)} \qquad (7.1)$$

where:

   C   = Concentration, mass/volume

I   = Intake Rate, volume/time

EF   = Exposure Frequency, time/time

ED   = Exposure Duration, time

A   = Absorption Factor, unitless

AT   = Averaging Time, time

BW = Body Weight, mass

Example Problem 7.10

Calculate the Lifetime Average Daily Dose (LADD) for an adult male from drinking water containing 10 μg/L of chloroform. Assume the individual is exposed for 30 years, drinks two L/day of water, 7 days/wk, and weighs 70 kg. LADDs are daily doses calculated over an average lifetime of 70 years. Assume 100% of the chloroform is absorbed.

Solution

$$\text{LADD} = \frac{\left\{(10 \ \mu g/L)(2L/day)(7d/wk)(30 \ yrs)(1)\right\}}{(1000 \ \mu g/mg)(70 \ kg)(70 \ yrs)}$$

$$\text{LADD} = 8.57 \times 10^{-4} \ \text{mg/kg-wk}$$

$$\text{LADD} = 1.22 \times 10^{-4} \ \text{mg/kg-d}$$

Table 7.7 Typical exposure and intake rates for risk assessment.

Variable	Value
Soil Ingestion Rate	200 mg/day (children 1 through 6 years old)
	100 mg/day (age groups greater than 6 years old)
Exposure Duration	30 years at one residence (national 90th percentile) (adult)
	6 years (child)
Body Weight	70 kg (adult)
	10 kg (child)
Averaging Time	for noncarcinogens, exposure duration
	for carcinogens, 70 years
Water Ingestion Rate	1 L/d (child, age 1–6)
	2 L/d (adult)
Inhalation Rate	0.83 m³/hr (adult)
	0.46 m³/hr (child)

The **risk characterization** estimates the probability of adverse incidence occurring under the conditions identified during the exposure assessment. For carcinogens, the daily dose (developed during the exposure assessment) is multiplied by the CSF developed from dose-response curves to calculate the added risk of cancer. Risks are additive for multiple carcinogenic contaminants. For noncarcinogens, a hazard index (HI) is calculated by dividing the daily dose by the RfD (developed from dose-response assessments). Where more than one contaminant is present a hazard quotient (HQ) is determined by summing all of the HIs. A HQ value greater than one indicates an unacceptable risk.

Example Problem 7.11

Calculate the added carcinogenic risk posed by the chloroform consumption of example problem 7.10 using a CSF for chloroform of 0.0061 $(mg/kg\text{-}day)^{-1}$.

Solution

Risk = Dose x Toxicity = Intake Rate x CSF

Risk = $(1.22 \times 10^{-4}$ mg/kg-day$)(0.0061$ $(mg/kg\text{-}day)^{-1})$

Risk = 7.4×10^{-7}

Risk = 1 additional cancer per million people

In example problem 7.11 a lifetime risk of an additional 1 cancer case per million exposed individuals was calculated as a result of consuming the contaminated water. This risk is in addition to background incidence which is estimated to be 300,000 per million individuals (Ames and Gold, 1992).

Example Problem 7.12

Chloroform also has a noncarcinogenic risk. Calculate the hazard index for chloroform exposure calculated in example problem 7.10 using a reference dose for chloroform of 0.010 mg/kg-day.

Solution

HI = Intake Rate/RfD

HI = $(1.22 \times 10^{-4}$ mg/kg-day$)/(0.010$ mg/kg-day$)$

HI = 0.0122

Since the HI is well below 1, little noncarcinogenic risk is present.

7.7.2 Risk Management

Risk management refers to making decisions using the results of the risk assessment. The risk is evaluated in light of the cost of action and such factors as the technical feasibility of reducing risk, the size of the population involved, the possibility of increasing risks during the response action (for example exposure to dust during soil excavation), statutory mandates, and public concern. The resulting decision may indicate the level of clean-up desired at an uncontrolled hazardous waste site, an acceptable drug dosage, a treatment standard for potable water, or an acceptable location for an industry. Generally, a carcinogenic risk between 10^{-4} and 10^{-6} is deemed acceptable by the U.S. EPA. For noncarcinogens, a HQ or HI greater than one indicates that an unacceptable risk exists.

7.7.3 Uncertainty in Evaluating Risks

The risk assessment process is extremely conservative in nature and utilizes measurements that are uncertain. In addition, insufficient data or information gaps often exist in characterizing the potential risk of an agent, necessitating the need for assumptions or educated guesses. Risk is generally expressed as a single number because it is easier to present to the public in that form. However, it would more properly be represented by a numerical range. For example, the risk calculated in example problem 7.11 to be 1 out of every million exposed individuals may be somewhere between zero and 1 (and perhaps closer to zero than 1). A recent trend has been to use computational tools from the field of decision analysis (for example, Monte Carlo simulations) to account for the uncertainties in the process. These tools allow risk to be expressed as a probability distribution rather than a single number that can then be used to make a more informed decision during risk management.

End-of-Chapter Problems

7.1   A city has a stable population of 250,000 and is running out of room at its MSW landfill. Estimate the land area (in acres) that must be set aside for a new landfill given the following conditions: MSW generation rate = 4.0 lb/cap/day, density of compacted MSW in the landfill is 800 lb/yd^3, average depth of waste in the landfill is 50 ft. Assume that 50% of the total land area must be used for roads, buildings, and buffer zones. Design for a landfill life of 20 years.

7.2   Estimate the potential annual electricity generation (in units of kwh per year) from a MSW incinerator with turbine/generator that burns all the MSW from the city in problem 7.1 above. Assume that the MSW has a heat content of 4500 Btu/lb and that an overall thermal efficiency of 30% can be achieved. If an average household uses 1500 kwh per month, how many households can be powered by this MSW incinerator?

7.3 Find the critical distance from a city to a landfill. The critical distance is that distance at which it becomes economical to buy and use long-haul trucks to take the MSW to the landfill (in this alternative, the city also must build and operate a transfer station in the city) rather than using the city's MSW collection trucks to drive the extra distance to the remote landfill. Use only the following data to make your estimate. Collection trucks hold 25 yd^3 of MSW and get 4 miles/gal of fuel. Long-haul trucks hold 100 yd^3 of waste and get 10 miles/gal. Fuel costs $1.25/gal. The city presently operates 30 MSW collection trucks. To build and operate a transfer station will cost the city $500,000 per year, and the annualized cost of owning and operating long-haul trucks is $75,000/yr per truck plus fuel costs. The city already owns the collection trucks, but assume that the extra labor, maintenance, and depreciation costs associated with driving them the extra distance to the remote landfill will add another $15,000 per year per truck to that alternative. Assume the city operates the trucks 6 days/week, 52 weeks/yr. Both types of trucks will make one trip per day to the landfill.

7.4 Use the Internet to identify and research a Superfund site. Summarize your findings: Is it still on the National Priority List? What were the costs? What impacts were there on the surrounding area? What technologies were used? How long was the process?

7.5 Discuss the pros and cons of waste recycling and reuse considering the environmental and economic implications.

7.6 Calculate the time required for 100,000 m^3 of TRU waste to decay to the point that only 1000 m^3 remains radioactive. Consider the waste to be composed of plutonium-239, which has a half-life of 24,000 years.

7.7 If you were required to start up a hazardous waste management program in a country that had no environmental regulations, would you follow the approach to define hazardous wastes used in the United States (i.e., list the waste and characteristics) or would you develop a different approach? Why or why not?

7.8 How would you convince your community to accept a MSW landfill, hazardous waste treatment facility, or other "undesirable" land use?

7.9 Consider the manufacture of the automobile. List possible sources of hazardous wastes, ways to reduce the toxicity of the waste, and ways to improve the recyclability of the auto after its useful life is over.

7.10 Determine the recycling efficiency of a community with the waste composition shown in Figure 7.2 and which recycles 10% of its plastics, 50% of metals, and 30% of its paper and paperboard products.

7.11 Compare the efficiency of a fleet of 20 yd^3 vs. 30 yd^3 collection vehicles to pick up a community's waste by calculating the total number of hours required to collect the waste per week for each type truck and the number of trucks of each type required. Use the following data: 25,000 locations to pick up, 2.4 min/loc to pick up (includes drive between location time), compaction ratio is 2 for the 30 yd³ truck and 2.5 for the 20 yd³ truck, collection once per week, 0.24 yd³/location, 20 min. to and 20 min. from landfill, 12 min. at the landfill, and a 40-hr work week.

7.12 A community with 100 homes is considering switching from its present hauled container system to a weekly pickup system by a compacting vehicle. Compare the total labor requirements for the two systems. Use the following data: five 9-yd³ containers clustered together are presently filled to 70% of their capacity on average. The hauling truck spends 12 minutes at the landfill, which is 20 minutes from the community. It takes 4 minutes to load each container. It takes 30 minutes to get to and from the community from the dispatch site. The compactor vehicle is 27 yd³, has a compaction ratio of 2, and it takes 1 min/container to load. The compactor truck spends 6 minutes at the landfill per trip. The compactor truck has a crew of 2, the haul truck only has a driver.

7.13 A groundwater is contaminated with 0.1 mg/L of trichloroethylene. Determine the lifetime cancer risk for an adult associated with utilization of this groundwater as a potable supply. Risk assessment data for trichloroethylene are provided in Table 7.6. The cancer slope factor for trichloroethylene is 0.011 (mg/kg-d)$^{-1}$.

7.14 For the groundwater described in problem 7.13 above, determine the removal efficiency required to achieve compliance with the maximum contaminant level (MCL). Assuming that treatment is provided to reduce the concentration to the MCL, determine the lifetime cancer risk associated with utilization of this groundwater as a potable supply.

7.15 Calculate the number of additional deaths per year occurring in a population of 250,000,000 from motor vehicle accidents. Calculate the number of deaths over an average lifetime.

7.16 Calculate the cumulative risk associated with exposure to the carcinogens chloroform, dioxin, and 1,1,2-trichloroethane in drinking water using the information provided below.

Carcinogen	Cancer Slope Factor, (mg/kg-d)$^{-1}$	Concentration, μg/L
Chloroform	0.0061	10
Dioxin	4.25 x 10⁵	0.001
1,1,2-Trichloroethane	0.057	100

7.17 Calculate the required action level in mg/L for treatment of drinking water containing the carcinogen aldrin if the acceptable risk level is 10^{-6}. The CSF for aldrin is 17 (mg/kg-d)$^{-1}$.

References

American Council for Capital Formation. 1992. "Study Calls U.S. Hazardous Waste Laws Much More Stringent." *Journal of the Air and Waste Management Association* 42(1): 10–11.

Ames, B. N., and L. S. Gold. 1992. "Environmental Pollution and Cancer: Some Misconceptions." In *Rational Readings on Environmental Concerns*, edited by J. H. Lehr. New York: Van Nostrand Reinhold.

Brunner, D. R., and D. J. Kelly. 1972. *Sanitary Landfill Design and Operation*. Washington, DC: Environmental Protection Agency, Office of Solid Waste Management.

Clay, Don R. 1991. "Ten Years of Progress in the Superfund Program." *Journal of the Air and Waste Management Association* 41(2): 144–147.

Fortuna, R. C., and D. J. Lennett. 1987. *Hazardous Waste Regulation: The New Era*. New York: McGraw Hill.

Franklin and Associates. 1998. "Characterization of Municipal Solid Waste in the United States—1996 Update." Prepared for the U.S. EPA, Municipal and Solid Waste Division, Office of Solid Waste. EPA 530-S-98-007.

Freilich, Irvin M. 1992. "Causation Becomes a Factor Under CERCLA." *Journal of the Air and Waste Management Association* 42(10): 1274–1275, 1392.

Harris, C., William L. Wart, and Morris A. Ward. 1987. *Hazardous Waste: Confronting the Challenge*. New York: Quorum Books.

King, D., and A. Mureebe. 1992. "Leachate Management Successfully Implemented at Landfill." *Water Environment & Technology* 4(9): 18.

Oppelt, E. Tim. 1987. "Incineration of Hazardous Waste: A Critical Review." *Journal of the Air Pollution Control Association* 37(5).

Reinhart, D. R., C. D. Cooper, and N. E. Ruiz. 1994. *Estimation of Landfill Gas Emissions at the Orange County, Fl Landfill*. Civil and Environmental Engineering Department, University of Central Florida.

Rowe, W. D. 1977. *An Anatomy of Risk*. New York: Wiley Interscience.

Tchobanoglous, G., H. Theisen, and S. Vigil. 1993. *Integrated Solid Waste Management Engineering Principles and Management Issues*. New York: McGraw-Hill.

U.S. Department of Energy, Office of Environmental Management. 1996. *Closing the Circle on the Splitting of the Atom*. DOE/EM-0266, Washington, DC.

U.S. Environmental Protection Agency. 1987. "The Hazardous Waste System." Office of Solid Waste and Emergency Response, Washington, DC.

U.S. Environmental Protection Agency. 1989. Integrated Risk Information (IRIS) database.

U.S. Environmental Protection Agency. 1995a. "The Preliminary Biennial RCRA Hazardous Waste Report-Executive Summary." Office of Solid Waste and Emergency Response, Washington, DC (March).

U.S. Environmental Protection Agency. 1995b. EPA RCRA "Hotline." "National Oversite Query for Counts of RCRA Subtitle C Handlers and Counts of Facilities with Combustion." Washington, DC. (June 16).

U.S. Environmental Protection Agency. 1995c. EPA Superfund "Hotline." Washington, DC. (June 23).

U.S. Environmental Protection Agency. 1995d. *U.S. Federal Register.* 60FR20330, pp. 20334.

Wentz, Charles A. 1989. *Hazardous Waste Management*. New York: McGraw Hill.

Wilson, W. R. and E. A. C. Crouch. 1987. "Risk Assessment and Comparisons: An Introduction." *Science* 236:267–270.

Woods, Randy. 1995. "Studies Show Number of Hazardous Waste Facilities on Decline." *Waste Age* 28(6): 35.

Chapter 8

```
┌─
│
│  Case Studies
│
└─
```

8.1   Introduction

Engineering is both an art and a science. As such, not everything is learned from theory and first principles; engineers learn much from past experiences. Through open and honest communication, engineers document and use their own accomplishments and mistakes, as well as the triumphs and failures of others, to advance the knowledge of the entire profession. Thus, other engineers avoid repeating similar mistakes and can use past successes to build even better systems in the future. This chapter strives to take advantage of this experiential learning process through case studies. The successful parts of these case studies should be used not as blueprints to be followed exactly, but as general guidelines for future creative efforts.

8.2   Biomedical Waste Incineration and Pollution Prevention in Hospitals

Background

From about 1950 through 1990, the number and size of full-service hospitals grew markedly as part of the explosive growth of the health care industry. Large hospitals produce a variety of medical wastes (such as pathological tissue samples, blood and body fluid wastes, various chemicals, used syringes, and broken glass) that, if released without being properly treated, could contaminate the environment and pose a threat to public health. In many hospitals, the generation of excessive paper and plastic wastes in addition to medical wastes gave rise to serious concerns about properly managing and disposing of those wastes. In the 1980s, the issue of disposal of medical wastes received national media attention when discarded syringes and other medical wastes washed up onto beaches in several states.

Hospitals generate three types of solid waste: ordinary wastes, biomedical or "red bag" wastes, and small amounts of hazardous wastes. Ordinary wastes include such things as paper, cardboard, plastic wraps, and food waste. The biomedical waste category includes items such as used syringes, used catheter tubing, bloody bandages or human blood and/or body fluids, nonliquid human tissue, and any waste from rooms that house patients with infectious diseases. Hazardous wastes from hospitals are items such as chemotherapy chemicals, formaldehyde, radionuclides, x-ray film chemicals, mercury, and certain solvents. At most hospitals in the 1980s, hazardous wastes were being handled separately and appropriately, but often the biomedical and ordinary wastes were mixed together, and handled inappropriately.

For many hospitals, the answer seemed to be incineration. Since hospitals typically use large amounts of steam, burning all the wastes on-site and recovering some of the heat to generate steam seemed to make a great deal of sense. Initially, there was little or no thought given to air pollution control, despite the fact that most hospitals were located very close to houses and businesses. Research on emissions from hospital waste incinerators published in the 1980s provided ample evidence that significant concentrations of a variety of serious air pollutants were present in incinerator exhaust gases (Walker and Cooper, 1992). This led to the eventual shutdown of a number of hospital incinerators, or their retrofit with modern air pollution control (APC) equipment.

Incinerator emissions are tied to (1) the type of materials entering the combustion chamber, (2) the design, operation, and maintenance of the incinerator, and (3) the type and efficiency of the APC equipment. Thus, control of certain air pollution emissions (e.g., mercury) can be effected by prevention; that is, by not putting certain things (e.g., mercury thermometers) into the incinerator, as well as by pollution control (e.g., a baghouse with lime and carbon injection). Because of the cost of sending wastes off-site for incineration and disposal, there was a significant effort at hospitals to minimize their wastes and to manage them better. Pollution prevention (carefully managing the purchase, use, and disposal of materials) is usually more cost effective than pollution control and certainly better than site remediation (U.S. EPA, 1990). Green (1992) expands on the concept of pollution prevention with respect to medical wastes.

The Case

General Hospital in Orlando, Florida, has about 1000 patient beds and is a typical full service hospital. It is located in the heart of a commercial and residential area, and is bounded on the east and west by two busy arterial roadways and by single family homes to the north. In 1991, General Hospital was operating a 15-year-old incinerator with a waste heat boiler. They burned all their nonhazardous waste, a mixed stream of trash and biomedical wastes, and generated steam for use in the hospital. The incinerator had no air pollution control equipment.

As of 1991, General Hospital was generating about 18,000 pounds per day of solid waste, about one-third of which was red bag waste. This amount strained their existing incinerator, and often resulted in 30 to 40 bags of waste piled in open-top dumpsters awaiting loading into the burner. There had been several complaints from people in the immediate vicinity about smoke during the daytime and noise from the incinerator at night.

As new air pollution regulations for medical waste incinerators were passed in Florida, hospital management was faced with the problem of what to do about the old incinerator and, consequently, the hospital's wastes. They could (1) cease waste incineration and pay for off-site disposal of all wastes, (2) burn only the true red bag wastes and haul the ordinary wastes to a land-fill, or (3) continue to burn all the wastes. The latter two options required the addition of a modern air pollution control system, and the third option might require a new incinerator, as well.

The Analysis

Management at General Hospital commissioned a consulting firm to study and delineate its options. The consultants started with a review of the liter-ature to see how similar hospitals across the country had solved their waste-handling problems. They spent considerable time characterizing the sources and amounts of all the waste streams at General Hospital, and found that much of the ordinary waste was getting mixed in with the red bag wastes. Workers were not doing a good job of segregating the waste streams, and supervisors were not doing a good job of training the workers. The existing incinerator was inspected and found to need considerable maintenance; the existing stack and surrounding area on the roof of the building were badly corroded.

The stack gases were tested and found to contain significant amounts of carbon monoxide, hydrogen chloride, particulate matter, and some small amounts of unburned hydrocarbons and heavy metals. On the plus side, there were no viable organisms, either in the gases or in the incinerator ash. There was an occasional problem with the incinerator producing visible smoke, and frequent problems with odor near the waste bag storage area.

Costs were obtained from vendors for landfill disposal of ordinary waste, and for transport and off-site incineration of biomedical waste. At that time, the cost for proper disposal of General's red bag wastes was estimated as $525,000 per year, with landfilling of their ordinary waste adding another $190,000 per year.

The Solution

The first option—dispose of everything off-site—would cost more than $700,000 per year and would leave the hospital vulnerable to future price increases and/or possible liability problems for accidents that could occur during waste transport. It would also mean purchasing more fuel gas to gen-erate the steam the hospital needed to run its other operations. Option three—burn everything—would require a new, larger incinerator with its consequent large capital cost. In addition, space was somewhat limited, so constructing the new incinerator while keeping the old one running was not possible. Thus, a new incinerator would impose a big disruption on hospital operations. Therefore, the second option—burning only the red bag waste and landfilling the ordinary waste—was selected.

At the same time, administrators at General Hospital developed a comprehensive waste management plan. They instituted a good training program to help workers minimize unnecessary waste generation, and to keep such things as lead, cadmium, and mercury wastes out of the incinerator. Office and kitchen workers were educated about the problem to help ensure that all office and kitchen wastes were kept with the ordinary solid wastes. An aluminum can recycling program was started. Furthermore, hospital purchasing practices were examined to try to find acceptable substitutes for things that were causing emissions problems in the incinerator. For example, mercury thermometers were replaced with electronic digital thermometers that did not contain mercury. PVC plastic containers were replaced with high-density polyethylene ones. (PVC contains a significant percentage of chlorine, which was contributing to the high HCl content of the stack gases and the corrosion problems around the stack.)

Despite the best management practices to reduce total waste generation and to burn only those wastes necessary, the exhaust gas emissions still contained enough pollutants to require a modern air pollution control system. The system chosen was a pulse-jet baghouse with lime (calcium oxide) and powdered carbon injection. The lime neutralized the HCl and the carbon adsorbed any residual mercury vapors as well as trace amounts of unburned hydrocarbons.

After several years of operation, the baghouse continues to operate well, capturing 99.9% of the particulate matter, including all the lime and carbon. Mercury emissions are often nondetectable during emission tests. The furnace ash and collected baghouse ash are mixed, and the resulting stabilized ash is sent to an ordinary landfill. Because of the reduced load on the incinerator, it operates better, thus reducing the amounts of CO and unburned hydrocarbons being emitted. In fact, the load is now low enough that the unit is shut down from 9:00 P.M. to 7:00 A.M., thus reducing the number of noise complaints from neighboring residences.

8.3   Landfill Leachate Treatment at a Publicly Owned Wastewater Treatment Plant

Background

Leachate is produced at municipal solid waste (MSW) landfills as a result of water (either from infiltrating precipitation or from the moisture contained in the waste) moving through disposed waste. As moisture comes into contact with waste, material dissolves or is entrained in the leachate, producing a highly contaminated liquid. With the promulgation of federal MSW landfill liner requirements, leachate collection and treatment has become necessary in many areas of the country. One management approach is to treat leachate at a nearby publicly owned treatment facility along with municipal wastewater.

The Case

A new cell at the Orange County (Florida) landfill was opened in 1991. Soon after it was opened, a series of very intense rainstorms created much larger leachate volumes than were expected during design of the landfill. To accommodate these large volumes, a force main was installed to deliver leachate to the Eastern Service Area Wastewater Treatment Facility (ESAWTF) operated by the Orange County Public Utilities Division. This facility uses an activated sludge process to remove organic pollutants as well as nutrients (phosphorous and nitrogen). Four attempts were made to treat this leachate during late 1991 and early 1992. Each time leachate was introduced, nitrogen concentration in the ESAWTF effluent increased significantly (to greater than 3.0 mg/L). Also, specific operational problems related to nitrification were observed when leachate was introduced. Due to the sequential nature of biological nitrification and denitrification, it was not possible to determine the leachate's impact on denitrification. Plant records, however, showed that the removal of BOD, suspended solids, and phosphorus was not affected by the addition of leachate. A study was then conducted to investigate factors that might have contributed to high levels of effluent nitrogen and to recommend an operating strategy which would help the plant meet effluent limitations while treating leachate.

The ESAWTF provides advanced treatment of municipal wastewater using a modified Bardenpho process for the biological removal of nitrogen and phosphorus, as shown in Figure 8.1. At the time of the study, the plant was permitted at a capacity of 51,100 m^3/d (13.5 MGD) with annual average effluent limits of 5.0 mg/L BOD$_5$, 5.0 mg/L total suspended solids, 3.0 mg/L nitrogen, and 1.0 mg/L phosphorus. The facility is operated as two parallel five-stage treatment trains, one with a capacity of 22,700 m^3/d (6.0 MGD) and the other with a capacity of 28,400 m^3/d (7.5 MGD). The two trains share a common pretreatment structure for bar screening and grit removal. Each of the two treatment trains includes a fermentation basin (no oxygen), first- and second-stage anoxic basins (nitrate serves as the electron acceptor), racetrack-type aeration tanks, reaeration tanks, secondary clarification, tertiary filters, and chlorination. Alum is added at the reaeration stage in order to polish the liquid prior to clarification and insure phosphorus levels in the effluent are maintained below 1.0 mg/L. ESAWTF effluent is disposed using one of three methods: recharge of groundwater using rapid infiltration basins, discharge to natural and artificial wetlands, or reuse as cooling water at a nearby coal-powered electrical generation facility. Tank capacities for the plant are provided in Table 8.1.

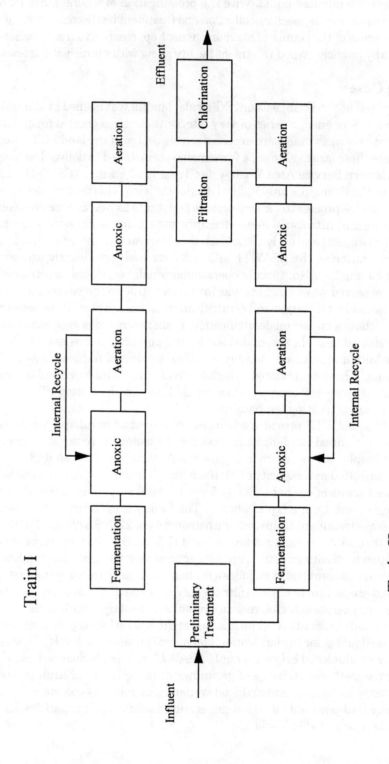

Figure 8.1
ESAWTF process flow diagram.

Table 8.1 ESAWTF unit process tank capacities.

Treatment Train	Basin	Number of Units	Unit Volume, m³	Hydraulic Detention[1], hr
I	Fermentation	2	946	2
I	First Anoxic	4	870	3.6
I	Aeration	4	2690	11
I	Second Anoxic	2	950	2
I	Reaeration	2	105	0.2
II	Fermentation	2	1110	1.9
II	First Anoxic	2	2030	3.4
II	Aeration	2	7140	12
II	Second Anoxic	2	1100	1.9
II	Reaeration	2	350	0.6

[1]at design capacity

Leachate generated from the Orange County landfill was sent to the ESAWTF for treatment on four occasions between September 1991 and January 1992. Following the addition of leachate, total nitrogen levels in the effluent increased almost immediately, reaching concentrations as high as 18 mg/L as nitrogen. Dissolved oxygen (DO) concentrations in the aeration basins fell dramatically during the first two leachate additions, but remained at adequate levels during the last two additions.

After reviewing ESAWTF and landfill records, it was concluded that early increases in effluent nitrogen levels may have been caused by inhibition of nitrification, most likely resulting from insufficient DO. However, during two of the treatment attempts the DO concentration was not low enough to cause inhibition. Nevertheless, during these two attempts, effluent nitrogen concentrations increased very rapidly and recovery required three to five days, even after leachate pumping stopped. In these cases, it was thought that not all of the ammonia was converted into nitrate because ammonia loadings were greater than the nitrifying capacity of the biological process.

Full-Scale Testing and Analysis

A full-scale testing program was designed so that plant performance with and without leachate addition could be measured. Leachate was added to Train I only; Train II served as a control (no leachate added). The two trains were operated as identically as possible so that any differences in performance would be due only to the effects of leachate addition.

Leachate was brought from the landfill in tanker trucks and introduced to Train I just downstream from the grit chambers. Leachate was initially added at a rate of 19 m³/d (5,000 gpd) then gradually increased to 95 m³/d (25,000 gpd) over a period of five weeks. The final leachate addition rate was

selected to be 1% of the plant influent flow rate, and was about equal to estimated normal leachate production rates. The five-week period was believed
to be necessary to allow the biological population to acclimate to the leachate.
This rate of introduction was much slower than previous additions.

Plant performance was measured by studying operating records (flows
and suspended solids concentrations) and by sampling and analyzing wastewater collected at several locations throughout the plant. Grab samples were
collected every two hours and combined over a 24-hour period for analysis.
Samples were collected from the influent (prior to leachate addition),
leachate tank, first anoxic basins, aeration basins, second anoxic basins, and
clarifier effluents. Leachate characteristics are shown in Table 8.2.

The Solution

Very high nitrification efficiencies were observed in both trains during full-
scale testing. However, nitrate concentration data were distinctly different in
Trains I and II. It was also noted that the nitrate concentration levels in the
Train I aeration basin exceeded concentrations in Train II even before
leachate was added to Train I. The difference between nitrate concentrations
increased as leachate flow rates increased. Nitrate concentrations at various
sampling locations are summarized in Table 8.3.

After investigation, two explanations were offered for the accumulation
of nitrate in Train I: (1) the increased nitrogen loading associated with the
leachate addition, and (2) an unintentional lower internal recycle ratio (3.78)
employed in Train I (compared to 4.20 for Train II). The reduced capacity to
recirculate nitrate for removal in the first anoxic basin, coupled with a
greater input of nitrogen from leachate, resulted in a greater accumulation
of nitrate in the Train I aeration basin.

Table 8.2 Leachate characteristics during full-scale testing.

Parameter[1]	Average	Range	Standard Deviation	Number of Samples
Ammonia-N	440	217-944	193	19
Total Kjeldahl nitrogen (TKN)	443	296-738	162	7
Total solids	11339	766-2040	444	10
BOD_5	513	60-950	376	7
COD	3298	2577-4497	721	5
BOD_5/COD	0.36	0.06-0.65	0.20	7

[1]All parameters in mg/L, except BOD/COD ratio.

Table 8.3   Intensive sampling nitrogen concentrations, mg/L.

Location	Train I TKN, mg/L	Train II TKN, mg/L	Train I NO_3^- & NO_2^- mg/L	Train II NO_3^- & NO_2^- mg/L
Influent	31.17	31.17	–	–
First Anoxic	6.75	5.78	0.20	0.31
Aeration	1.72	1.80	4.17	2.18
Second Anoxic	–	–	2.45	0.55
Effluent	1.35	1.06	1.84	0.62

Ammonia removal ability of the plant during earlier leachate introduction was estimated using a mass balance approach. Depending on the volume of leachate treated at that time, ammonia effluent concentrations ranging from 4.0 to 38 mg/L were predicted. Thus, effects of early leachate treatment can be accounted for by considering plant capacity for nitrogen removal. Treatment of leachate at the ESAWTF appeared to be an acceptable means of disposal provided that (1) leachate loadings were not too high, (2) a period of acclimation was provided, and (3) operational adjustments for high nitrogen loading were made. Leachate addition to the entire plant, at volumetric loading well below 1%, has continued since the fall of 1992 without adversely impacting plant operations.

8.4   Refinery Operations—Catalytic Hydrodesulfurization of Oil Products to Recover Elemental Sulfur

Background

During the period following World War II, the United States experienced a tremendous economic boom that lasted for more than twenty years. However, in 1974–1975, a severe economic recession occurred that was triggered by the oil embargo on the United States, and the subsequent tripling of energy costs. Prior to that time, oil prices were very low (about $3 per barrel), and several new, large oil fields had been discovered so that supply exceeded demand for an extended period. However, demand for oil and gasoline continued to grow steadily, so the increase in oil supply was largely filled by "sour" crudes from west Texas, Saudi Arabia, Venezuela, and other places. Sour crude has much higher sulfur content than sweet crude and results in more SO_2 emissions from within the refinery, as well as more SO_2 emissions from other industries that burn fuel products with sulfur in them.

Because sulfur in gasoline and other fuels is not desirable, oil refineries began desulfurizing the intermediate oil products by a process called catalytic hydrodesulfurization. This process used hydrogen and a catalyst to remove the sulfur from the oil as H_2S. H_2S is a toxic and odorous gas, but will burn and produce significant amounts of heat. Thus, the H_2S usually was sent to the refinery fuel gas system (a network of pipes collecting refinery light-ends—methane, ethane, ethylene, and other combustible gases—and mixing them with natural gas). The refinery fuel gas was burned in the numerous furnaces and combustion devices throughout the refinery. This reduced sulfur in the refinery products being sold, but increased SO_2 emissions from the refinery itself. In 1970, the Clean Air Act Amendments required reductions in the emissions of SO_2 and other pollutants, so refiners were faced with the prospect of adding expensive pollution-control equipment to capture the SO_2 from furnaces. However, a much better plan was devised.

The Case

The general situation just described was experienced by many companies at many locations; one such location was the Exxon refinery at Baytown, Texas. The Baytown refinery in 1974 was one of the largest in the world. It processed about 400,000 barrels (a barrel is 42 gallons) of crude oil every day, turning it into such products as gasoline, heating oil, diesel fuel, kerosene, lubricating oils, paint solvents, and more. For decades, the Baytown refinery had processed sweet crude oil from the east Texas area, but as that crude oil supply had dwindled, and the demand for petroleum products had grown, the Baytown refinery had gradually been converted into running more and more sour crudes, especially pipelined oil from west Texas, and large tankers of oil from Saudi Arabia.

Among other processes, the refinery made high-octane gasoline stocks by catalytic reforming—a process that changed aliphatic hydrocarbons (like hexane and heptane) into aromatics (such as benzene and toluene). This process produced hydrogen gas as a by-product. The excess hydrogen gas was blended into the refinery fuel gas system and burned in the refinery's furnaces as a fuel. As the sulfur content of the crude oil increased in the mid-1960s to early 1970s, more catalytic hydrodesulfurization units were built, which used some of the hydrogen to produce "sweeter" gasolines or heating oils, but which also produced large quantities of H_2S.

At the same time, throughout the country, many industries were using more H_2SO_4 in the expanding U.S. economy in the late 1960s. Sulfuric acid was and is such a widely used industrial chemical that its level of use is often interpreted as a barometer of the degree of industrialization of a country. Sulfuric acid is produced by burning elemental sulfur in air and then further oxidizing the SO_2 to SO_3, then absorbing the SO_3 in water. Elemental sulfur production in the United States in 1970 was about 9 million long tons (Manderson and Cooper, 1982), about 80% of which was by the Frasch process (a specialized mining of underground sulfur deposits). In 1970, sulfur was selling for about $25 per long ton (2240 pounds).

The Analysis

Recognizing the problem with SO_2 emissions, and the value of elemental sulfur, engineers and chemists worked together to develop and commercialize a new process—the Claus process. The process partially oxidizes the H_2S to SO_2, and then reacts the two compounds together to produce elemental sulfur and water. The reactions are as follows:

$$H_2S + 1.5\ O_2 \rightarrow SO_2 + H_2O$$

$$2\ H_2S + SO_2 \rightarrow 3\ S + 2\ H_2O$$

There were two significant advantages of the Claus process: (1) SO_2 emissions were reduced, and (2) sulfur was recovered in elemental form, and could be sold as a product.

The Solution

In the early 1970s, the Exxon refinery invested millions of dollars in a Claus plant. It was designed to produce some 150 long tons per day of elemental sulfur. Although company officials expected this pollution-control process to help pay for itself, they mainly thought of this as a way to help them comply with environmental regulations to reduce SO_2 emissions from the refinery. However, they got a very pleasant surprise. Sulfur prices escalated rapidly, and by 1979, elemental sulfur was selling for about $95/long ton. So the process originally built to handle an air pollutant and waste product was now generating revenues of more than $5,000,000 per year for the refinery! In this case, a pollutant was turned into a saleable product, and became a source of significant revenues for the company. Of course, not all pollution problems turn out to have such a nice solution, but because of the ingenuity of engineers and chemists, this one did!

8.5   The Clean-up of Love Canal

Background

The name Love Canal has become almost synonymous with uncontrolled hazardous waste. It is a classic example of waste management practiced according to the standards and knowledge of the time—with tragic results. The history of Love Canal actually dates back to 1892 when William T. Love purchased land to construct a hydroelectric power plant. He was attempting to build a series of canals that connected the Niagara River with Lake Ontario. He abandoned his effort after beginning the construction of one canal, due to financial difficulties attributed to the advent of alternating current. The canal, which was 60-feet wide, ten-feet deep, and three blocks long, was used as a local swimming hole for almost 50 years until it was purchased by the Hooker Electrochemical Company for waste disposal. The site was well suited for waste disposal because the clay soil in which the canal was dug has a very low hydraulic conductivity. Hooker Electrochemical placed some 21,800 tons of chemical wastes over a ten-year period in the north and south sector of the canal. Municipal solid wastes were placed in the central sector. Table 8.4 provides a list of chemicals placed in the ground. The site was closed in 1953, at which time Hooker installed a clay cap.

Shortly after closing the site, Hooker was pressured by the City of Niagara Falls Board of Education to deed the land to the city. Hooker sold 16 acres to the city for one dollar, under threat of seizure, while warning city officials that hazardous chemicals had been buried in the ground. In addition, the deed stipulated that hazardous chemicals were present on the site.

In 1954, the Board of Education constructed the 99th St. Elementary School in the center of the site, with associated sewers and roads. Surrounding land was sold for residential development. The LaSalle Freeway was constructed on the southern end of the site, which changed the natural flow of groundwater and disrupted the clay cap over the wastes.

As early as 1958, children were complaining of chemical burns while playing in road-building debris. Vegetation blackened and died. At that time, ruptured drums were excavated and removed, and the area was refilled, apparently solving the problem.

In 1976 residents began complaining of odors and black oozing material in their basements. Medical problems were associated with this material. Lois Gibbs established a homeowner's association that actively pursued governmental intervention in the situation and brought media attention to the problem. In 1978 the State of New York Commissioner of Health declared a health emergency and relocated 236 families. In 1980, an additional set of 800

Table 8.4 Chemicals found at Love Canal.

Chemical	Amount Found, tons
Acidic chlorides	400
Benzene hexachloride	6900
Benzoic chlorine	800
Benzyl chlorides	2400
Chlorobenzene	2000
Dodecylmercaptans	2400
Liquid disulfides, monochlorotoluene	700
Metal chlorides	400
Sulfides	2100
Thionyl chloride	500
Trichlorophenyl	200
Miscellaneous chlorinated compounds	1000
Miscellaneous other compounds	2000

Source: Adapted from Feinberg, 1982.

families living in the second ring around the canal were relocated. It soon became apparent that clean-up of the site was going to be quite expensive. There were no provisions at the time for federal funding (like the Superfund of today); therefore, in 1981, President Jimmy Carter declared the Love Canal a disaster area to make federal funds available through the Federal Emergency Management Act. Another 2500 families were relocated at a cost of $30 million.

Site investigation and clean-up continued over the next few years. By 1988, four of seven areas were found to be habitable according to air quality monitoring. Homes north of the site were sold at 20 percent below market value. As of June 1994, 193 of 280 homes were sold. Remaining areas were zoned commercial and light industrial and now house an industrial park and middle school.

Site Investigation

The site was subjected to an exstensive and costly investigation. The magnitude of the investigation and technology used made history. Nearly 100 wells were dug in and around the site to determine subsurface hydrologic and geologic conditions and to monitor the fate and transport of contaminants. Seismic investigations were used to delineate the canal edge and to locate buried drums.

The site investigation helped to identify the cause of the contamination. Because the original cap was disrupted by construction, precipitation could infiltrate into the wastes. The clay surrounding the wastes restricted the flow out of the canal, causing it to fill with water that became contaminated with chemicals. Before long the water depth was sufficient to reach silty soil at the surface of the area which had a much higher hydraulic conductivity. Contaminated leachate flowed out of the canal, through these highly permeable soil layers and along sewer trenches, and entered basements of houses and schools north, east, and west of the site. In addition, contaminated soils from the canal area were used as fill in other parts of the site.

A study of health effects in 1976 revealed animals and children with chemical burns, increased incidences of miscarriages, cancer, blood and liver defects, low birth weight, and congenital defects. In 1978 over 400 chemicals were identified, some at concentrations 5,000 times maximum safe levels. Although there was no evidence of acute illnesses, pregnant women and children under two were asked to leave. Cytogenic assessments performed in 1979 documented chromosomal abnormalities and other damage. In 1980, a study of 900 children found high rates of occurrences of seizures, learning disabilities, eye and skin irritation, incontinence, and severe abdominal pain.

The Solution

The immediate response to the Love Canal situation was to relocate families, install a new 22-acre clay cap, construct a drainage system to divert groundwater, close the elementary school, and fence the area. Long-term remediation of the site was aimed at hydraulic containment and included a four-prong approach: (1) contain, collect, and treat leachate, (2) dewater the canal, (3) cover the site with an impermeable cap, and (4) institute a monitoring program. Removal of the waste was considered but not implemented. Consequently, the site must be monitored in perpetuity, for all practical purposes.

Collection of leachate and drainage of the canal was accomplished using 7,000 feet of French drains installed 15 to 20 feet below land surface along the outer boundaries of the site (see Figure 8.2). The drains are composed of two feet of gravel, 8-inch perforated vitrified clay pipe, and sand backfill. The drains direct leachate to wells that are equipped with pumps to transfer leachate to an on-site treatment facility. The treatment train consists of pH adjustment, clarification, and activated carbon. The treated wastewater is discharged to the city sewer. Ultimately, the site was closed with a 40-acre cap consisting of two three-foot layers of clay overlain by geosynthetic membranes. The total cost of site investigation and cleanup was $325 million. Ironically, the cost of locating and constructing a secure landfill in 1953 would have been $4 million.

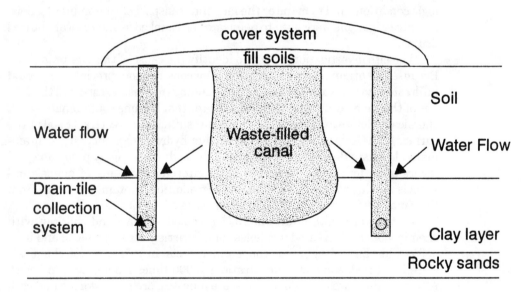

Figure 8.2
Leachate draining plan at Love Canal.

Liability and Costs

In order to recover costs related to the clean-up of Love Canal, both the State of New York and the U.S. government sued the Occidental Chemical Company (which had purchased Hooker in 1968). After 14 years of legal battle,

Occidental settled with the state for $99 million in 1994. In December of 1995, Occidental agreed to reimburse the Superfund $102 million and to pay $27 million to FEMA. In response to a countersuit, the U.S. government reimbursed the Superfund $8 million, reflecting its share of responsibility attributed to the disposal of U.S. Army wastes in Love Canal. Finally, some 900 Love Canal residents sued Occidental over health-related problems. These claims were denied because it had been more than three years since the plaintiffs had been exposed to toxic chemicals when the lawsuits were filed.

8.6   Metal Finishing Industry— Cleaning and Rinsing Operations

Background

Electroplating operations require frequent cleaning and rinsing stages in order to prepare the surface for deposition of a metal. A schematic representation of a nickel-chromium plating line is shown in Figure 8.3. The overall process includes alkaline cleaning, acid pickling, nickel plating, and chrome plating. Each of these four surface operations is followed by a rinsing stage.

As any individual piece or "part" leaves any process tank, a residual film of process tank solution remains on the surface of the part. This residual film, known as "drag-out," travels with the part into the next tank in the processing sequence. The drag-out represents the principal source of pollutants that require treatment in the electroplating industry. Minimization of drag-out is an integral component in any pollution prevention program. The following discussion focuses on reducing rinse water quantities, and thus reducing the quantity of wastewater that requires treatment. In addition, minimizing drag-out reduces the consumption of fresh water at these plants.

The Case

Manufacturing specifications dictate standards of cleanliness that must be achieved by the rinsing operation. In many cases, these criteria address contaminants on the surface of the part that would interfere with subsequent electroplating processes. In other cases, residual chemicals are associated with accelerated corrosion, spotting, or appearance problems. It also may be necessary to achieve a high degree of cleaning (by rinsing) in order to prevent drag-in of chemicals that would poison the chemistry in downstream process tanks. For any of these reasons, use of substantial quantities of water for rinsing may be necessary in order to assure satisfactory surface finishing operations.

Rinse water requirements may be determined through a series of material balances, provided that various operating factors can be defined. These factors include drag-out volumes, drag-out concentrations, source water concentrations, and required water quality in final rinsing operations. Development of staged, counterflow rinsing tanks (Figure 8.4) has been very effective in reducing the amount of water required to achieve rinsing criteria (degree of cleanliness).

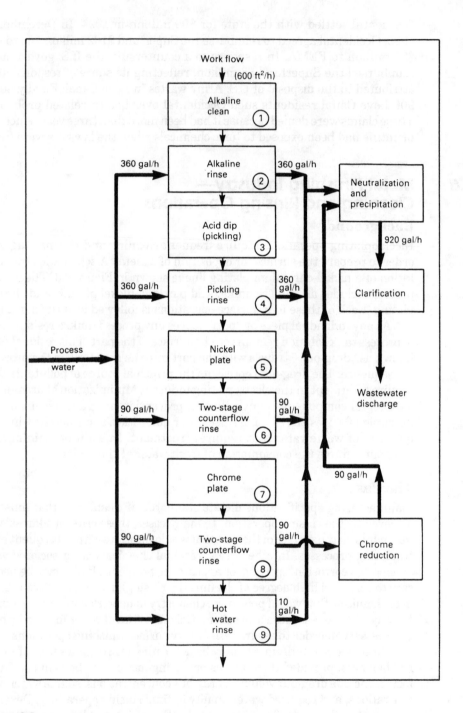

Figure 8.3
Nickel-chrome plating line. (*U.S. EPA, 1985*)

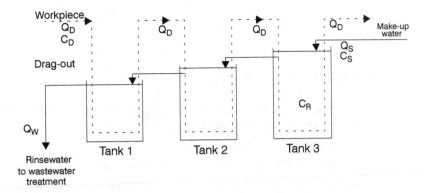

Figure 8.4
Counterflow rinsing configuration. (*U.S. EPA, 1985*)

The Analysis

Steady-state material balances for a constituent in drag-out from a process tank can be written for each stage of a rinsing operation. For a **single-stage** rinse tank, a mass balance yields the following equation:

$$0 = Q_D\, C_D + Q_S\, C_S - Q_D\, C_R - Q_S\, C_R \tag{8.1}$$

where: Q_D = drag-out flow rate from upstream process tank
C_D = concentration in upstream process tank
Q_S = rinse water (make-up) flow rate
C_S = concentration in rinse water supply
C_R = concentration in rinse tank

Development of equation (8.1) assumes steady state, equal drag-in and drag-out flow rates to and from the rinsing tank, complete mixing in the rinse tank, negligible evaporation losses, and no reaction in the rinse tank. If criteria for the final rinse tank concentration are established, the required rinse water supply rate can be determined for this single-stage rinsing operation with equation (8.2):

$$Q_S = \frac{Q_D\, C_D - Q_D\, C_R}{C_R - C_S} \tag{8.2}$$

The Solution

A similar equation can be developed to define the wastewater generation rate for n-stage rinsing operations (see Figure 8.4) by simultaneous solution of n mass balance equations to yield equation (8.3):

$$Q_W \cong \left[\frac{C_D - C_S}{C_R - C_S} \right]^{1/n} Q_D \qquad (8.3)$$

where: $Q_w = Q_S$ = wastewater flow rate

Various strategies for reducing wastewater flow rates are evident from examination of equation (8.3):

1. Increase the number of stages.
2. Reduce the drag-out flow rate from the process tank.
3. Improve the quality of the source water used as rinse water make-up.
4. Reduce the drag-out concentration.

Many examples of successful implementation of these strategies have been documented in the literature (U.S. EPA, 1985); some are summarized in the following example problems.

Example Problem 8.1

A typical Watts-type nickel plating solution contains 270,000 mg/L of total dissolved solids (C_D) and the process requires a final rinse concentration of less than 37 mg/L (C_R). Assuming that deionized water is used as the source for rinse water make-up with negligible total dissolved solids (C_S), determine the wastewater generation rates (gallons per hour) for a system with a drag-out of one gallon per hour for a single-stage, two-stage, and three-stage counterflow rinsing operation.

Solution

Equation (8.3) is used to calculate required rinse water volumes. The volume of wastewater is equal to the rinse water volume. The results are summarized below:

Number of Stages	Wastewater Flow Rate (gal/hr)
1	7300
2	86
3	20

The flow reductions that can be achieved by counterflow rinsing are impressive.

Example Problem 8.2

Implementation of improved parts-holder racks in a chrome-plating process is projected to increase the effectiveness of drainage of parts with return of more process solution to the process tank. If this modification is anticipated to achieve an 85% reduction in drag-out volume, determine the associated reduction in wastewater flow rate.

Solution

Examination of equation (8.3) reveals that the wastewater quantity is directly proportional to the drag-out rate. Thus, an 85% reduction in drag-out volume would yield an identical reduction, 85%, in wastewater generation rate.

Example Problem 8.3

The numerical analysis in example problem 8.1 utilized an ultra-high-purity source water as make-up for rinsing. Repeat the analysis assuming the source water contains 25 mg/L of total dissolved solids. [This source (with 25 mg/L of dissolved solids) would still be considered a very high quality water.] Compare the results with wastewater generation rates for the deionized water source.

Solution

Rinse water requirements are again calculated with equation (8.3). The results are summarized below:

	Wastewater Flow Rate (gal/hr)	
Number of Stages	$C_S = 0$ mg/L	$C_S = 25$ mg/L
1	7300	22500
2	86	150
3	20	28

Example Problem 8.4

Various strategies are available for reducing the concentration of drag-out, including still rinses, spray rinses, and reduction in process bath concentration. For purposes of this analysis, repeat example problem 8.1 assuming that a spray rinse is employed prior to the counterflow rinsing system which achieves a 75% reduction in drag-out concentration by return of the spray rinse waters to the process tank. Compare these results with wastewater volumes for the system without spray rinsing.

Solution

Rinse water requirements are again calculated with equation (8.3). The drag-out concentration (C_D) is reduced from 270,000 mg/L to 67,500 mg/L. The results are summarized below:

	Wastewater Flow Rate (gal/hr)	
Number of Stages	Without Spray Rinse	With Spray Rinse
1	7300	1824
2	86	43
3	20	12

8.7 Halogenated Solvent Degreasing Operations— Replacement with Alternative Cleaning Methods

In many cases, it is necessary to machine parts prior to application of a metal coating by electroplating. The machining operations require cutting or lubricating oils, and these oils must be removed prior to electro-deposition of a metal surface coating. Historically, the method of choice for removing oil from these surfaces involved use of halogenated solvents. Similar surface preparation (cleaning) processes throughout the surface-finishing industry employed halogenated solvents, including trichloroethane (methyl chloroform), CFCs, HCFCs, trichloroethylene, tetrachloroethylene, and methylene chloride.

The preceding solvents possess certain undesirable environmental characteristics. Many of the solvents are ozone-depleting substances (chapter 5), including trichloroethane, CFCs, and HCFCs. The remaining compounds are potential carcinogens: trichloroethylene, tetrachloroethylene, and methylene chloride. In view of these adverse environmental characteristics, the surface-finishing industry has aggressively developed alternative cleaning methods to reduce or eliminate dependence on these solvents.

The most common alternative solution has relied on aqueous cleaning methods, typically alkaline detergents (Vidmar, 1991; Willis, 1992; Smith, 1994). Other replacement solvents include alcohols (for example, isopropanol), terpenes (for example, d-limonene), and aromatic compounds (for example, naphtha) (Peterson, 1994; Stone, 1994).

Methylene chloride has been the long-standing chemical of choice for paint-stripping operations. Use of abrasive methods (for example, sand blasting) has been examined, but the operations generate large quantities of hazardous debris (blasting media plus hazardous paint residues). Development of an alternative blasting technology using carbon dioxide pellets as the blasting media is reported to greatly reduce the quantity of hazardous waste (Ina, 1997). The CO_2 pelletizing turbine process achieves excellent metal surface preparation for adhesion of organic coatings (paint).

References

Feinberg, D. W. 1982. "Denial of a Remedy: Former Residents of Hazardous Waste Sites and New York's Statute of Limitations." *Columbia Journal of Environmental Law* (Winter): 161–74.

Green, Alex E. S., ed. 1992. *Medical Waste Incineration and Pollution Prevention*. New York: Van Nostrand Reinhold.

Ina, Paul. 1997. "CO_2 Turbine Coatings Removal System." EPA Green Technologies Conference, November, Atlanta, GA.

Manderson, M. C., and C. David Cooper. 1982. "Sulfur Supply and Demand and its Relationship to New Energy Sources." In *Sulfur: New Sources and Uses,* ACS Symposium Series 183, edited by Michael E. D. Raymont. Washington, DC: American Chemical Society.

Peterson, D. 1994. "Industrial Lubricants and Cleaning Processes." Proceedings of the Fifteenth AESF/EPA Conference on Environmental Control for the Surface Finishing Industry, January, Kissimmee, FL.

Smith, C.A. 1994. "Elimination of Ozone Depleting Chemicals and Implementation of New Cleaning Methods." Proceedings of the Fifteenth AESF/EPA Conference on Environmental Control for the Surface Finishing Industry, January, Kissimmee, FL.

Stone, H. J. 1994. "Chemical and Mechanical Factors in the Aqueous Cleaning of Manufactured Parts." Proceedings of the Fifteenth AESF/EPA Conference on Environmental Control for the Surface Finishing Industry, January, Kissimmee, FL.

U.S. Environmental Protection Agency. 1990. "Guides to Pollution Prevention—Selected Hospital Waste Streams." EPA/625/7-90/009. Cincinnati: U.S. EPA.

U.S. Environmental Protection Agency. 1985. "Environmental Pollution Control Alternatives: Reducing Water Pollution Control Costs in the Electroplating Industry." EPA/625/5-85/016 (revised).

Vidmar, K. P. 1991. "Elimination of 1,1,1 Trichloroethane through Critical Evaluation and Aqueous Cleaning." Proceedings of the Twelfth AESF/EPA Conference on Environmental Control for the Surface Finishing Industry, January, Kissimmee, FL.

Walker, Barry L., and C. David Cooper. 1992. "Air Pollution Emission Factors for Medical Waste Incinerators." *Journal of Air & Waste Management Association* 42(8).

Willis, D. G. 1992. "Decorative Plater Eliminates Vapor Degreasing." Proceedings of the Thirteenth AESF/EPA Conference on Environmental Control for the Surface Finishing Industry, January, Kissimmee, FL.

Appendix A: Conversion Factors

How to use these tables:

1 unit from column A = **table entry** units from row B

Examples: 1 lb_m = 453.6 g; 1 kg = 2.205 lb_m

Table A.1 Mass

| | Row B | | | | | |
Column A	lb_m	g	gr	kg	ton	tonne
lb_m	1.0	453.6	7,000	0.4536	0.00050	0.000454
g	0.002205	1.0	15.43	0.001	$1.10 (10)^{-6}$	$1.0 (10)^{-6}$
gr	0.000143	0.0648	1.0	$6.48 (10)^{-5}$	$7.14 (10)^{-8}$	$6.48 (10)^{-8}$
kg	2.205	1,000	$1.54 (10)^4$	1.0	0.0011	0.001
ton	2,000	$9.07 (10)^5$	$1.40 (10)^7$	907	1.0	0.907
tonne (metric ton)	2,205	$(10)^6$	$1.54 (10)^7$	1,000	1.102	1.0

Table A.2 Length

| | Row B | | | | | |
Column A	m	ft	in.	μm	km	miles
m	1.0	3.281	39.37	10^6	0.001	$6.21 (10)^{-4}$
ft	0.3048	1.0	12	$3.05 (10)^5$	$3.05 (10)^{-4}$	$1.894 (10)^{-4}$
in.	0.0254	0.0833	1.0	$2.54 (10)^4$	$2.54 (10)^{-5}$	$1.578 (10)^{-5}$
μm	10^{-6}	$3.28 (10)^{-6}$	$3.94 (10)^{-5}$	1.0	$1.0 (10)^{-9}$	$6.22 (10)^{-10}$
km	1,000	3,281	$3.94 (10)^4$	$1.0 (10)^9$	1.0	0.6215
miles	1,609	5,280	$6.336 (10)^4$	$1.61 (10)^9$	1.609	1.0

Table A.3 Volume

Column A	Row B			
	ft^3	L	gal	m^3
ft^3	1.0	28.32	7.481	0.02832
L	0.03531	1.0	0.2642	0.001
gal	0.1337	3.785	1.0	0.003785
m^3	35.31	1,000	264.2	1.0

Table A.4 Force

Column A	Row B			
	N	lb$_f$	kg-m/s^2	lb$_m$-ft/sec^2
N	1.0	0.2248	1.0	7.232
lb$_f$	4.448	1.0	4.448	32.17
kg-m/s^2	1.0	0.2248	1.0	7.232
lb$_m$-ft/s^2	0.1383	0.03108	0.1383	1.0

NOTE: 1 dyne = 10.0 μN

Table A.5 Pressure

Column A	Row B					
	atm	psi	mm Hg	in. H$_2$O	mbar	Pa (N/m^2)
atm	1.0	14.70	760	406.8	1,013	101,300
psi	0.068	1.0	51.7	27.67	68.9	6,891
mm Hg	1.316 (10)$^{-3}$	0.0193	1.0	0.535	1.333	133.3
in. H$_2$O	0.002458	0.03614	1.868	1.0	2.49	249
mbar	9.87 (10)$^{-4}$	0.0145	0.750	0.4016	1.0	100
Pa	9.87 (10)$^{-6}$	1.45 (10)$^{-4}$	0.0075	0.00402	0.01	1.0

NOTE: 1 Pa = 1 N/m^2

Table A.6   Energy

Column A	Btu	kJ	cal	ft.-lb$_f$	kwh	liter-atm
			Row B			
Btu	1.0	1.055	252	778	$2.93\,(10)^{-4}$	10.41
kJ	0.948	1.0	239	737.5	$2.778(10)^{-4}$	98.62
cal	0.00397	0.004184	1.0	3.087	$1.163\,(10)^{-6}$	0.0413
ft-lb$_f$	0.001285	0.001356	0.3239	1.0	$3.766\,(10)^{-7}$	0.01338
kwh	3,412	3,600	$8.60\,(10)^5$	$2.66\,(10)^6$	1.0	$3.55\,(10)^4$
liter-atm	0.0961	0.01014	24.22	74.74	$2.82\,(10)^{-5}$	1.0

NOTE: 1 J = 1 N-m

Table A.7   Power

Column A	W	kw	ft-lb$_f$/sec	hp	Btu/hr
			Row B		
W	1.0	0.001	0.737	0.00134	3.412
kw	1,000	1.0	737.6	1.341	3,412
ft-lb$_f$/sec	1.356	0.001356	1.0	0.001818	4.63
hp	745.5	0.7455	550	1.0	2,545
Btu/hr	.293	$2.93\,(10)^{-4}$	0.216	$3.93\,(10)^{-4}$	1.0

NOTE: 1W = 1 J/s

Table A.8   Speed

Column A	ft/sec	m/s	mi/hr	ft/min
		Row B		
ft/sec	1.0	0.3048	0.6818	60.0
m/s	3.281	1.0	2.237	196.8
mi/hr	1.467	0.447	1.0	88.0
ft/min	0.01667	0.00508	0.01136	1.0

Table A.9 Viscosity

Column A	Row B			
	cp	g/cm-s	lb_m/ft-hr	kg/m-hr
cp	1.0	0.01	2.42	3.61
g/cm-s	100	1.0	242	361
lb_m/ft-hr	0.413	0.00413	1.0	1.492
kg/m-hr	0.277	0.00277	0.670	1.0

Appendix B: Selected Physical and Chemical Properties

Selected properties of air
Values of the Ideal Gas Law constant R
Selected properties of liquid water
Solubility constants of selected solids
Ionization constants of selected acids and bases
Henry's law constants for selected gases in water
Enthalpies of saturated steam and water
Standard heats of combustion for various organic compounds

Table B.1 Selected Properties of Air

Approximate Composition (% by moles or volume)		Molecular Weight of Dry Air
(2 gas approximation)	79% N_2, 21% O_2	M.W.=28.85*
(3 gas approximation)	78% N_2, 21% O_2, 1% Ar	M.W.=28.96
(4 gas approximation)	78.09% N_2, 20.94% O_2, 0.93% Ar, 0.04% CO_2	M.W.=28.97

*M.W. calculated as $\Sigma y_i M.W._i$ where y_i = mole fraction of gas I; $M.W._i$ = molecular weight of gas I

Temp, °F	Density*		Specific Heat (C_p)
	lb/ft³	kg/m³	Btu/lb–°F or cal/g–°C
77	0.0740	1.185	0.240
150	0.0650	1.041	0.240
300	0.0521	0.834	0.241
500	0.0412	0.660	0.242
1000	0.0275	0.440	0.246
2000	0.0161	0.258	0.260

*at standard pressure of 1.00 atm.

Table B.2   Values of the Ideal Gas Law Constant R [$R = (PV/nT)$]

	Units
R	(Pressure-Volume)/(Matter-Temperature)
0.08206	atm-L/gmol-K
82.06	atm-cm^3/gmol-K
62.36	mm Hg-L/gmol-K
1.314	atm-ft^3/lbmol-K
0.08314	bar-L/gmol-K
998.9	mm Hg-ft^3/lbmol-K
0.7302	atm-ft^3/lbmol-R
21.85	in. Hg-ft^3/lbmol-R
555.0	mm Hg-ft^3/lbmol-R
10.73	psia-ft^3/lbmol-R
1545.	psfa-ft^3/lbmol-R
8.314	Pa-m^3/gmol-K
1.987	cal/gmol-K
8314.	J/kgmol-K

NOTE: 1 pascal (Pa) = 1 N/m^2; 1 N-m = 1 joule (J)

Table B.3   Selected Properties of Liquid Water

Temperature °F	Density lb$_m$/ft^3	Viscosity cp	Vapor Pressure psi	Specific Heat Btu/lb-°F or cal/g-°C
40	62.43	1.55	0.122	1.004
70	62.30	0.98	0.363	0.9993
100	62.00	0.68	0.949	0.9986
200	60.13	0.30	11.53	1.005

Adapted from Cooper, C. D., and F. C. Alley. 1994. *Air Pollution Control: A Design Approach*, 2nd ed. Prospect Heights, IL: Waveland Press.

Table B.4   Solubility Constants of Selected Solids

Bromides		Chlorides	
$PbBr_2$	4.6×10^{-6}	$PbCl_2$	1.6×10^{-5}
Hg_2Br_2	1.3×10^{-22}	Hg_2Cl_2	1.1×10^{-18}
$AgBr$	5.0×10^{-13}	$AgCl$	1.7×10^{-10}
Carbonates		**Chromates**	
$BaCO_3$	1.6×10^{-9}	$BaCrO_4$	8.5×10^{-11}
$CdCO_3$	5.2×10^{-12}	$PbCrO_4$	2×10^{-16}
$CaCO_3$	4.7×10^{-9}	Hg_2CrO_4	2×10^{-9}
$CuCO_3$	2.5×10^{-10}	Ag_2CrO_4	1.9×10^{-12}
$FeCO_3$	2.1×10^{-11}	$SrCrO_4$	3.6×10^{-5}
$PbCO_3$	1.5×10^{-15}		
$MgCO_3$	1×10^{-15}	**Phosphates**	
$MnCO_3$	8.8×10^{-11}	$Ba_3(PO_4)_2$	6×10^{-39}
Hg_2CO_3	9.0×10^{-17}	$Ca_3(PO_4)_2$	1.3×10^{-32}
$NiCO_3$	1.4×10^{-7}	$Pb_3(PO_4)_2$	1×10^{-54}
Ag_2CO_3	8.2×10^{-12}	Ag_3PO_4	1.8×10^{-18}
$SrCO_3$	7×10^{-10}	$Sr_3(PO_4)_2$	1×10^{-31}
$ZnCO_3$	2×10^{-10}		
		Sulfates	
Fluorides		$BaSO_4$	1.0×10^{-10}
BaF_2	2.4×10^{-5}	$CaSO_4$	2.4×10^{-5}
CaF_2	3.9×10^{-11}	$PbSO_4$	1.3×10^{-8}
PbF_2	4×10^{-8}	Ag_2SO_4	1.2×10^{-5}
MgF_2	8×10^{-8}	$SrSO_4$	7.6×10^{-7}
SrF_2	7.9×10^{-10}		
		Sulfides	
Hydroxides		Bi_2S_3	1.6×10^{-72}
$Al(OH)_3$	5×10^{-33}	CdS	1.0×10^{-28}
$Ba(OH)_2$	5.0×10^{-3}	CoS	5×10^{-22}
$Cd(OH)_2$	2.0×10^{-14}	CuS	8×10^{-37}
$Ca(OH)_2$	1.3×10^{-6}	FeS	4×10^{-19}
$Cr(OH)_3$	6.7×10^{-31}	PbS	7×10^{-29}
$Co(OH)_2$	2.5×10^{-16}	MnS	7×10^{-16}
$Co(OH)_3$	2.5×10^{-43}	HgS	1.6×10^{-54}
$Cu(OH)_2$	1.6×10^{-19}	NiS	3×10^{-21}
$Fe(OH)_2$	1.8×10^{-15}	Ag_2S	5.5×10^{-51}
$Fe(OH)_3$	6×10^{-38}	SnS	1×10^{-26}
$Pb(OH)_2$	4.2×10^{-15}	ZnS	2.5×10^{-22}
$Mg(OH)_2$	8.9×10^{-12}		
$Mn(OH)_2$	2×10^{-13}	**Miscellaneous**	
$Hg(OH)_2(HgO)$	3×10^{-26}	$NaHCO_3$	1.2×10^{-3}
$Ni(OH)_2$	1.6×10^{-16}	$KClO_4$	8.9×10^{-3}
$AgOH(Ag_2O)$	2.0×10^{-8}	$K_2[PtCl_6]$	1.4×10^{-6}
$Sr(OH)_2$	3.2×10^{-4}	$AgC_2H_3O_2$	2.3×10^{-3}
$Sn(OH)_2$	3×10^{-27}	$AgCN$	1.6×10^{-14}
$Zn(OH)_2$	4.5×10^{-17}	$AgCNS$	1.0×10^{-12}
Iodides			
PbI_2	8.3×10^{-9}		
Hg_2I_2	4.5×10^{-29}		
AgI	8.5×10^{-17}		

Adapted from Wanielista, M. P., Y. A. Yousef, J. S. Taylor, and C. D. Cooper. 1984. *Engineering and the Environment.* Monterey, CA: Brooks Cole.

Table B.5   Ionization Constants of Selected Acids and Bases

MONOPROTIC ACIDS

acetic	$HC_2H_3O_2 \leftrightarrows H^+ + C_2H_3O_2^-$	2.5×10^{-5}
benzoic	$HC_7H_5O_2 \leftrightarrows H^+ + C_7H_5O_2^-$	6.0×10^{-5}
chlorous	$HClO_2 \leftrightarrows H^+ + ClO_2^-$	1.1×10^{-2}
cyanic	$HOCN \leftrightarrows H^+ + OCN^-$	1.2×10^{-4}
formic	$HCHO_2 \leftrightarrows H^+ + CHO_2^-$	1.8×10^{-4}
hydrazoic	$HN_3 \leftrightarrows H^+ + N_3^-$	1.9×10^{-5}
hydrocyanic	$HCN \leftrightarrows H^+ + CN^-$	4.0×10^{-10}
hydrofluoric	$HF \leftrightarrows H^+ + F^-$	6.7×10^{-4}
hypobromous	$HOBr \leftrightarrows H^+ + OBr^-$	2.1×10^{-9}
hypochlorous	$HOCl \leftrightarrows H^+ + OCl^-$	3.2×10^{-8}
nitrous	$HNO_2 \leftrightarrows H^+ + NO_2^-$	4.5×10^{-4}

POLYPROTIC ACIDS

arsenic	$H_3AsO_4 \leftrightarrows H^+ + H_2AsO_4^-$	$K_1 = 2.5 \times 10^{-4}$
	$H_2AsO_4^- \leftrightarrows H^+ + HAsO_4^{-2}$	$K_1 = 5.6 \times 10^{-8}$
	$HAsO_4^{-2} \leftrightarrows H^+ + AsO_4^{-3}$	$K_3 = 3 \times 10^{-13}$
carbonic	$CO_2 + H_2O \leftrightarrows H^+ + HCO_3^-$	$K_1 = 4.5 \times 10^{-7}$
	$HCO_3^- \leftrightarrows H^+ + CO_3^{-2}$	$K_2 = 4.7 \times 10^{-11}$
hydrosulfuric	$H_2S \leftrightarrows H^+ + HS^-$	$K_1 = 1.1 \times 10^{-7}$
	$HS^- \leftrightarrows H^+ + S^{-2}$	$K_2 = 1.0 \times 10^{-14}$
oxalic	$H_2C_2O_4 \leftrightarrows H^+ + HC_2O_4^-$	$K_1 = 5.9 \times 10^{-2}$
	$HC_2O_4^- \leftrightarrows H^+ + C_2O_4^{-2}$	$K_2 = 6.4 \times 10^{-5}$
phosphoric	$H_3PO_4 \leftrightarrows H^+ + H_2PO_4^-$	$K_1 = 7.5 \times 10^{-3}$
	$H_2PO_4^- \leftrightarrows H^+ + HPO_4^{-2}$	$K_2 = 6.2 \times 10^{-8}$
	$HPO_4^{-2} \leftrightarrows H^+ + PO_4^{-3}$	$K_3 = 1 \times 10^{-12}$
phosphorous	$H_3PO_3 \leftrightarrows H^+ + H_2PO_3^-$	$K_1 = 1.6 \times 10^{-2}$
(diprotic)	$H_2PO_3^- \leftrightarrows H^+ + PO_3^{-2}$	$K_2 = 7 \times 10^{-7}$
sulfuric	$H_2SO_4 \leftrightarrows H^+ + HSO_4^-$	strong
	$HSO_4^- \leftrightarrows H^+ + SO_4^{-2}$	$K_2 = 1.3 \times 10^{-2}$
sulfurous	$SO_2 + H_2O \leftrightarrows H^+ + HSO_3^-$	$K_1 = 1.3 \times 10^{-2}$
	$HSO_3^- \leftrightarrows H^+ + SO_3^{-2}$	$K_2 = 5.6 \times 10^{-8}$

BASES

ammonia	$NH_3 + H_2O \leftrightarrows NH_4 + OH^-$	1.8×10^{-5}
aniline	$C_6H_5NH_2 + H_2O \leftrightarrows C_6H_5NH_3^+ + OH^-$	4.6×10^{-10}
dimethylamine	$(CH_3)_2NH + H_2O \leftrightarrows (CH_3)_2NH_2^+ + OH^-$	7.4×10^{-4}
hydrazine	$N_2H_4 + H_2O \leftrightarrows N_2H_5^+ + OH^-$	9.8×10^{-7}
methylamine	$CH_3NH_2 + H_2O \leftrightarrows CH_3NH_3^+ + OH^-$	5.0×10^{-4}
pyridine	$C_5H_5N + H_2O \leftrightarrows C_5H_5NH^+ + OH^-$	1.5×10^{-9}
trimethylamine	$(CH_3)_3N + H_2O \leftrightarrows (CH_3)_3NH^+ + OH^-$	7.4×10^{-5}

Adapted from Wanielista, M. P., Y. A. Yousef, J. S. Taylor, and C. D. Cooper. 1984. *Engineering and the Environment*. Monterey, CA: Brooks Cole.

Table B.6 Henry's Law Constants for Selected Gases in Water

where $\bar{P}_a = H_a x_a$

$\bar{P}_a$ = partial pressure of the solute a in the gas phase, atm

x_a = mole fraction of solute a in the liquid phase, mole fraction

H_a = Henry's law constant, atm/mole fraction

$H_a \times 10^{-4}$, atm/mole fraction

Temp °C	Air	CO_2	CO	C_2H_6	H_2	H_2S	CH_4	NO	N_2	O_2
0	4.32	0.0728	3.52	1.26	5.79	0.0268	2.24	1.69	5.29	2.55
10	5.49	0.104	4.42	1.89	6.36	0.0367	2.97	2.18	6.68	3.27
20	6.64	0.142	5.36	2.63	6.83	0.0483	3.76	2.64	8.04	4.01
30	7.71	0.186	6.20	3.42	7.29	0.0609	4.49	3.10	9.24	4.75
40	8.70	0.233	6.96	4.23	7.51	0.0745	5.20	3.52	10.4	5.35
50	9.46	0.283	7.61	5.00	7.65	0.0884	5.77	3.90	11.3	5.88
60	10.1	0.341	8.21	5.65	7.65	0.103	6.26	4.18	12.0	6.29
70	10.5		8.45	6.23	7.61	0.119	6.66	4.38	12.5	6.63
80	10.7		8.45	6.61	7.55	0.135	6.82	4.48	12.6	6.87
90	10.8		8.46	6.87	7.51	0.144	6.92	4.52	12.6	6.99
100	10.7		8.46	6.92	7.45	0.148	7.01	4.54	12.6	7.01

Adapted from Foust, Wenzel, Clump, Maus, and Anderson. 1960. *Principles of Unit Operations.* New York: John Wiley & Sons.

Table B.7 Enthalpies of Saturated Steam and Water

T, °F	Vap. Press atm	Enthalpy, Btu/lb$_m$			T, °F	Vap. Press atm	Enthalpy, Btu/lb$_m$		
		Sat. Liq.	ΔH_v	Sat. Vapor			Sat. Liq.	ΔH_v	Sat. Vapor
32	0.0060	0	1,075.1	1,075.1	180	0.511	147.91	990.2	1,138.1
40	0.0083	8.05	1,070.5	1,078.6	190	0.635	157.95	984.1	1,142.1
50	0.0121	18.07	1,064.8	1,082.9	200	0.784	167.99	977.8	1,145.8
60	0.0174	28.07	1,059.1	1,087.2	210	0.961	178.06	971.5	1,149.6
70	0.0247	38.05	1,053.4	1,091.5	212	1.000	180.07	970.3	1,150.4
80	0.0345	48.02	1,047.8	1,095.8	220	1.170	188.14	965.2	1,153.3
90	0.0475	58.00	1,042.1	1,100.1	230	1.414	198.22	958.7	1,156.9
100	0.0646	67.97	1,036.4	1,104.4	240	1.699	208.34	952.1	1,160.4
110	0.0867	77.94	1,030.9	1,108.8	250	2.029	218.48	945.3	1,163.8
120	0.115	87.91	1,025.3	1,113.2	260	2.411	228.65	938.6	1,167.3
130	0.151	97.89	1,019.5	1,117.4	270	2.848	238.84	931.8	1,170.6
140	0.196	107.88	1,013.7	1,121.6	280	3.348	249.06	924.6	1,173.7
150	0.253	117.87	1,007.8	1,125.7	290	3.916	259.31	917.4	1,176.7
160	0.322	127.87	1,002.0	1,129.9	300	4.560	269.60	910.1	1,179.7
170	0.408	137.89	996.1	1,134.02					

Adapted from Cooper, C. D., and F. C. Alley. 1994. *Air Pollution Control: A Design Approach*, 2nd ed. Prospect Heights, IL: Waveland Press.

Table B.8   Standard Heats of Combustion for Various Organic Compounds
(Products are H_2O(g) and CO_2(g) at 25°C)

Compound	Formula	M.W.	ΔH_c, kJ/kg
n-Alkanes			
Methane	CH_4	16.0	50,150
Ethane	C_2H_6	30.1	47,440
Propane	C_3H_8	44.1	46,350
n-Butane	C_4H_{10}	58.1	45,730
n-Pentane	C_5H_{12}	72.2	45,320
n-Hexane	C_6H_{14}	86.2	45,090
1-Alkenes			
Ethylene	C_2H_4	28.1	47,080
Propylene	C_3H_6	42.1	45,760
1-Butene	C_4H_8	56.1	45,300
1-Pentene	C_5H_{10}	70.1	45,020
1-Hexene	C_6H_{12}	84.2	44,780
Miscellaneous			
Acetylene	C_2H_2	26.0	48,290
Benzene	C_6H_6	78.1	40,580
1,3-Butadiene	C_4H_6	54.1	44,540
Cyclohexane	C_6H_{12}	84.2	43,810
Ethylbenzene	C_8H_{10}	106.2	41,310
Methylcyclohexane	C_7H_{14}	98.2	43,710
Styrene	C_8H_8	104.2	40,910
Toluene	C_7H_8	92.1	40,950
Carbon monoxide	CO	28.0	10,110
Hydrogen	H_2	2.016	120,900
Ethanol	C_2H_5OH	46.1	25,960
Methanol	CH_3OH	32.0	18,790
Water (liq. to gas)	H_2O	18.0	2,445

Adapted from Cooper, C. D., and F. C. Alley. 1994. *Air Pollution Control: A Design Approach*, 2nd ed. Prospect Heights, IL: Waveland Press.

Appendix C: Computer Solutions to Selected Example Problems

BASIC Solution

```
100 REM     Example Problem 3.7.  Simulation of Holding Pit
110 REM     Define System Parameters
120         Q = 100000
130         V = 1000000
140         CI = 100
150 REM     Define Simulation Parameters
160         DELTIME = .1
170         ENDTIME = 10
180 REM     Define Initial Conditions
190         T = 0
200         C = 10
210 REM     Enter Time Domain Simulation
220         FOR I = DELTIME TO ENDTIME STEP DELTIME
230         DCDT =  Q*CI/V - Q*C/V
240         C = C + DCDT*DELTIME
250         T = T + DELTIME
260         NEXT I
270 REM     Print Output
280         PRINT T,C
```

BASIC Solution

```
100 REM     Example Problem 3.13.  First-Order Batch Reactor Simulation
110 REM     Define System Parameters
120         K = 0.1
130 REM     Define Simulation Parameters
140         DELTIME = 1
150         ENDTIME = 30
160 REM     Define Initial Conditions
```

```
170          T = 0
180          C = 100
190 REM      Enter Time Domain Simulation
200          FOR I = DELTIME TO ENDTIME STEP DELTIME
210          DCDT = - K*C
220          C = C + DCDT*DELTIME
230          T = T + DELTIME
240          NEXT I
250 REM      Print Output
260          PRINT T,C
```

BASIC Solution
```
100 REM      Example Problem 3.14. Numerical Solution for Single CSTR
110 REM      First-Order Kinetics
120 REM      Define System Parameters
130          K = .4
140          Q = 100
150          V = 1000
160          C0 = 100
170 REM      Define Simulation Parameters
180          X = .1
190 REM      Define Initial Estimate of Unknown
200          C = 100
210 REM      Begin Iterations
220          FOR I = 1 TO 25
230          CTEMP = C0 - V*K*C/Q
240          C = CTEMP*X + C*(1 - X)
250          PRINT I,C
260          NEXT I
```

BASIC Solution
```
100 REM      Example 3.15. Numerical Solution for Single CSTR
110 REM      Variable-Order Kinetics
120 REM      Define System Parameters
130          K = 45
140          KC = 40
150          Q = 100
160          V = 1000
170          C0 = 100
180 REM      Define Simulation Parameters
190          X = .1
200 REM      Define Initial Estimate of Unknown
210          C = 100
220 REM      Begin Iterations
```

```
230        FOR I = 1 TO 15
240        CTEMP = C0 - V*K*C/(Q*(KC + C))
250        C = CTEMP*X + C*(1 - X)
260        PRINT I,C
270        NEXT I
```

BASIC Solution
```
100 REM    Example Problem 3.16. Numerical Solution 3 CSTRs in Series
110 REM    First-Order Kinetics
120 REM    Define System Parameters
130        K = .4
140        Q = 100
150        C3 = 20
160        C0 = 100
170 REM    Define Simulation Parameters
180        X = .1
190 REM    Define Initial Estimates of Unknowns
200        C1 = 100
210        C2 = 100
220        V = 100
230 REM    Begin Iterations
240        FOR I = 1 TO 250
250        C1TEMP = C0/(1 + K*V/(3*Q))
260        C1 = C1TEMP*X + C1*(1 - X)
270        C2TEMP = C1/(1 + K*V/(3*Q))
280        C2 = C2TEMP*X + C2*(1 - X)
290        VTEMP = (Q*C2 - Q*C3)*3/(K*C3)
300        V = VTEMP*X + V*(1 - X)
310        PRINT I,C1,C2,V
320        NEXT I
```

BASIC Solution
```
100 REM    Example Problem 3.17. Determination of Neutralization pH
110 REM    Define System Parameters
120        KW = 10^-14
130        KA = 10^-4.6
140        CT = 5/1050
150        NA = 5/1050
160 REM    Define Simulation Parameters
170        X = .01
180 REM    Define Initial Estimates of Unknowns
190        H = 10^-7
200        OH = 10^-7
210        AC = CT
```

220		HAC = 0
230	REM	Begin Iterations
240		FOR I = 1 TO 2000
250		OHTEMP = KW/H
260		OH = OHTEMP*X + OH*(1-X)
270		HACTEMP = H*AC/KA
280		IF HACTEMP > CT THEN HACTEMP = CT
290		HAC = HACTEMP*X + HAC*(1-X)
300		AC = CT - HAC
310		HTEMP = AC + OH - NA
320		IF HTEMP < 0 THEN HTEMP = 0
330		H = HTEMP*X + H*(1-X)
340		NEXT I
350	REM	Print Output
360		PH = -LOG(H)/LOG(10)
370		PRINT I,H,OH,HAC,AC,PH

Spreadsheet Solution

Example Problem 3.7

	A	B	C	D	E	F	G	H
1	Example Problem 3.7							
2								
3	Q	100000	DCDT = $B$3*$B$5/$B$4 - $B$3*B10/$B$4					
4	V	1000000	CNEW = B10 + C10*$B$4					
5	CI	100	TNEW = A10 + $B$6					
6	DELTIME	0.1						
7	ENDTIME	10						
8								
9	T	C	DCDT	CNEW	TNEW			
10	0.00	10.00	9.00	10.90	0.10			
11	0.10	10.90	8.91	11.79	0.20			
12	0.20	11.79	8.82	12.67	0.30			
13	0.30	12.67	8.73	13.55	0.40			
14	0.40	13.55	8.65	14.41	0.50			
15	0.50	14.41	8.56	15.27	0.60			
16	0.60	15.27	8.47	16.11	0.70			
17	0.70	16.11	8.39	16.95	0.80			
18	0.80	16.95	8.30	17.78	0.90			
19	0.90	17.78	8.22	18.61	1.00			
20	1.00	18.61	8.14	19.42	1.10	Note: Rows 21-99 hidden.		
100	9.00	63.57	3.64	63.94	9.10			
101	9.10	63.94	3.61	64.30	9.20			
102	9.20	64.30	3.57	64.66	9.30			
103	9.30	64.66	3.53	65.01	9.40			
104	9.40	65.01	3.50	65.36	9.50			
105	9.50	65.36	3.46	65.71	9.60			
106	9.60	65.71	3.43	66.05	9.70			
107	9.70	66.05	3.40	66.39	9.80			
108	9.80	66.39	3.36	66.72	9.90			
109	9.90	66.72	3.33	67.06	10.00			
110	10.00	67.06						

Spreadsheet Solution   Example Problem 3.13

	A	B	C	D	E
1	Example Problem 3.13				
2					
3	K	0.1	DCDT = -$B$3*B8		
4	DELTIME	1	CNEW = B8 + C8*$B$4		
5	ENDTIME	30	TNEW = A8 + $B$4		
6					
7	T	C	DCDT	CNEW	TNEW
8	0.00	100.00	-10.00	90.00	1.00
9	1.00	90.00	-9.00	81.00	2.00
10	2.00	81.00	-8.10	72.90	3.00
11	3.00	72.90	-7.29	65.61	4.00
12	4.00	65.61	-6.56	59.05	5.00
13	5.00	59.05	-5.90	53.14	6.00
14	6.00	53.14	-5.31	47.83	7.00
15	7.00	47.83	-4.78	43.05	8.00
16	8.00	43.05	-4.30	38.74	9.00
17	9.00	38.74	-3.87	34.87	10.00
18	10.00	34.87	-3.49	31.38	11.00
19	11.00	31.38	-3.14	28.24	12.00
20	12.00	28.24	-2.82	25.42	13.00
21	13.00	25.42	-2.54	22.88	14.00
22	14.00	22.88	-2.29	20.59	15.00
23	15.00	20.59	-2.06	18.53	16.00
24	16.00	18.53	-1.85	16.68	17.00
25	17.00	16.68	-1.67	15.01	18.00
26	18.00	15.01	-1.50	13.51	19.00
27	19.00	13.51	-1.35	12.16	20.00
28	20.00	12.16	-1.22	10.94	21.00
29	21.00	10.94	-1.09	9.85	22.00
30	22.00	9.85	-0.98	8.86	23.00
31	23.00	8.86	-0.89	7.98	24.00
32	24.00	7.98	-0.80	7.18	25.00
33	25.00	7.18	-0.72	6.46	26.00
34	26.00	6.46	-0.65	5.81	27.00
35	27.00	5.81	-0.58	5.23	28.00
36	28.00	5.23	-0.52	4.71	29.00
37	29.00	4.71	-0.47	4.24	30.00
38	30.00	4.24	-0.42		

Spreadsheet Solution   Example Problem 3.14

	A	B	C	D	E	F
1	Example Problem 3.14					
2						
3	K	0.4	CTEMP = $B$6 - $B$5*$B$3*B10/$B$4			
4	Q	100	CNEW = C10*$B$7 + B10*(1-$B$7)			
5	V	1000				
6	C0	100				
7	X	0.1				
8						
9	I	C	CTEMP	CNEW		
10	0.00	100.00	-300.00	60.00		
11	1.00	60.00	-140.00	40.00		
12	2.00	40.00	-60.00	30.00		
13	3.00	30.00	-20.00	25.00		
14	4.00	25.00	0.00	22.50		
15	5.00	22.50	10.00	21.25		
16	6.00	21.25	15.00	20.63		
17	7.00	20.63	17.50	20.31		
18	8.00	20.31	18.75	20.16		
19	9.00	20.16	19.38	20.08		
20	10.00	20.08	19.69	20.04		
21	11.00	20.04	19.84	20.02		
22	12.00	20.02	19.92	20.01		
23	13.00	20.01	19.96	20.00		
24	14.00	20.00	19.98	20.00		
25	15.00	20.00	19.99	20.00		
26	16.00	20.00	20.00	20.00		
27	17.00	20.00	20.00	20.00		
28	18.00	20.00	20.00	20.00		
29	19.00	20.00	20.00	20.00		
30	20.00	20.00	20.00	20.00		
31	21.00	20.00	20.00	20.00		
32	22.00	20.00	20.00	20.00		
33	23.00	20.00	20.00	20.00		
34	24.00	20.00	20.00	20.00		
35	25.00	20.00	20.00	20.00		

Spreadsheet Solution

	A	B	C	D	E	F	G
1	Example Problem 3.15						
2							
3	K	45	CTEMP = $B$7 - $B$6*$B$3*B11/($B$5*($B$4+B11))				
4	KC	40	CNEW = C11*$B$8 + B11*(1-$B$8)				
5	Q	100					
6	V	1000					
7	C0	100					
8	X	0.1					
9							
10	I	C	CTEMP	CNEW			
11	0.00	100.00	-221.43	67.86			
12	1.00	67.86	-183.11	42.76			
13	2.00	42.76	-132.50	25.23			
14	3.00	25.23	-74.07	15.30			
15	4.00	15.30	-24.52	11.32			
16	5.00	11.32	0.73	10.26			
17	6.00	10.26	8.12	10.05			
18	7.00	10.05	9.65	10.01			
19	8.00	10.01	9.94	10.00			
20	9.00	10.00	9.99	10.00			
21	10.00	10.00	10.00	10.00			
22	11.00	10.00	10.00	10.00			
23	12.00	10.00	10.00	10.00			
24	13.00	10.00	10.00	10.00			
25	14.00	10.00	10.00	10.00			
26	15.00	10.00	10.00	10.00			

Spreadsheet Solution

Example Problem 3.16

	A	B	C	D	E	F	G
1	Example Problem 3.16						
2							
3	K		0.4	C1TEMP = $B$6- D11*$B$3*B11/(3*$B$4))			
4	Q		100	C1NEW = E11*$B$7 + B11*(1-$B$7)			
5	C3		20	C2TEMP = B11-D11*$B$3*C11/(3*$B$4))			
6	C0		100	C2NEW = F11*$B$7 + C11*(1-$B$7)			
7	X		0.1	VTEMP = ($B$4*C11 - $B$4*$B$5)*3/($B$3*$B$5)			
8				VNEW = G11*$B$7 + D11*(1-$B$7)			
9							
10	I	C1	C2	V	C1TEMP	C2TEMP	VTEMP
11	0.00	100.00	100.00	100.00	88.24	88.24	3000.00
12	1.00	98.82	98.82	390.00	65.79	65.02	2955.88
13	2.00	95.52	95.44	646.59	53.70	51.30	2829.10
14	3.00	91.34	91.03	864.84	46.44	42.42	2663.55
15	4.00	86.85	86.17	1044.71	41.79	36.29	2481.28
16	5.00	82.34	81.18	1188.37	38.69	31.86	2294.25
17	6.00	77.98	76.25	1298.96	36.60	28.54	2109.30
18	7.00	73.84	71.48	1379.99	35.21	26.00	1930.41
19	8.00	69.98	66.93	1435.03	34.32	24.02	1759.87
20	9.00	66.41	62.64	1467.52	33.82	22.46	1598.96
21	10.00	63.15	58.62	1480.66	33.62	21.23	1448.29
71	60.00	59.62	35.62	557.10	57.38	34.21	585.89
72	61.00	59.40	35.48	559.98	57.25	34.01	580.59
73	62.00	59.18	35.34	562.04	57.16	33.83	575.06
74	63.00	58.98	35.18	563.34	57.11	33.68	569.42
75	64.00	58.79	35.03	563.95	57.08	33.56	563.79
76	65.00	58.62	34.89	563.94	57.08	33.46	558.26
77	66.00	58.47	34.74	563.37	57.11	33.39	552.92
78	67.00	58.33	34.61	562.32	57.15	33.34	547.84
79	68.00	58.21	34.48	560.87	57.21	33.31	543.07
80	69.00	58.11	34.36	559.09	57.29	33.29	538.66
81	70.00	58.03	34.26	557.05	57.38	33.30	534.65
251	240.00	58.48	34.20	532.48	58.48	34.20	532.48
252	241.00	58.48	34.20	532.48	58.48	34.20	532.48
253	242.00	58.48	34.20	532.48	58.48	34.20	532.48
254	243.00	58.48	34.20	532.48	58.48	34.20	532.48
255	244.00	58.48	34.20	532.48	58.48	34.20	532.48
256	245.00	58.48	34.20	532.48	58.48	34.20	532.48
257	246.00	58.48	34.20	532.48	58.48	34.20	532.48
258	247.00	58.48	34.20	532.48	58.48	34.20	532.48
259	248.00	58.48	34.20	532.48	58.48	34.20	532.48
260	249.00	58.48	34.20	532.48	58.48	34.20	532.48
261	250.00	58.48	34.20	532.48	58.48	34.20	532.48

Spreadsheet Solution *(continued)*

	H	I	J	K	L	M
1						
2						
3						
4						
5						
6						
7						
8						
9						
10	C1NEW	C2NEW	VNEW			
11	98.82	98.82	390.00			
12	95.52	95.44	646.59			
13	91.34	91.03	864.84			
14	86.85	86.17	1044.71			
15	82.34	81.18	1188.37			
16	77.98	76.25	1298.96			
17	73.84	71.48	1379.99			
18	69.98	66.93	1435.03			
19	66.41	62.64	1467.52			
20	63.15	58.62	1480.66			
21	60.20	54.88	1477.42			
71	59.40	35.48	559.98	Note: Rows 22-70 hidden.		
72	59.18	35.34	562.04			
73	58.98	35.18	563.34			
74	58.79	35.03	563.95			
75	58.62	34.89	563.94			
76	58.47	34.74	563.37			
77	58.33	34.61	562.32			
78	58.21	34.48	560.87			
79	58.11	34.36	559.09			
80	58.03	34.26	557.05			
81	57.97	34.16	554.81			
251	58.48	34.20	532.48	Note: Rows 82-250 hidden.		
252	58.48	34.20	532.48			
253	58.48	34.20	532.48			
254	58.48	34.20	532.48			
255	58.48	34.20	532.48			
256	58.48	34.20	532.48			
257	58.48	34.20	532.48			
258	58.48	34.20	532.48			
259	58.48	34.20	532.48			
260	58.48	34.20	532.48			
261	58.48	34.20	532.48			

Appendix D: Solutions to Odd-Numbered Problems

1.1	500 tires; 16,000 lb of tires
1.3	$7.28\ (10)^4$ tons/yr
1.5	$1.44\ (10)^5$ MT/yr collected; $1.46\ (10)^3$ MT/yr emitted
1.7	0.231 years or 12.0 weeks
1.9	8.47 billion; 13.3 billion
2.3	70.8 g of oxygen
2.5	$Al(OH)_3$ precipitates first
2.7	(a) pH = 12; (b) pH = 2; (c) pH = 4.8; (d) pH = 6
2.9	$5.07\ (10)^{-6}$ mole/L
2.11	2520 mg/L as $CaCO_3$
2.13	106 L of CO_2
2.15	0.0355 mole fraction
2.19	855 min
2.21	111 hours; zero-order is reasonable, first-order is not
2.23	3.41 hours
3.1	1800 kg/day
3.3	$Q_C = 55$ L/s; $Q_E = 5$ L/s; $Q_{A'} = 150$ L/s
	$C_D = 628.5$ mg/L; $C_E = 59{,}715$ mg/L; $C_{A'} = 2209$ mg/L
3.5	818.5 gmole/day of CO_2
3.7	728 kg/day
3.9	rainy season C = 18.2 mg/L; dry season C = 100 mg/L
3.11	687,500 L/day
3.13	3103 gal of straight-run and 6897 gal of high-octane
3.15	(a) 13,333 L; (b) 3600 mole B/day

3.17   0.042 mole/L

3.19   all in million kg/day;
pipeline slurry: 20.31 coal, 20.31 water, 40.62 total
separated water: 0.31 coal, 15.31 water, 15.62 total
separated coal: 20.0 coal, 5.0 water, 25.0 total

3.21   1286 L

4.1   722 mg/L

4.3   COD = 0 mg/L; $CBOD_u$ = 0 mg/L; NBOD = 213 mg/L

4.5   COD = 107 mg/L; $CBOD_u$ = 107 mg/L; NBOD = 0 mg/L

4.7   COD = 0 mg/L; $CBOD_u$ = 0 mg/L; NBOD = 80 mg/L

4.9   236 mg/L

4.11   466 mg/L

4.13   949 mg/L

4.15   No. Also need to know either k or BOD_u.

4.17   Eutrophication.

4.19   Eutrophication.

4.21   There will be higher DO, a lower level of ammonia, and reduced algae growth.

4.23   Allowable BOD_5 = 5.6 mg/L

4.25   Allowable BOD_5 = 1430 mg/L

4.27   Variable order;   from nonlinear, least-squares fit:
k = 1.97 mg BOD_5/mg VSS-L; K_s = 9.3 mg/L;
Y_{max} = 0.60 mg VSS/mg BOD_5 ; K_e = 0.04 day^{-1}

5.1   187.5 ppm

5.3   12 μg/m^3

5.5   0.50 ppm

5.7   98.25%

5.9   520 g/s of NO_x and 26 g/s of particulate matter

5.11   3.4 (10)^5 g/yr of CO;   40,000 g/yr of NO_x;   41,000 g/yr of VOC

5.13   480 kg/day of SO_2

5.15   No

5.17   235 pCi/ft^3

5.21   74.4 dB(A); L_{eq}

5.23   63.3 dB(A); L_{eq}

5.25   42 μg/m^3 at 2 km downwind, and 3.9 μg/m^3 at 5 km

6.1   0.20 year^{-1}

6.3   No

6.5   33.6 cfs

6.7   advantage: decrease peak flow and pollutant loading; disadvantage: consume land

6.9   94.0 cfs

6.11   pH 7 is better

6.13   (1) remove iron and H_2S; (2) disinfect

6.15   raise pH to precipitate Mg

6.17   all as $CaCO_3$: CO_2 = 80 mg/L; Ca = 500 mg/L; Mg = 247 mg/L; Na = 76 mg/L; total hardness = 747 mg/L as $CaCO_3$; softening is needed

6.19   lime =4180 lb/day; soda =1740 lb/day; CO_2 =183 lb/day; residuals =8930 lb/day

6.21   L = 20 ft, W = 7.5 ft, D = 7.5 ft; HRT = 48.5 min

6.23   L = 4.17 ft, W = 4.17 ft, D = 7 ft; HLR = 5.0 gpm/ft^2

6.25   154 min

6.27   0.18 lb BOD_5/capita-day; 0.18 lb SS/capita-day

6.29   7.2 mg BOD_5/L; 1120 lb residuals/day

6.31   0.29 lb BOD_5/lb VSS-day; 62.6 lb BOD_5/1000 ft^3-day; 5.98 hour; 67%; 531 gpd/ft^2; 7.48 days

6.33   64,100 gpd

6.37   160,300 ft^3

7.1   226 acres

7.3   107 miles

7.11   20 yd^3 trucks are better

7.13   1.47 (10)$^{-5}$

7.15   50,000 deaths per year; 3.5 million deaths per 70-year "lifetime"

7.17   0.0048 µg/L

Index

Acid deposition (acid rain), 51–53, 151
Acidity, 51
Acids, 39–53
Activated sludge process
    description, 253–260
    design equations, 255–259
    kinetic models, 257
    organic loading, 251
    recycled solids (X_r), 256
    solids retention time, 250
Aeration, 243
Aerobic sludge digestion, 261
Aerobic treatment, 138, 261
Air pollutants
    carbon monoxide, 150, 173–175, 177, 296, 319–320
    causes, sources, and effects, 152
    emissions of, 152, 296
    HAPs, 150, 319
    lead, 174, 176, 320
    nitrogen oxides, 149, 151, 173–174, 296
    ozone, 150
    particulate matter, 149, 162–168, 296, 319
    primary, 149
    secondary, 149
    sulfur oxides, 149, 151, 170–173, 325–327
    volatile organic compounds, 150, 151, 169–170, 173–175, 177, 300–302
Air pollution episodes, 11
Air resource management, 147–148
Alkalinity, 51, 238

Alum, 230
Ambient air quality standards, 159–160
Ammonia (in water), 129, 247, 321–325
Anaerobic sludge digestion, 138, 261
Anoxic (denitrification), 138, 321–323
Arrhenius equation, 72
Atmospheric dispersion
    box model, 178–179
    Gaussian model, 184–187
    stability categories, 178–181
Average global temperature, 157–158
Avogadro's number, 31

Bacterial growth
    biomass, 137–139, 247–251, 257–259
    cell yield, 139, 248
    half-saturation constant, 67, 140
    kinetics, 139–140, 247–248, 251–252
    reactions, 137, 247–248
    specific growth rate (μ), 139, 248
    specific substrate removal rate (q), 139, 247–248
    yield (Y), 139, 248
Baghouse, 165, 320
Bases, 39–53
Biochemical oxygen demand (BOD)
    carbonaceous oxygen demand, 129
    definition, 122
    deoxygenation rate constant, k_d, 124
    measurements, 122
    nitrification, 129, 247, 321–325

Names, symbols, atomic numbers, and atomic weights of selected elements

	Sorted by Name				Sorted by Symbol		
Name	Symbol	Atomic Number	Atomic Weight	Symbol	Name	Atomic Number	Atomic Weight
Aluminum	Al	13	26.98	Ag	Silver	47	107.87
Antimony	Sb	51	121.75	Al	Aluminum	13	26.98
Argon	Ar	18	39.95	Ar	Argon	18	39.95
Arsenic	As	33	74.92	As	Arsenic	33	74.92
Barium	Ba	56	137.34	Au	Gold	79	196.97
Beryllium	Be	4	9.01	B	Boron	5	10.81
Bismuth	Bi	83	208.98	Ba	Barium	56	137.34
Boron	B	5	10.81	Be	Beryllium	4	9.01
Bromine	Br	35	79.90	Bi	Bismuth	83	208.98
Cadmium	Cd	48	112.40	Br	Bromine	35	79.90
Calcium	Ca	20	40.08	C	Carbon	6	12.01
Carbon	C	6	12.01	Ca	Calcium	20	40.08
Cerium	Ce	58	140.12	Cd	Cadmium	48	112.40
Cesium	Cs	55	132.90	Ce	Cerium	58	140.12
Chlorine	Cl	17	35.45	Cl	Chlorine	17	35.45
Chromium	Cr	24	52.00	Co	Cobalt	27	58.93
Cobalt	Co	27	58.93	Cr	Chromium	24	52.00
Copper	Cu	29	63.54	Cs	Cesium	55	132.90
Fluorine	F	9	19.00	Cu	Copper	29	63.54
Gold	Au	79	196.97	F	Fluorine	9	19.00
Helium	He	2	4.00	Fe	Iron	26	55.85
Hydrogen	H	1	1.01	H	Hydrogen	1	1.01
Iodine	I	53	126.90	He	Helium	2	4.00
Iron	Fe	26	55.85	Hg	Mercury	80	200.59
Krypton	Kr	36	83.80	I	Iodine	53	126.90
Lead	Pb	82	207.20	K	Potassium	19	39.10
Lithium	Li	3	6.94	Kr	Krypton	36	83.80
Magnesium	Mg	12	24.30	Li	Lithium	3	6.94
Manganese	Mn	25	54.94	Mg	Magnesium	12	24.30
Mercury	Hg	80	200.59	Mn	Manganese	25	54.94
Molybdenum	Mo	42	95.94	Mo	Molybdenum	42	95.94
Neon	Ne	10	20.18	N	Nitrogen	7	14.01
Nickel	Ni	28	58.70	Na	Sodium	11	23.00
Nitrogen	N	7	14.01	Ne	Neon	10	20.18
Oxygen	O	8	16.00	Ni	Nickel	28	58.70
Palladium	Pd	46	106.40	O	Oxygen	8	16.00
Phosphorus	P	15	30.97	P	Phosphorus	15	30.97
Platinum	Pt	78	195.09	Pb	Lead	82	207.20
Plutonium	Pu	94	242.00	Pd	Palladium	46	106.40
Potassium	K	19	39.10	Pt	Platinum	78	195.09
Radium	Ra	88	226.03	Pu	Plutonium	94	242.00
Radon	Rn	86	222.00	Ra	Radium	88	226.03
Selenium	Se	34	78.96	Rn	Radon	86	222.00
Silicon	Si	14	28.09	S	Sulfur	16	32.06
Silver	Ag	47	107.87	Sb	Antimony	51	121.75
Sodium	Na	11	23.00	Se	Selenium	34	78.96
Strontium	Sr	38	87.62	Si	Silicon	14	28.09
Sulfur	S	16	32.06	Sn	Tin	50	118.69
Tin	Sn	50	118.69	Sr	Strontium	38	87.62
Titanium	Ti	22	47.90	Ti	Titanium	22	47.90
Tungsten	W	74	183.85	U	Uranium	92	238.03
Uranium	U	92	238.03	V	Vanadium	23	50.94
Vanadium	V	23	50.94	W	Tungsten	74	183.85
Xenon	Xe	54	131.30	Xe	Xenon	54	131.30
Zinc	Zn	30	65.38	Zn	Zinc	30	65.38
Zirconium	Zr	40	91.22	Zr	Zirconium	40	91.22